AF361899

Deep Learning and Edge Computing for Internet of Things

Deep Learning and Edge Computing for Internet of Things

Guest Editors

Shaohua Wan
Yirui Wu

Basel • Beijing • Wuhan • Barcelona • Belgrade • Novi Sad • Cluj • Manchester

Guest Editors

Shaohua Wan
National Engineering
Laboratory for Big Data System
Computing Technology
Shenzhen University
Shenzhen
China

Yirui Wu
College of Computer Science
and Software Engineering
Hohai University
Nanjing
China

Editorial Office
MDPI AG
Grosspeteranlage 5
4052 Basel, Switzerland

This is a reprint of the Special Issue, published open access by the journal *Applied Sciences* (ISSN 2076-3417), freely accessible at: https://www.mdpi.com/journal/applsci/special_issues/ CD7Q35B72J.

For citation purposes, cite each article independently as indicated on the article page online and as indicated below:

Lastname, A.A.; Lastname, B.B. Article Title. *Journal Name* **Year**, *Volume Number*, Page Range.

ISBN 978-3-7258-2995-8 (Hbk)
ISBN 978-3-7258-2996-5 (PDF)
https://doi.org/10.3390/books978-3-7258-2996-5

Contents

applied sciences

Editorial

Editorial: Deep Learning and Edge Computing for Internet of Things

Shaohua Wan [1,2,3] **and Yirui Wu** [4,*]

[1]   National Engineering Laboratory for Big Data System Computing Technology, Shenzhen University, Shenzhen 518060, China; shaohua.wan@uestc.edu.cn
[2]   Key Laboratory of AI and Information Processing, Hechi University, Hechi 546300, China
[3]   Shenzhen Institute for Advanced Study, University of Electronic Science and Technology of China, Shenzhen 518110, China
[4]   College of Computer Science and Software Engineering, Hohai University, Nanjing 211100, China
[*]   Correspondence: wuyirui@hhu.edu.cn

Citation: Wan, S.; Wu, Y. Editorial: Deep Learning and Edge Computing for Internet of Things. *Appl. Sci.* **2024**, *14*, 11063. https://doi.org/10.3390/app142311063

Received: 20 November 2024
Accepted: 26 November 2024
Published: 28 November 2024

The evolution of 5G and Internet of Things (IoT) technologies is leading to ubiquitous connections among humans and their environment, such as autopilot transportation, mobile e-commerce, unmanned vehicles, and healthcare applications, bringing revolutionary changes to our daily lives. Moreover, the computing environment requires support to an increasing range of functionality, including multi-sensory data processing and analysis, complex systems control strategies, and, ultimately, artificial intelligence. After several years of development, edge computing for deep learning has shown incomparable practical value in the IoT environment. Pushing computing resources to the edge, closer to devices, enables low-latency service delivery for both safety and applications. However, edge computing still has abundant untapped potential for deep learning. Systems should leverage awareness of the surrounding environment and attach more importance to edge–edge intelligence collaboration and edge–cloud communication. This Special Issue aims to explore recent advances in edge computing technologies.

This issue includes eleven peer-reviewed papers that focus on deep learning and edge computing for the Internet of Things. "Techniques for detecting the start and end points of sign language utterances to enhance recognition performance in mobile environments" written by Kim et al. [1] proposes a technique to dynamically adjust the sampling rate based on the number of frames extracted in real-time during sign language utterances in mobile environments to accurately detect the start and end points of the sign language. Since partitioning and offloading are important for delay-sensitive CNN inference in edge computing, Zha et al. [2] wrote "Energy-efficient joint partitioning and offloading for delay-sensitive CNN inference in edge computing", which proposes a parallel partitioning method based on matrix convolution to partition foundation model inference tasks. Inspired by the widespread use of mobile and IoT devices, Sanabria et al. [3] wrote "connection-aware heuristics for scheduling and distributing jobs under dynamic dew computing environments", which proposes integrating the "reliability" concept into cutting-edge human-design job distribution heuristics, adapting to the dynamic and ever-changing network conditions caused by nodes' mobility.

Subsequently, deoxyribonucleic acid (DNA) computing has demonstrated great potential in data encryption due to its capability of parallel computation, minimal storage requirement, and unbreakable cryptography. Li et al. [4] wrote "hash-based DNA computing algorithm for image encryption" in which the experimental results suggest that the proposed method performs better than most comparative methods in key space, histogram analysis, pixel correlation, information entropy, and sensitivity measurements. Basori et al. [5] wrote "Hybrid Deep Convolutional Generative Adversarial Network (DCGAN) and Xtreme Gradient Boost for X-ray Image Augmentation and Detection", which provides a technique for the automated analysis of X-ray pictures using server processing with a deep convolutional generative adversarial network (DCGAN), thus improving the overall image quality of X-ray scans. Wang et al. [6] wrote "A Combined Multi-Classification Network Intrusion Detection System Based on Feature Selection and Neural Network Improvement", which uses 23 subframes and a mixer for multi-classification work, thus improving the parallelism of NIDS and is more adaptable to edge networks.

Furthermore, to improve the detection and recognition accuracy of small-sized, occluded, or truncated objects in complex scenes, Sheng et al. [7] wrote "Faster RCNN Target Detection Algorithm Integrating CBAM and FPN", which incorporates the convolutional block attention module in the feature extraction network, linking high- and bottom-level feature data to obtain high-resolution and strong semantic data. Yu et al. [8] wrote "Rainfall Similarity Search Based on Deep Learning by Using Precipitation Images". They proposed a rainfall similarity research method based on deep learning by using precipitation images, discovering similar rainfall processes, and providing new ideas for hydrological forecasting. To decrease the cost of data annotation and improve segmentation accuracy, Fu et al. [9] wrote "Attention-Based Active Learning Framework for Segmentation of Breast Cancer in Mammograms", which consists of a basic breast cancer segmentation model, an attention-based sampling scheme, and an active learning strategy for labeling. Wang et al. [10] wrote "An Adaptive Dynamic Channel Allocation Algorithm Based on a Temporal–Spatial Correlation Analysis for LEO Satellite Networks" in which they propose an adaptive dynamic channel allocation algorithm based on temporal–spatial correlation analysis for LEO satellite networks, reducing the rejection probability of the handoff call and then improving the total performance of the LEO satellite network.

Finally, Yan et al. [11] wrote a review "UAV Detection and Tracking in Urban Environments Using Passive Sensors: A Survey" in which they provide an overview of the existing military and commercial anti-UAV systems, proposing several suggestions for developing general-purpose UAV-monitoring systems tailored for urban environments.

Overall, we released eleven excellent papers in our Special Issue, which show promising developments in deep learning and edge computing for the Internet of Things. However, there are still many open challenges. For example, the complete architecture under heterogeneous networks has not been well addressed yet. Moreover, the difficulty of the optimization problem is further exacerbated due to the dynamic changes in the network environments; and though the edge computing architecture brings some benefits to the network applications, it raises unforeseen security and privacy issues at the same time. To further explore this discipline, we hope researchers and practitioners from academia and related industries can investigate this topic for further developments. Finally, we would like to thank the authors of the papers published in this Special Issue and the editorial team of *Applied Sciences*.

Conflicts of Interest: The authors declare no conflict of interest.

References

1. Kim, T.; Kim, B. Techniques for Detecting the Start and End Points of Sign Language Utterances to Enhance Recognition Performance in Mobile Environments. *Appl. Sci.* **2024**, *14*, 9199. [CrossRef]
2. Zha, Z.; Yang, Y.; Xia, Y.; Wang, Z.; Luo, B.; Li, K.; Ye, C.; Xu, B.; Peng, K. Energy-Efficient Joint Partitioning and Offloading for Delay-Sensitive CNN Inference in Edge Computing. *Appl. Sci.* **2024**, *14*, 8656. [CrossRef]
3. Sanabria, P.; Montoya, S.; Neyem, A.; Toro Icarte, R.; Hirsch, M.; Mateos, C. Connection-Aware Heuristics for Scheduling and Distributing Jobs under Dynamic Dew Computing Environments. *Appl. Sci.* **2024**, *14*, 3206. [CrossRef]
4. Li, H.; Zhang, L.; Cao, H.; Wu, Y. Hash Based DNA Computing Algorithm for Image Encryption. *Appl. Sci.* **2023**, *13*, 8509. [CrossRef]
5. Basori, A.H.; Malebary, S.J.; Alesawi, S. Hybrid Deep Convolutional Generative Adversarial Network (DCGAN) and Xtreme Gradient Boost for X-ray Image Augmentation and Detection. *Appl. Sci.* **2023**, *13*, 12725. [CrossRef]
6. Wang, Y.; Liu, Z.; Zheng, W.; Wang, J.; Shi, H.; Gu, M. A Combined Multi-Classification Network Intrusion Detection System Based on Feature Selection and Neural Network Improvement. *Appl. Sci.* **2023**, *13*, 8307. [CrossRef]
7. Sheng, W.; Yu, X.; Lin, J.; Chen, X. Faster rcnn target detection algorithm integrating cbam and fpn. *Appl. Sci.* **2023**, *13*, 6913. [CrossRef]

8. Yu, Y.; He, X.; Zhu, Y.; Wan, D. Rainfall Similarity Search Based on Deep Learning by Using Precipitation Images. *Appl. Sci.* **2023**, *13*, 4883. [CrossRef]
9. Fu, X.; Cao, H.; Hu, H.; Lian, B.; Wang, Y.; Huang, Q.; Wu, Y. Attention-Based Active Learning Framework for Segmentation of Breast Cancer in Mammograms. *Appl. Sci.* **2023**, *13*, 852. [CrossRef]
10. Wang, J.; Sun, L.; Zhou, J.; Han, C. An Adaptive Dynamic Channel Allocation Algorithm Based on a Temporal–Spatial Correlation Analysis for LEO Satellite Networks. *Appl. Sci.* **2022**, *12*, 10939. [CrossRef]
11. Yan, X.; Fu, T.; Lin, H.; Xuan, F.; Huang, Y.; Cao, Y.; Hu, H.; Liu, P. UAV Detection and Tracking in Urban Environments Using Passive Sensors: A Survey. *Appl. Sci.* **2023**, *13*, 11320. [CrossRef]

 *applied sciences*

Article

An Adaptive Dynamic Channel Allocation Algorithm Based on a Temporal–Spatial Correlation Analysis for LEO Satellite Networks

Juan Wang [1,2], Lijuan Sun [1,2,*], Jian Zhou [1,2,*] and Chong Han [1,2]

1   School of Computer Science, Nanjing University of Posts and Telecommunications, Nanjing 210023, China
2   Jiangsu High Technology Research Key Laboratory for Wireless Sensor Networks, Nanjing 210023, China
*   Correspondence: sunlijuan_nupt@163.com (L.S.); zhoujian@njupt.edu.cn (J.Z.)

Abstract: Low Earth orbit (LEO) satellites that can be used as computing nodes are an important part of future communication networks. However, growing user demands, scarce channel resources and unstable satellite–ground links result in the challenge to design an efficient channel allocation algorithm for the LEO satellite network. Edge computing (EC) provides sufficient computing power for LEO satellite networks and makes the application of reinforcement learning possible. In this paper, an adaptive dynamic channel allocation algorithm based on a temporal–spatial correlation analysis for LEO satellite networks is proposed. First, according to the user mobility model, the temporal–spatial correlation of handoff calls is analyzed. Second, the dynamic channel allocation process in the LEO satellite network is formally described as a Markov decision process. Third, according to the temporal–spatial correlation, a policy for different call events is designed and online reinforcement learning is used to solve the channel allocation problem. Finally, the simulation results under different traffic distributions and different traffic intensities show that the proposed algorithm can greatly reduce the rejection probability of the handoff call and then improve the total performance of the LEO satellite network.

Keywords: dynamic channel allocation; temporal–spatial correlation analysis; LEO satellite network; reinforcement learning; edge computing

Citation: Wang, J.; Sun, L.; Zhou, J.; Han, C. An Adaptive Dynamic Channel Allocation Algorithm Based on a Temporal–Spatial Correlation Analysis for LEO Satellite Networks. *Appl. Sci.* **2022**, *12*, 10939. https://doi.org/10.3390/app122110939

Academic Editors: Shaohua Wan and Yirui Wu

Received: 28 September 2022
Accepted: 26 October 2022
Published: 28 October 2022

Publisher's Note: MDPI stays neutral with regard to jurisdictional claims in published maps and institutional affiliations.

1. Introduction

Satellite networks not only provide a call admission service to terminal users at any time and anywhere, but also provide reliable communication in many scenes such as natural disasters and emergency rescues. Therefore, it has become a favorable supplement to terrestrial networks [1]. Low Earth orbit (LEO) satellite networks have advantages such as global coverage, real-time communication and small terminals, which makes them a research hotspot of satellite networks [2]. Several researchers have combined LEO satellite networks with edge computing (EC) to deploy EC servers on LEO satellites [3–5]. As edge computing nodes, LEO satellites are an important part of future communication networks [6]. The effectiveness of the application of an EC framework has been verified in existing systems [7–9]. A reasonable channel allocation algorithm can improve the utilization of communication resources and the performance of satellite networks. Edge computing provides sufficient computing power for LEO satellites and makes it possible to apply reinforcement learning to channel allocations [10].

LEO satellites operate at a low altitude with a high speed, which makes its coverage area prone to handoff. Therefore, the visibility time of LEO satellites to terminal users is very short. By using multiple antennas and a satellite-fixed cell (SFC) mode, the coverage area is divided into multi-beams. Each beam is called as a cell [11]. A call for the terminal user will hand over between multiple cells or multiple satellites during the whole communication process. Each handoff will cause a reallocation of the channel resources. An efficient channel allocation algorithm can reduce the rejection probability of calls and improve the total performance of LEO satellite networks [12].

In recent years, scholars have researched channel allocation algorithms in LEO satellite networks [13–18]. A fixed channel allocation (FCA) algorithm can allocate unchanged channels in each specified cell. Del Re et al. [19] analyzed the performance of an FCA for LEO satellite networks. An FCA is simple to implement, but it has a poor adaptability to variations in the demands of terminal users. A dynamic channel allocation (DCA) algorithm is superior to an FCA in performance. Li et al. [20] used a DCA to improve the resource utilization in a satellite network. However, the computational complexity of a DCA is higher than that of an FCA. Reinforcement learning (RL) is suitable for solving the DCA problem [21]. Nie et al. [22] used Q-learning to solve the DCA problem and reduce the computational complexity. Hu et al. [23] proposed a deep RL framework to solve the DCA problem and further improve the resource utilization in a satellite network. Liu et al. [24] considered the temporal correlation of a satellite network and used deep RL to further improve the resource utilization. Zheng et al. [25] extracted the state features through a convolution neural network and used deep learning to solve the DCA problem. The above scholars optimized a DCA with RL, which effectively improved the resource utilization in LEO satellite networks. However, LEO satellite networks have the problem of frequent handoffs. The existing dynamic channel allocation algorithms based on RL rarely evaluate the performance of handoff calls.

The channel reservation technique is an effective way to resolve the problem of frequent handoffs [26]. Maral et al. [27] designed a channel locking mechanism for handoff users with a successful handoff. Del Re et al. [28] proposed different handoff queueing strategies with dynamic and fixed channel allocation techniques. However, it is difficult to balance the complexity and performance. A channel allocation algorithm combined with RL technology can be implemented more flexibly and efficiently [29,30]. The traffic prediction of the calls of terminal users plays a decisive role in the resource allocation [31]. Due to LEO satellites moving along their orbit regularly and periodically, a handoff call has the characteristic of a temporal–spatial correlation. The temporal correlation means that the departure call of the current cell and the new call of the adjacent cell occur at the same time. The spatial correlation means that the adjacent cell and the current cell have a neighbor relationship in space. We made full use of the temporal–spatial correlation of handoff calls in LEO satellite networks to propose an adaptive DCA algorithm based on RL. This algorithm not only considered the problems caused by a frequent handoff, but also improved the total performance of the network. The main efforts of this paper were:

- The temporal–spatial correlation of handoff calls was analyzed and the Markov decision process (MDP) was used to formally describe the channel allocation process so that the channel allocation could be dynamically adjusted according to the environment.
- A policy for different call events was designed. Afterwards, an online RL algorithm—namely, SARSA—was used to solve the DCA problem. SARSA iteratively updated the policy from the performed actions so that the channel allocation could be adjusted in real-time according to the environment.
- The effectiveness of the proposed algorithm was verified by simulation experiments under different traffic distributions and different traffic intensities.

The remainder of this paper is structured as follows. Section 2 introduces the related technologies. The proposed algorithm is presented in Section 3. Section 4 presents and discusses the simulation results. Finally, conclusions are drawn in Section 5.

2. Related Technologies

2.1. Markov Decision Process

The MDP, which is a discrete time stochastic control process, provides a mathematical framework for modeling the decision process [32,33]. Typically, the MDP is defined as a tuple (S, A, P, R), where S represents a finite set of states and A represents a finite set of actions. P represents the probability of the state transition from s ($s \in S$) to s' ($s' \in S$) after performing an action a ($a \in A$). R represents the immediate reward obtained after performing this action. In the MDP, policy π is defined as the mapping from a state to an action. The ultimate goal of the MDP is to find the optimal policy π^* to maximize the benefit that the performed actions can cumulatively obtain from the environment.

2.2. SARSA

RL, an important branch of machine learning, is an effective way to solve the MDP. The state-action value is very important in RL, which is called the Q-value. The Q-value refers to the average value of the benefits cumulatively obtained from the environment when a policy π performs an action from the current state to the final state. It is generally expressed as the expectation of the sum of the immediate reward and the subsequent rewards. The Q-value is a measure of the quality of the policy and it is calculated by:

$$Q(s_t, a_t) = E_\pi \left(R_t + \gamma R_{t+1} + \gamma^2 R_{t+2} + \gamma^3 R_{t+3} + \ldots \middle| s_t = s, a_t = a \right) \qquad (1)$$

where s_t is the environmental state at time t, a_t is the performed action at time t and π is the policy. In particular, R_t is the immediate reward after performing action a_t in the state s_t and $R_{t+1}, R_{t+2} \ldots$ are the subsequent rewards after time t. $\gamma \in [0,1)$ is the discount factor, which weighs the immediate reward and subsequent rewards obtained after performing the current action.

Finding the optimal policy means that an optimal action is performed at each state. When the space of the environmental state is very large or the probability of the current state reaching the final state is very small, it is difficult for Equation (1) to update the Q-value. Q-learning can update the Q-value in one step with the learning rate.

SARSA can calculate the Q-value of a given policy without a state transition probability and complete the state sequence so it can find the optimal policy without requiring knowledge of the environment [34,35]. Different from Q-learning, SARSA performs the actions with the same policy in real-time at each iteration time step [36,37]. Its Q-value is iteratively updated at each time step by:

$$Q_{t+1}(s_t, a_t) = Q_t(s_t, a_t) + \alpha_t [R_t(s_t, a_t) + \gamma Q_t(s_{t+1}, a_{t+1}) - Q_t(s_t, a_t)] \qquad (2)$$

where α_t is the learning rate at time t and $Q_t(s_{t+1}, a_{t+1})$ is the value of the next state and the next action. According to $\alpha_{t+1} = \alpha_t * \delta$, the learning rate decreases with an increase in the iteration time t. δ is the decay factor of α and $\alpha_t \in (0,1]$.

3. The Proposed Algorithm

3.1. Temporal–Spatial Correlation Analysis

The mobility model of the users is shown in Figure 1. The LEO satellite used the SFC coverage mode. The user mobility was simplified to a linear motion. Compared with the satellite movement speed, the speed of the terminal users had the same value and the opposite direction [38].

The high-speed movement of LEO satellites makes calls frequently hand over between cells or satellites. The departure call of the current cell and the new call of the adjacent cell occur at the same time. A handoff call occurs in the only neighbor cell that is in the opposite direction to the satellite movement. Therefore, a handoff call has the characteristic of a temporal–spatial correlation in LEO satellite networks.

A call of the current cell hands over at a certain time because of the high-speed movement of the LEO satellite. The call of the current cell departs; meanwhile, a new call of the adjacent cell occurs. The adjacent cell is the only neighbor cell where the call of the current cell hands over. As shown in Figure 1, cell 19 is the adjacent cell of cell 18. In other words, the call of cell 18 will hand over to cell 19.

When a handoff call occurs, the channel allocation algorithm not only releases the occupied channel for the departure call, but also allocates the channel for the handoff call of the adjacent cell. Generally speaking, one channel resource is usually allocated for one call in LEO satellite networks. The allocated channel must also meet any electromagnetic compatibility constraints to avoid co-channel interference [39].

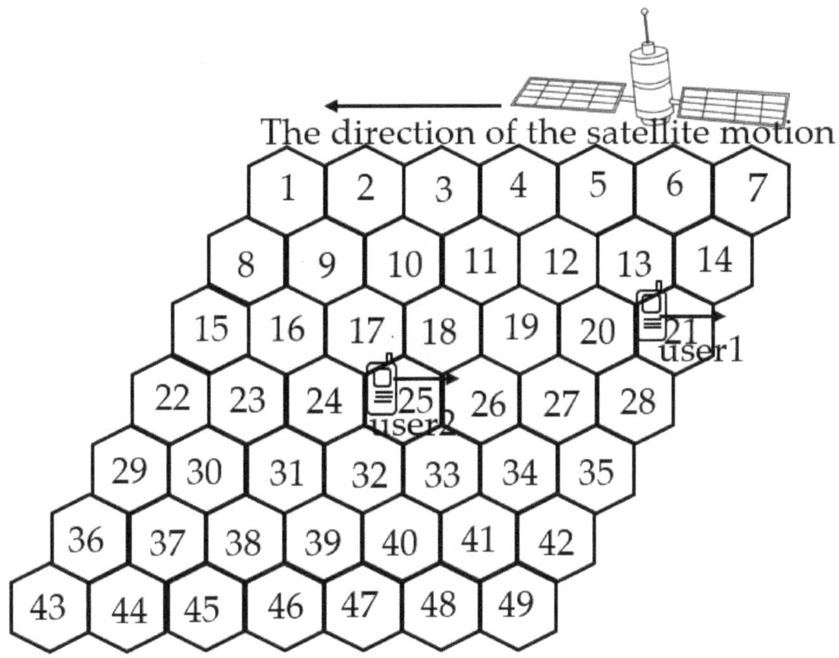

Figure 1. The mobility model of the users.

We assumed that the LEO satellite network had N cells $= \{1, 2, 3, \ldots \ldots N\}$ and K channels $K = \{1, 2, 3, \ldots \ldots K\}$. The conflicting cell $I(n)$ of cell n referred to the cells whose distances from themselves to cell n were less than the minimal reuse distance. This could be expressed by:

$$I(n) = \{m \in, dist(n, m) < d\} \tag{3}$$

where $dist(n, m)$ represents the distance between cell n and cell m and d represents the minimal reuse distance. As shown in Figure 1, the 18 cells in gray were conflicting the cell of cell 25 if $d = 3$.

The same channel cannot simultaneously be allocated to the calls of the current cell and its conflicting cell. The eligible channel $\tilde{A}(n)$ of cell n refers to the idle channel in both cell n itself and its conflicting cell and can be expressed by:

$$\tilde{A}(n) = \{k \in K, \sum_{m \in I(n)} x(m, k) = 0\} \tag{4}$$

where $x(m, k)$ represents the status of channel k of cell m. The value 0 means idle and 1 means occupied.

3.2. Markov Decision Process for the Dynamic Channel Allocation Process

In this paper, the channel status and the call event formed the environment in the channel allocation problem. The transition of the environmental state had the Markov property. The whole channel allocation process was described as per the following MDP model to dynamically adapt to the environmental changes.

State S: $s_t = (X_t, e_t)$ represents the environmental state at time t. X_t is the channel status of the LEO satellite network at time t and e_t is the call event for the admission service to the terminal users at time t. They are expressed, respectively, by:

$$X_t = \begin{bmatrix} x_t(1,\,1) & x_t(1,\,2) & x_t(1,\,3) & \cdots & x_t(1,\,k) & \cdots & x_t(1,K) \\ x_t(2,\,1) & x_t(2,\,2) & x_t(2,,,,3) & \cdots & x_t(2,\,k) & \cdots & x_t(2,K) \\ \cdots & \cdots & \cdots & \cdots & \cdots & \cdots & \cdots \\ x_t(n,\,1) & x_t(n,\,2) & x_t(n,3) & \cdots & x_t(n,\,k) & \cdots & x_t(n,K) \\ x_t(N,1) & x_t(N,2) & x_t(N,3) & \cdots & x_t(N,\,k) & \cdots & x_t(N,K) \end{bmatrix} \tag{5}$$

$$e_t \in \left\{ e_{n,t}^{new},\ e_{n,m,k,t}^{handoff},\ e_{n,k,t}^{end} \right\} \tag{6}$$

where $x_t(n,\,k) = 0$ represents that channel k of cell n is idle at time t and $x_t(n,\,k) = 1$ represents that this channel is occupied. $e_{n,t}^{new}$ represents a new call event in cell n at time t. $e_{n,m,k,t}^{handoff}$ represents that the handoff call event on channel k in cell n will hand over to the adjacent cell m at time t according to its temporal–spatial correlation. $e_{n,k,t}^{end}$ represents that the departure call event on channel k in cell n occurs at time t.

Action *A*: to accept or reject the call event. If a new call arrives, a channel should be occupied. If the call departs, a channel should be free. If the call hands over, the related new call and the departure call should be handled. In this paper, the actions of different call events were expressed by:

$$A((x_t, e_t)) = \begin{cases} (n),\ e_t = e_{n,t}^{new} \\ \{l \in K, x_t(n,l) = 1\},\ e_t = e_{n,k,t}^{end} \\ A\left(\left(x_t, e_{n,k,t}^{end}\right)\right) \cup A\left((x_{t+1}, e_{m,t}^{new})\right)\ e_t = e_{n,m,k,t}^{handoff} \end{cases} \tag{7}$$

Immediate reward *R*: the reward obtained from the environment after performing action a_t in the current state s_t. In this paper, the immediate reward referred to the total number of calls served in the current network. It was expressed by:

$$R_t = (s_t,\,)\ \sum_{n=1}^{N} \sum_{k=1}^{K} x_{t+1}\,(n,\,k) \tag{8}$$

where $x_{t+1}(n,\,k)$ represents the transformed status of channel k in cell n after performing action a_t.

Policy *π*: the mapping from an environmental state to the performed action. The performed action in the current state corresponds with the selected channel for the current call. The relationship between action, state and policy is expressed by:

$$a_t = \pi(s_t = (X_t, e_t)) \tag{9}$$

3.3. SARSA for Solving the MDP Model

RL can discover an optimal policy and obtain the maximal benefit from the environment. As an online RL algorithm, SARSA was used in this paper to solve the MDP for the channel allocation process. SARSA selected the optimal actions at each time step in real-time and directly updated the policy by the performed actions through interacting with the environment so that the channel allocation algorithm could be adjusted in real-time with the environmental changes.

First, parameters such as the channel status, the Q-value, the learning rate and the discount factor were initialized. Second, by using the temporal–spatial correlation of handoff calls, action a_t was performed according to the policy π for different call events. Subsequently, the immediate reward R_t was obtained after performing a_t and then the environment reached a new state. Third, with the same policy, an action was selected for the next call event and the current Q-value was updated. Last, the iteration continued in the new state until the ending conditions were satisfied. SARSA for solving the MDP model is shown in Algorithm 1.

Algorithm 1: SARSA for solving the MDP model

Input: α, δ, γ
Output: $Q^*(s, a)$
$X \leftarrow x_0(n, k) = 0$ $\square\square$ //initialize the channel status
$Q(s, a) = 0$ //initialize the Q-value
while *list { call event }* $\neq \phi$ do $\square\square\square\square$
    if $e_t == e_{n,t}^{new}$
        $a_t \leftarrow \pi\left(s_t = \left(X_t, e_{n,t}^{new}\right)\right)$ //perform a_t for the new call event
    else if $e_t == e_{n,k,t}^{end}$
        $a_t \leftarrow \pi\left(s_t = \left(X_t, e_{n,k,t}^{end}\right)\right)$ //perform a_t for the departure call event
    else if $e_t == e_{n,m,k,t}^{handoff}$
        $a_t \leftarrow \pi\left(s_t = \left(X_t, e_{n,m,k,t}^{handoff}\right)\right)$ //perform a_t for the handoff call event
    end if
    $X_{t+1} \leftarrow z(s_t, a_t)$ //transform the channel status
    $R_t = \sum_{n=1}^{N} \sum_{k=1}^{K} x_{t+1}(n, k)$ //calculate the immediate reward
    Update $Q(s_t, a)$
    $\alpha \leftarrow \alpha * \delta$ //decline the learning rate with the decay factor δ
    $s_t \leftarrow s_{t+1}$
    $t = t + 1$
end

In view of different call events, the proposed algorithm had the following policy. For a new call event and a departure call event, the actions were performed with ε-greedy and then the Q-value was updated. For a handoff call event, two actions were successively performed with the temporal–spatial correlation of the handoff calls and then the Q-value was updated. The policy is described in detail below.

3.3.1. The New Call Event

When a new call event occurred ($e_t = e_{n,t}^{new}$), the current state was $s_t = (X_t, e_{n,t}^{new})$. If the eligible channel $\tilde{A}(n)$ was empty, the new call was rejected. Otherwise, the channel was allocated for the new call in the current state. The policy for the new call event was expressed by:

$$\pi\left(s = \left(X_t, e_{n,t}^{new}\right)\right) \begin{cases} random\ (n), & \varepsilon \\ arg\ max\ Q(s_t, a), & 1 - \varepsilon \end{cases} \tag{10}$$

where ε is the exploration factor. ε could control selected actions according to a stochastic scheme. ε was used to explore a greater state space, which yielded other benefits. A channel from $\tilde{A}(n)$ was randomly selected with ε. Otherwise, a channel represented by the action with the maximal Q-value was selected with $1 - \varepsilon$. The policy for the new call event is shown in Algorithm 2.

3.3.2. The Departure Call Event

When a departure call event occurred ($e_t = e_{n,k,t}^{end}$), the current state was $s_t = (X_t, e_{n,k,t}^{end})$. The channel represented by the performed action with the minimal Q-value was selected and then the channel occupied by the departure call was released. The policy for the departure call event was expressed by:

$$\pi\left(s_t = \left(X_t, e_{n,k,t}^{end}\right)\right) = a_t = arg\ min\ Q\ (s_t, a). \tag{11}$$

If a_t was equal to k, the channel occupied by the departure call was directly released. If a_t was not equal to k, channel k was reallocated to the call occupying the channel represented by a_t. The policy for the departure call event is shown in Algorithm 3.

Algorithm 2: Policy for a new call event

Input: $e_t = e_{n,t}^{new}$, $s_t = (X_t, e_t)$
Output: a_t
if $\tilde{A}(n) == \phi$
    rejected   //reject the current new call event
else
    if $rand(\ ) < \varepsilon$ //select the eligible channel with ε
        $a_t = random(\tilde{A}(n))$ //select the eligible channel for the performed action
    else
        for $a \in \tilde{A}(n)$ **do**
        if $Q(s_t, a) > Q(s_t, a_t)$
        $a_t = a$ //select the channel represented by the selected action
        end if
        $X_{t+1} \leftarrow x_t(n, a_t) = 1$ //transform the channel status
end if

Algorithm 3: Policy for a departure call event

Input: $e_t = e_{n,k,t}^{end}$, $s_t = (X_t, e_t)$
Output: a_t
for $a \in \{a_j, x_t(n, j) = 1\}$ **do**
    if $Q(s_t, a) < Q(s_t, a_t)$
        $a_t = a$
        //select the channel represented by the action with the minimal Q-value
    end if
if$(a_t == k)$ //judge whether the value of the selected action is k
    $x_t(n, k) = 0$ //release the channel occupied by the current departure call event
else
    reallocation( )   //reallocate channel k to the call occupying the channel a_t
end if

3.3.3. The Handoff Call Event

If the policy could ensure an eligible channel for the handoff call of its adjacent cell, the rejection probability of the handoff call was effectively reduced. We assumed that there were 5 channels (*ch1–ch5*) in the LEO satellite network and the minimal reuse distance was 3. The served calls and the channel status are shown in Figure 2a. When the call on channel 1 of cell 25 handed over, the following three options were available for handling the handoff call event: (1) As shown in Figure 2b, without a reallocation, the call on channel 1 of cell 25 departed. The status of channel 1 of cell 25 was transformed from *1* to *0* and then the channel was released. According to the relative definitions and the channel status in $I(26)$, $\tilde{A}(26)$ was none. Thus, adjacent cell 26 had no eligible channel. As a result, the handoff call was rejected. (2) As shown in Figure 2c, channel 1 (occupied by the departure call) was reallocated to the call on channel 3 and then channel 3 was released. According to the current channel status in $I(26)$, $\tilde{A}(26)$ was none. Thus, adjacent cell 26 still had no eligible channel. As a result, the handoff call was rejected. (3) As shown in Figure 2d, channel 1 (occupied by the departure call) was reallocated to the call on channel 4 and then channel 4 was released. According to the current channel status in $I(26)$, $\tilde{A}(26) = \{ch4\}$. Thus, adjacent cell 26 had an eligible channel *ch4*. As a result, the handoff call was successfully accepted. If the policy selected option 3, as shown in Figure 2d, the policy ensured an eligible channel for the handoff call and reduced the rejection possibility of the handoff call.

When a handoff call event occurred $(e_t = e_{n,m,k,t}^{handoff})$, the current state was $s_t = (X_t, e_{n,m,k,t}^{handoff})$. The departure call event was handled first and then the related new call event was handled. The currently performed action was adjusted by the action for the new call event in the adjacent cell. The policy for the handoff call event was expressed by Equation (12) and Equation (13), respectively.

$$\pi\left(s_t = \left(X_t, e_{n,m,k,t}^{handoff}\right)\right) = \pi\left(s_t = \left(X_t, e_{n,k,t}^{end}\right)\right) = \mathrm{argmin}Q(s_t, a) \tag{12}$$

$$\pi\left(s_t = \left(X_t, e_{n,m,k,t}^{handoff}\right)\right) = \pi\left(s_{t+1} = \left(X_{t+1}, e_{m,t+1}^{new}\right)\right) = \mathrm{argmax}Q(s_{t+1}, a) \tag{13}$$

where the first action for the departure call event is expressed by a_t, corresponding with the minimal state-action value of $Q(s_t, a)$ and the second action for the new call event in the adjacent cell is expressed by a_{t+1}, corresponding with the maximal state-action value of $Q(s_{t+1}, a)$. The maximum of $Q(s_{t+1}, a_{t+1})$ and $Q(s_t, a_t)$ was selected to update the current Q-value. The channel occupied by the departure call event was reallocated to the call on the channel represented by the action with the maximum Q-value. The policy for the handoff call event is shown in Algorithm 4.

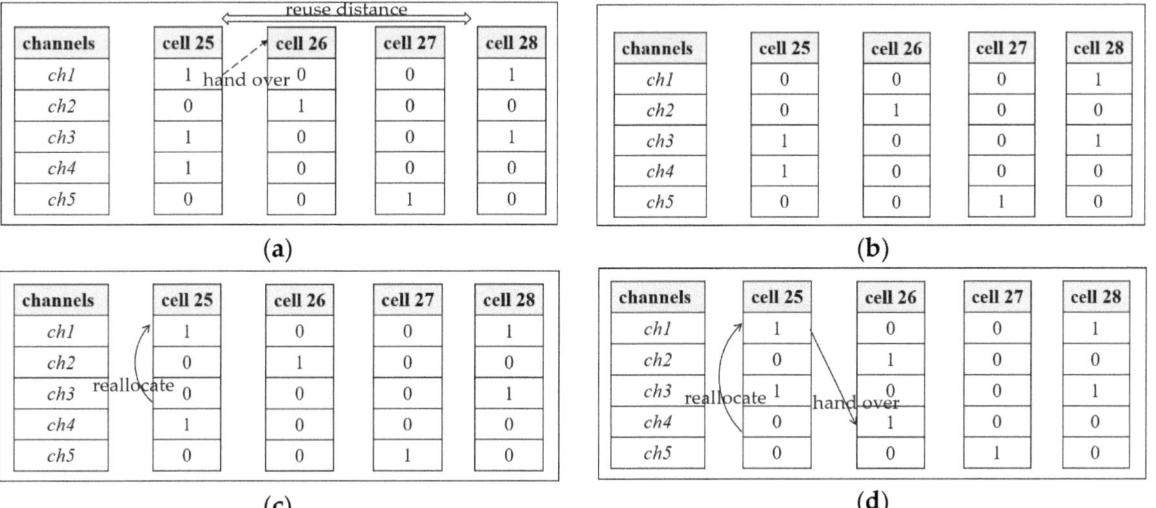

(a)

channels	cell 25	cell 26	cell 27	cell 28
ch1	1	0	0	1
ch2	0	1	0	0
ch3	1	0	0	1
ch4	1	0	0	0
ch5	0	0	1	0

(b)

channels	cell 25	cell 26	cell 27	cell 28
ch1	0	0	0	1
ch2	0	1	0	0
ch3	1	0	0	1
ch4	1	0	0	0
ch5	0	0	1	0

(c)

channels	cell 25	cell 26	cell 27	cell 28
ch1	1	0	0	1
ch2	0	1	0	0
ch3	0	0	0	1
ch4	1	0	0	0
ch5	0	0	1	0

(d)

channels	cell 25	cell 26	cell 27	cell 28
ch1	1	0	0	1
ch2	0	1	0	0
ch3	1	0	0	1
ch4	0	1	0	0
ch5	0	0	1	0

Figure 2. Reallocation for a handoff call event. (**a**) The handoff call occurs; (**b**) option 1: no reallocation; (**c**) option 2: reallocate to the call on *ch3*; (**d**) option 3: reallocate to the call on *ch4*.

Algorithm 4: Policy for a handoff call event

Input: $e_t = e_{n,m,k,t}^{handoff}$, $s_t = (X_t, e_t)$
Output: a_t, $Q(s_t, a_t)$
  for $a_t \in \pi((X_t, e_{n,k,t}^{end}))$ **do**   //handle the departure call event
    $k = a_t$ //perform the current action corresponding to channel k
      $X_{t+1} \leftarrow x_{t+1}(n, k) = 0$ //transform the status of channel k in cell n
      for $a_{t+1} \in \pi(X_{t+1}, e_{m,t+1}^{new})$ **do**   //handle the new call event in adjacent cell
        $X_{t+2} \leftarrow x_{t+2}(m, a_{t+1}) = 1$ //transform the channel status for the new call event
        if $Q(s_{t+1}, a_{t+1}) > Q(s_t, a_t)$
          $a_t = a_{t+1}$
          $Q(s_t, a_t) = Q(s_{t+1}, a_{t+1})$ // update the current Q-value
      end if
    end for
  end for

---

4. Simulation

4.1. Simulation Settings

The discrete events were programmed using Python to simulate the calls of the terminal users. The call arrival was assumed to follow a Poisson distribution with a mean rate λ. The call duration was assumed to follow an exponential distribution with a mean value $1/\mu$. According to [15], several parameters were set as shown in Table 1. The adaptive dynamic channel allocation algorithm based on the temporal–spatial correlation analysis (TSCA) was compared with SARSA [40], a random DCA algorithm (RDCA) [37] and an FCA [19] under a uniform traffic distribution and a nonuniform traffic distribution. The performance of the channel allocation algorithm was measured by the rejection probabilities of the new call, the handoff call and the total call [41]. The rejection probability of the total call was the ratio of the sum of the rejected new call and the handoff call to the total call. It could evaluate the total performance of the LEO satellite network. Without considering the temporal–spatial correlation, SARSA dynamically allocated channels according to the environment. The RDCA randomly selected a channel from the eligible channels. The FCA allocated 10 fixed channels to each cell.

Table 1. Parameter settings of the simulation.

Name	Description	Value
α	Learning rate	0.019389
δ	Decay factor of α	0.999999
γ	Discount factor	0.845
ε	Exploration factor	0.8
N	The number of cells	49
K	The number of channels	70
r	The radius of cell	450 km
v_s	The velocity of satellites	7 k/s
$1/\mu$	Call duration	3 min

4.2. Results and Analysis

Two simulations were carried out, including the cases of uniform and nonuniform traffic distributions. Under a uniform traffic distribution, the traffic intensity of each cell was the same. Under a nonuniform traffic distribution, the traffic intensity of each cell was different.

Figure 3 shows the comparison results of the four channel allocation algorithms under a uniform traffic distribution with different traffic intensities. It can be seen from Figure 3 that the performances of the different channel allocation algorithms decreased with an increase in the traffic intensity. In terms of the total performance, the RDCA was better than the FCA because the RDCA dynamically allocated eligible channels. The TSCA and SARSA were better than the RDCA. The reason was that RL can learn how to select and perform the actions from a continuous interaction with the environment; thus, the TSCA and SARSA could allocate more appropriate channels to the calls. The rejection probability of a new call with the TSCA was higher than SARSA. That was because the TSCA allocated more eligible channels to the handoff calls and the remaining eligible channels for the new calls were reduced. The rejection probability of a handoff call with the TSCA was much lower than that of the other algorithms. That was because the TSCA made full use of the temporal–spatial correlation of the handoff calls and allocated the eligible channels to the handoff calls. The rejection probabilities of a total call of the TSCA and SARSA were almost the same because they all performed optimal actions in real-time to allocate appropriate channels.

Figure 4 shows a case of a nonuniform traffic distribution. The numbers in the hexagons were the traffic distribution proportions of each cell. Value 1 represented the standard traffic distribution proportion and its traffic intensity was equal to 5.

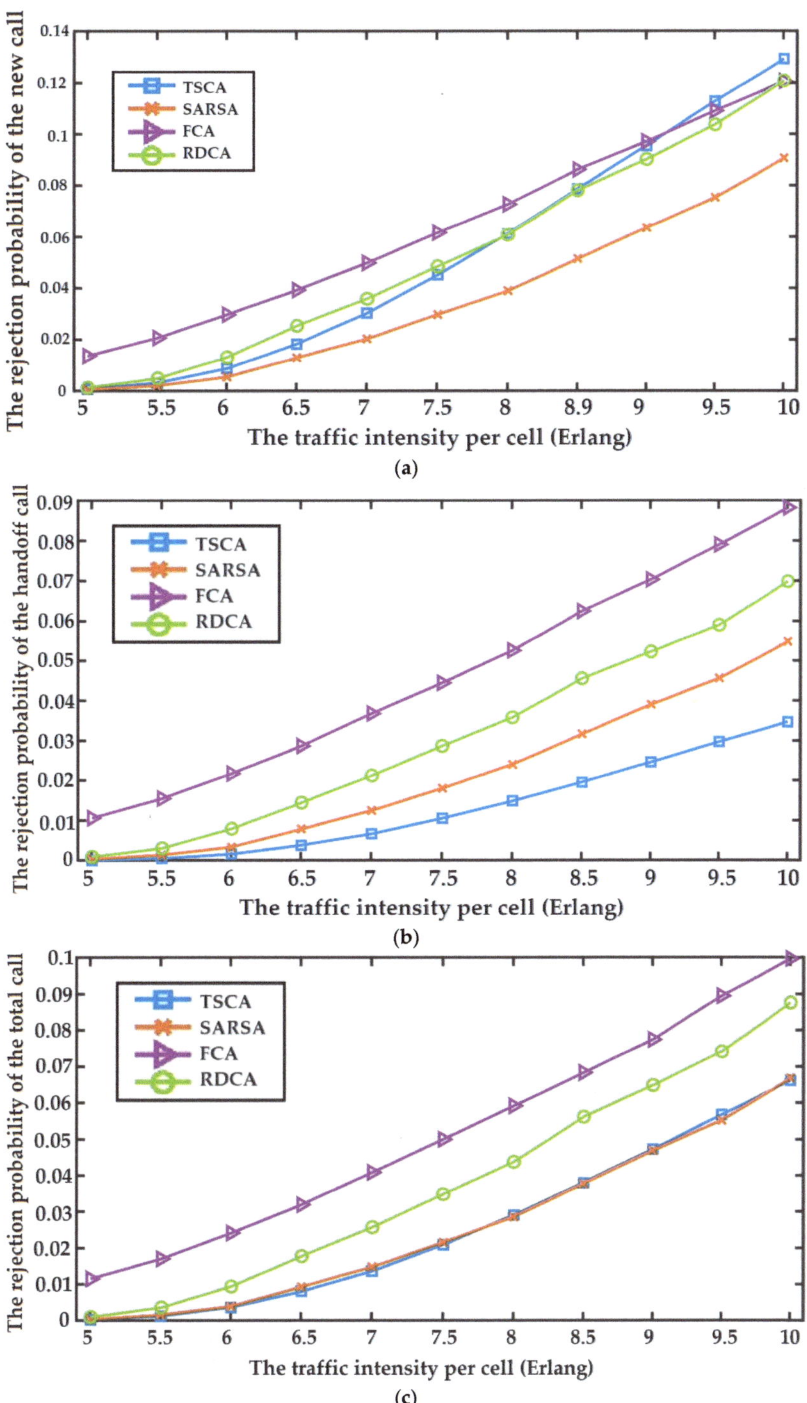

Figure 3. Performance comparison under uniform traffic distributions. (**a**) The rejection probability of the new call; (**b**) the rejection probability of the handoff call; (**c**) the rejection probability of the total call.

Figure 5 shows the comparison results of the four channel allocation algorithms under a nonuniform traffic distribution with different traffic intensities. It can be seen from Figure 5 that the performances of the different channel allocation algorithms decreased with an increase in the average traffic intensity. The performances of the four algorithms under a nonuniform traffic distribution were generally worse than that under a uniform traffic distribution.

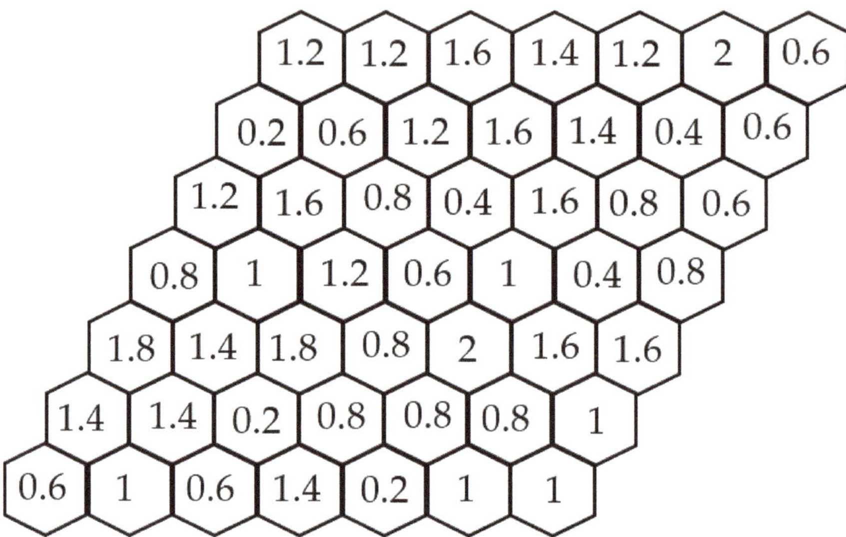

Figure 4. Nonuniform traffic distribution in cells.

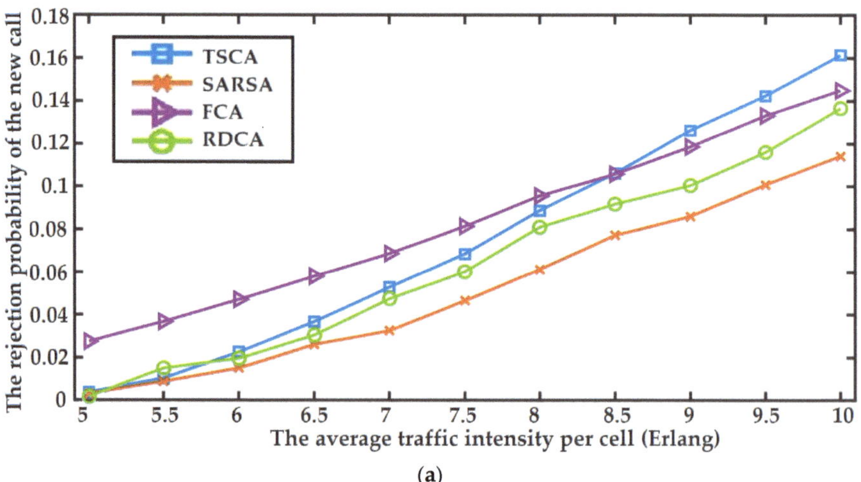

(**a**)

Figure 5. *Cont.*

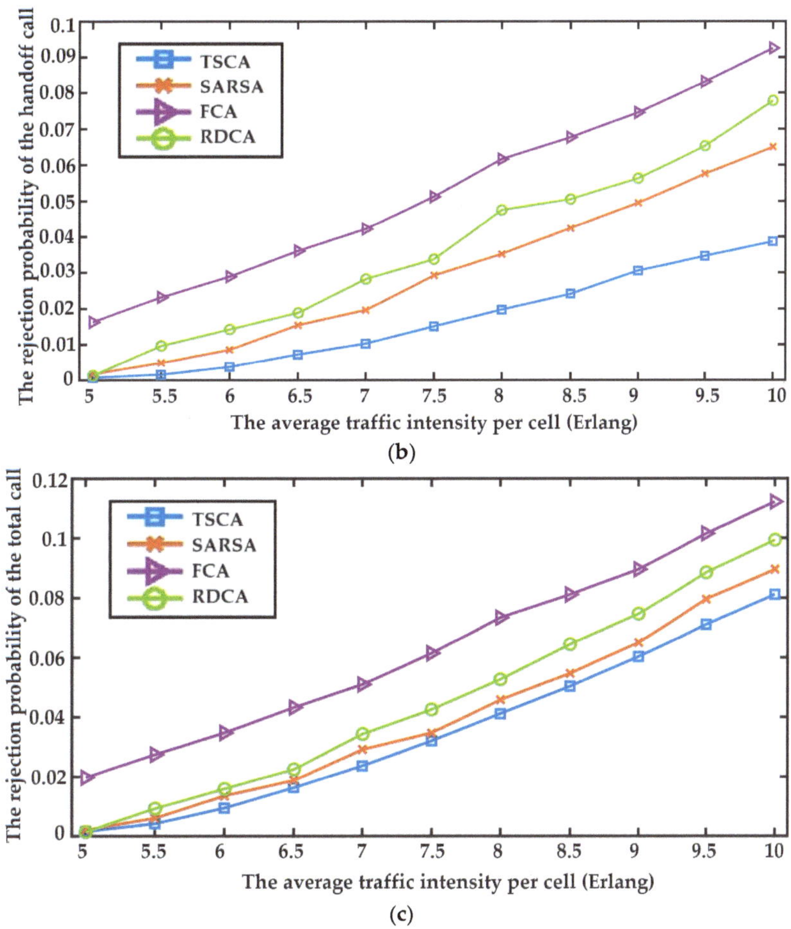

Figure 5. Performance comparison under nonuniform traffic distribution. (**a**) The rejection probability of the new call; (**b**) the rejection probability of the handoff call; (**c**) the rejection probability of the total call.

The performance of the same algorithm under different traffic distributions can be seen in Figures 3 and 5. It can be seen from Figures 3a and 5a that the advantage of a dynamic channel allocation was clearly reflected under a nonuniform traffic distribution. For instance, the rejection probability of a new call of the FCA was 2.4% higher than that of a uniform traffic distribution and the RDCA was 1.5% higher than that of a uniform traffic distribution at 10 Erlangs. In other words, the rejection probability of a new call of the FCA under a nonuniform traffic distribution was 19.98% higher than that under a uniform traffic distribution at 10 Erlangs and the rejection probability of a new call of the RDCA under a nonuniform traffic distribution was 12.74% higher. It can be seen from Figures 3b and 5b that the rejection probability of a handoff call of SARSA under a nonuniform traffic distribution was 18.17% higher than that under a uniform traffic distribution at 10 Erlangs and the rejection probability of a handoff call of the TSCA under a nonuniform traffic distribution was 11.04% higher. The TSCA not only had the lowest rejection probability of a handoff call, but also played a significant role in the performance, especially under a nonuniform traffic distribution and a high traffic intensity. As the TSCA considered the temporal–spatial correlation of the handoff calls, it dynamically allocated more appropriate channels for the handoff calls than SARSA.

The performance of the different a"gori'hms under the same traffic distribution can be seen in Figures 3 and 5. It can be seen from Figures 3b and 5b that the rejection probability of a handoff call of the TSCA was 40.59% lower than that of SARSA under a nonuniform traffic distribution at 10 Erlangs. In the case of a uniform traffic distribution, the rejection probability of a handoff call of the TSCA was 36.68% lower than that of SARSA at 10 Erlangs. It can be seen from Figures 3c and 5c that the rejection probability of a total call of the TSCA was 9.3% lower than that of SARSA at 10 Erlangs under a nonuniform traffic distribution. In the case of a uniform traffic distribution, the rejection probability of a total call of the TSCA was 0.8% lower than that of SARSA at 10 Erlangs. The TSCA improved the performance of the handoff calls, especially under a nonuniform traffic distribution and a high traffic intensity. The TSCA was better than SARSA, especially under a nonuniform traffic distribution.

The RDCA was better than the FCA because the FCA did not have enough channels in high traffic cells and then rejected the calls. The TSCA and SARSA were better than the RDCA. The reason was that RL performs actions in real-time to obtain more benefits from the environment during the iterative process. The TSCA was better than SARSA because the TSCA had a policy considering the temporal–spatial correlation of the handoff calls.

4.3. Parameter Discussion

First, the effect of the change of the parameter ε on the performance of the TSCA was discussed. Five values of the parameter (0.5, 0.6, 0.7, 0.8 and 0.9) were selected. Figure 6 shows the result after using the above values for the proposed algorithm. Simulations were carried out under uniform and nonuniform traffic distributions. It could be seen that the various rejection probabilities fluctuated in the range of one thousandth and the fluctuation was irregular. Therefore, we concluded that the proposed algorithm was insensitive to a parameter value of epsilon.

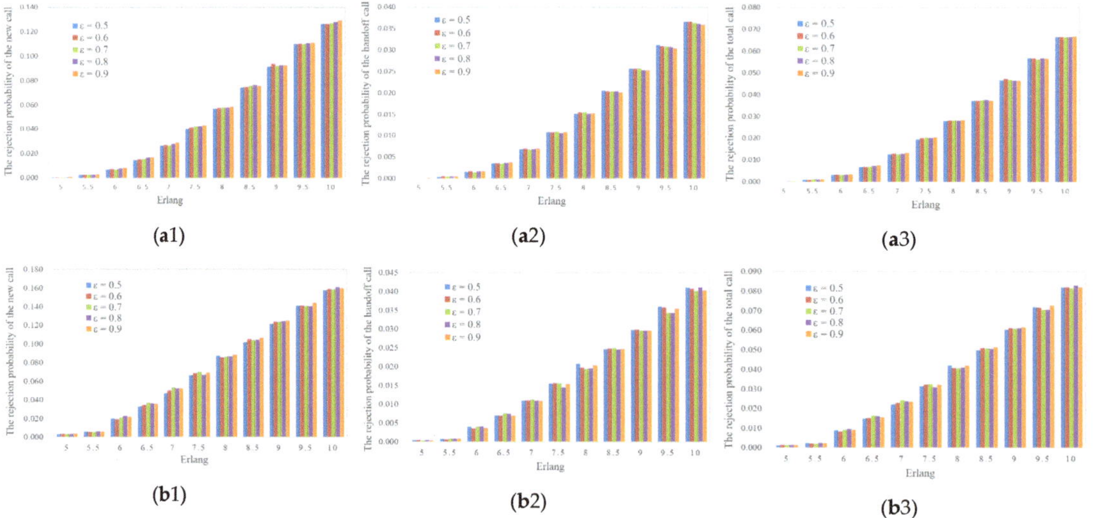

Figure 6. Performance comparison of TSCA with different values of epsilon. (**a**1) The rejection probability of the new call under uniform distribution; (**a**2) the rejection probability of the handoff call under uniform distribution; (**a**3) the rejection probability of the total call under uniform distribution; (**b**1) the rejection probability of the new call under nonuniform distribution; (**b**2) the rejection probability of the handoff call under nonuniform distribution; (**b**3) the rejection probability of the total call under nonuniform distribution.

We then discussed the effect of the change of the parameter γ on the performance of the TSCA. In addition to the previous experience value of 0.845, five different values of the parameter (0.95, 0.85, 0.75, 0.65 and 0.55) were selected. Figure 7 shows the result after using the above values for the proposed algorithm. Simulations were carried out under uniform and nonuniform traffic distributions. The results related to the various rejection probabilities fluctuated in the range of one thousandth and the fluctuation was also irregular. Therefore, we concluded that the proposed algorithm was insensitive to a parameter value of gamma.

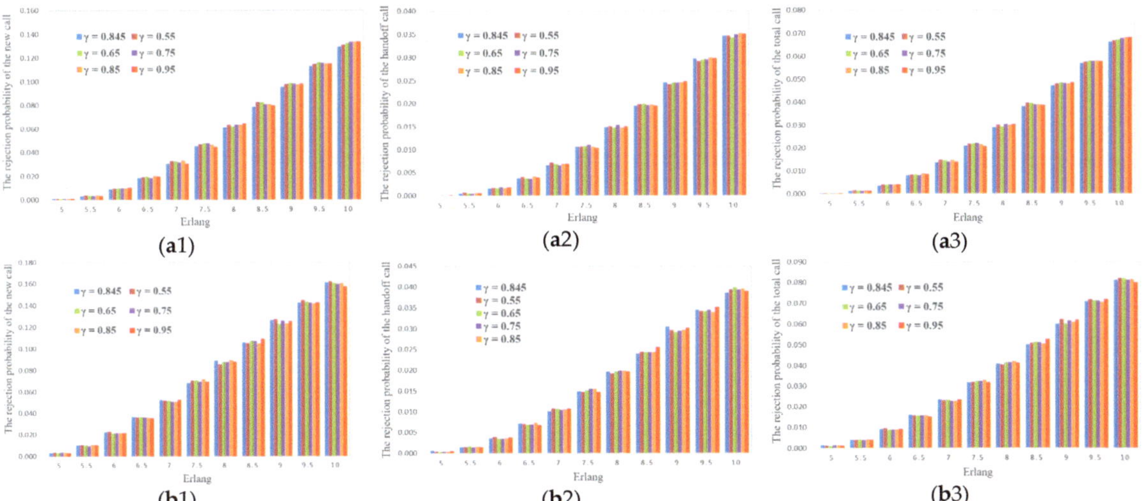

Figure 7. Performance comparison of TSCA with different values of gamma. (**a**1) The rejection probability of the new call under uniform distribution; (**a**2) the rejection probability of the handoff call under uniform distribution; (**a**3) the rejection probability of the total call under uniform distribution; (**b**1) the rejection probability of the new call under nonuniform distribution; (**b**2) the rejection probability of the handoff call under nonuniform distribution; (**b**3) the rejection probability of the total call under nonuniform distribution.

5. Conclusions

Frequent handoffs bring a new challenge of designing a reasonable channel allocation algorithm for LEO satellite networks. EC provides sufficient computing power for LEO satellite networks and makes it possible for the application of reinforcement learning. An adaptive DCA algorithm based on a temporal–spatial correlation analysis was proposed in this paper. First, the temporal–spatial correlation of handoff calls was analyzed. Second, the DCA process was formally described as an MDP. Third, a policy for different call events was designed and the MDP model was solved based on SARSA. Finally, simulation experiments were carried out under different traffic distributions and different traffic intensities. The simulation results showed that a TSCA could greatly reduce the rejection probability of handoff calls and improve the total performance of LEO satellite networks. However, the storage space of the Q-values in the TSCA rapidly increased with an increase in the network scale. In the future, we will optimize the required storage space of the proposed algorithm.

Author Contributions: Conceptualization, J.W. and J.Z.; methodology, J.W. and J.Z.; investigation, J.W. and J.Z.; software, J.W. and C.H.; supervision, J.W., L.S. and C.H.; writing—original draft preparation, J.W. and J.Z.; writing—review and editing, J.W., J. Z. and C.H.; project administration, J.W., L. S. and J.Z. All authors have read and agreed to the published version of the manuscript.

Funding: This research was funded by the National Natural Science Foundation of China (grant numbers 62272242, 61902237, 61972210) and the Innovation Project for Postgraduates of Jiangsu Province (grant number KYCX17 0782).

Institutional Review Board Statement: Not applicable.

Informed Consent Statement: Not applicable.

Data Availability Statement: Not applicable.

Acknowledgments: The authors would like to thank the editor and the anonymous reviewer whose constructive comments will help to improve the presentation of this paper.

Conflicts of Interest: The authors declare no conflict of interest.

References

1. Niephaus, C.; Kretschmer, M.; Ghinea, G. QoS provisioning in converged satellite and terrestrial networks: A survey of the state-of-the-art. *IEEE Commun. Surv. Tuts.* **2016**, *18*, 2415–2441. [CrossRef]
2. Su, Y.T.; Liu, Y.Q.; Zhou, Y.Q.; Yuan, J.H.; Cao, H.; Shi, J.L. Broadband leo satellite communications: Architectures and key technologies. *IEEE Wirel. Commun.* **2019**, *26*, 55–61. [CrossRef]
3. Zhang, Z.J.; Zhang, W.Y.; Tseng, F.H. Satellite mobile edge computing: Improving QoS of high-speed satellite-terrestrial networks using edge computing techniques. *IEEE Netw.* **2019**, *33*, 70–76. [CrossRef]
4. Wang, Y.X.; Yang, J.; Guo, X.Y.; Qu, Z. Satellite edge computing for the internet of things in aerospace. *Sensors* **2019**, *19*, 3607. [CrossRef] [PubMed]
5. Wei, J.Y.; Han, J.R.; Cao, S.Z. Satellite IoT edge intelligent computing: A research on architecture. *Electronics* **2019**, *8*, 1247. [CrossRef]
6. Wang, F.; Jiang, D.D.; Qi, S.; Qiao, C.; Shi, L. A dynamic resource scheduling scheme in edge computing satellite networks. *Mobile Netw. Appl.* **2021**, *26*, 597–608. [CrossRef]
7. Wang, S.H.; Ding, S.T.; Chen, C. Edge computing enabled video segmentation for real-time traffic monitoring in internet of vehicles. *Patten Recogn.* **2022**, *121*, 108146.
8. Wu, Y.R.; Guo, H.F.; Chakraborty, C. Edge Computing Driven Low-Light Image Dynamic Enhancement for Object Detection. *IEEE Trans. Netw. Sci. Eng.* **2022**. [CrossRef]
9. Wu, Y.R.; Zhang, L.L.; Berretti, S. Medical Image Encryption by Content-aware DNA Computing for Secure Healthcare. *IEEE Trans. Industr. Inform.* **2022**, 1–9. [CrossRef]
10. Xu, F.M.; Yang, F.; Zhao, C.L.; Wu, S. Deep reinforcement learning based joint edge resource management in maritime network. *China Commun.* **2022**, *5*, 211–222. [CrossRef]
11. Zhou, J.; Ye, X.G.; Pan, Y.; Xiao, F.; Sun, L.J. Dynamic channel reservation scheme based on priorities in LEO satellite systems. *J. Syst. Eng. Electron.* **2015**, *26*, 1–9. [CrossRef]
12. Moscholios, L.D.; Vassilakis, V.G.; Sagias, N.C.; Logothetis, M.D. On channel sharing policies in LEO mobile satellite systems. *IEEE T. Aerosp. Electron. Syst.* **2018**, *54*, 1628–1640. [CrossRef]
13. Chen, S.Z.; Sun, S.H.; Kang, S.L. System integration of terrestrial mobile communication and satellite communication–the trends, challenges and key technologies in B5G and 6G. *China Commun.* **2020**, *17*, 156–171. [CrossRef]
14. Liu, J.J.; Shi, Y.P.; Fadlullah, Z.M.; Kato, N. Space-air-ground integrated network: A survey. *IEEE Commun. Surv. Tutor.* **2018**, *20*, 2714–2741. [CrossRef]
15. Mnih, V.; Kavukcuoglu, K.; Silver, D.; Rusu, A.A.; Veness, J.; Bellemare, M.G.; Graves, A.; Riedmiler, M. Human-level control through deep reinforcement learning. *Nature* **2015**, *518*, 529–533. [CrossRef]
16. Du, J.; Jiang, C.X.; Wang, J.; Ren, Y.; Yu, S.; Han, Z. Resource allocation in space multiaccess systems. *IEEE Trans. Aerosp. Electron. Syst.* **2017**, *53*, 598–618. [CrossRef]
17. He, D.J.; You, P.; Yong, S.W. Mobility management in LEO satellite communication networks. *Chines Space Sci. Technol.* **2016**, *36*, 1–14.
18. Kato, N.; Fadlullah, Z.M.; Tang, F.X.; Mao, B.M.; Tani, S.; Okamura, A.; Liu, J.J. Optimizing space-air-ground integrated networks by artificial intelligence. *IEEE Wirel. Commun.* **2019**, *26*, 140–147. [CrossRef]
19. Del Re, E.; Fantacci, R.; Giambene, G. Handover queuing strategies with dynamic and fixed channel allocation techniques in low earth orbit mobile satellite systems. *IEEE Trans. Commun.* **1999**, *47*, 89–102. [CrossRef]
20. Li, Y.T.; Wang, S.; Zhou, W.Y. A novel dynamic resource optimization method in LEO-MSS downlink with multi-service based on handover forecasting. In Proceedings of the 5th International Conference Computer and Communications (ICCC), Chengdu, China, 9 June 2019; pp. 809–814.
21. Liu, S.J. *The Research on Dynamic Resource Management Techniques for Satellite Communication System*; Beijing University of Posts and Telecommunications: Beijing, China, 2018.
22. Nie, J.H.; Haykin, S. A Q-learning-based dynamic channel assignment technique for mobile communication systems. *IEEE Trans. Veh. Technol.* **1999**, *48*, 1676–1687.

23. Hu, X.; Liu, S.J.; Chen, R.; Wang, W.D.; Wang, C.T. A deep reinforcement learning-based framework for dynamic resource allocation in multibeam satellite systems. *IEEE Commun. Lett.* **2018**, *22*, 1612–1615. [CrossRef]
24. Liu, S.J.; Hu, X.; Wang, W.D. Deep reinforcement learning based dynamic channel allocation algorithm in multibeam satellite systems. *IEEE Access* **2018**, *6*, 15733–15742. [CrossRef]
25. Zheng, F.; Pi, Z.; Zhou, Z.; Wang, K.Z. Leo satellite channel allocation scheme based on reinforcement learning. *Mob. Inf. Syst.* **2020**, *2020*, 8868888. [CrossRef]
26. Wang, J.; Sun, L.J.; Zhou, J.; Han, C. A dynamic channel reservation strategy based on priorities of multi-traffic and multi-user in LEO satellite networks. *J. Circuit. Syst. Comp.* **2020**, *29*, 2050082. [CrossRef]
27. Maral, G.; Restrepo, J.; Del Re, E. Performance analysis for a guaranteed handover service in an LEO constellation with a 'satellite-fixed cell' system. *IEEE Trans. Veh. Technol.* **1998**, *47*, 1200–1214. [CrossRef]
28. Del Re, E. Different queuing policies for handover requests in low earth orbit mobile satellite systems. *IEEE Trans. Veh. Technol.* **1999**, *48*, 448–458. [CrossRef]
29. Deng, B.Y.; Jiang, C.X.; Yao, H.P. The next generation heterogeneous satellite communication of resource management and deep reinforcement learning. *IEEE Wirel. Commun.* **2020**, *27*, 105–111. [CrossRef]
30. Shi, G.C.; Wu, Y.R.; Liu, J. Incremental Few-Shot Semantic Segmentation via Embedding Adaptive-Update and Hyper-class Representation. *arXiv* **2022**. [CrossRef]
31. Zhou, J.; Han, T.T.; Xiao, F. Multi-scale network traffic prediction method based on deep echo state network for internet of things. *IEEE Internet Things J.* **2022**, *9*, 21862–21874. [CrossRef]
32. Luong, N.C.; Hoang, D.T.; Gong, S.M.; Niyato, D.; Wang, P.; Liang, Y.C.; Kim, D. Applications of deep reinforcement learning in communications and networking: A survey. *IEEE Commun. Surv. Tutor.* **2019**, *21*, 3133–3174. [CrossRef]
33. Puterman, M.L. *Markov Decision Processes: Discrete Stochastic Dynamic Programming*; John Wiley & Sons: New York, USA, 2014.
34. Alfakin, T.; Hassan, M.M.; Gumae, A.; Savaglio, C.; Fortino, G. Task offloading and resource allocation for mobile edge computing by deep reinforcement learning based on SARSA. *IEEE Access* **2020**, *8*, 54074–54084. [CrossRef]
35. Liu, C.F.; Bennis, M.; Debbah, M.; Vincent, P. Dynamic task offloading and resource allocation for ultra-reliable low-latency edge computing. *IEEE Trans. Commun.* **2019**, *67*, 4132–4150. [CrossRef]
36. Lilith, N.; Dogancay, K. Reduced-state SARSA featuring extended channel reassignment for dynamic channel allocation in mobile cellular networks. *LNCS* **2005**, *3421*, 531–542.
37. Torstein, S. *Contributions to Centralized Dynamic Channel Allocation Reinforcement Learning Agents*; Norwegian University of Science and Technology: Trondheim, Norway, 2018.
38. Zou, Q.Y.; Zhu, L.D. Dynamic channel allocation strategy of satellite communication systems based on grey prediction. In Proceedings of the International Symposium on Networks, Computers and Communications (ISNCC), Istanbul, Turkey, 18–20 June 2019; pp. 1–5.
39. Lima, M.A.; Araujo, A.F.; Cesar, A.C. Adaptive genetic algorithms for dynamic channel assignment in mobile cellular communication systems. *IEEE Trans. Veh. Technol.* **2007**, *56*, 2685–2696. [CrossRef]
40. Lilith, N.; Dogancay, K. Dynamic channel allocation for mobile cellular traffic using reduced-state reinforcement learning. In Proceedings of the Wireless Communications & Networking Conference (WCNC), Atlanta, GA, USA, 21–25 March 2004; pp. 2195–2200.
41. Wang, Z.P.; Mathiopoulos, P.T.; Schober, R. Performance analysis and improvement methods for channel resource management strategies of leo-mss with multiparty traffic. *IEEE Trans. Veh. Technol.* **2008**, *57*, 3832–3842. [CrossRef]

Article

Attention-Based Active Learning Framework for Segmentation of Breast Cancer in Mammograms

Xianjun Fu [1], Hao Cao [2], Hexuan Hu [2], Bobo Lian [1], Yansong Wang [2], Qian Huang [2] and Yirui Wu [2,3,*]

1 School of Artificial Intelligence, Zhejiang College of Security Technology, Wenzhou 325016, China
2 College of Computer and Information, Hohai University, Nanjing 210093, China
3 Key Laboratory of Symbolic Computation and Knowledge Engineering of Ministry of Education, Jilin University, Changchun 130015, China
* Correspondence: wuyirui@hhu.edu.cn

Abstract: Breast cancer is one of most serious malignant tumors that affect women's health. To carry out the early screening of breast cancer, mammography provides breast cancer images for doctors' efficient diagnosis. However, breast cancer lumps can vary in size and shape, bringing difficulties for the accurate recognition of both humans and machines. Moreover, the annotation of such images requires expert medical knowledge, which increases the cost of collecting datasets to boost the performance of deep learning methods. To alleviate these problems, we propose an attention-based active learning framework for breast cancer segmentation in mammograms; the framework consists of a basic breast cancer segmentation model, an attention-based sampling scheme and an active learning strategy for labelling. The basic segmentation model performs multi-scale feature fusion and enhancement on the basis of UNet, thus improving the distinguishing representation capability of the extracted features for further segmentation. Afterwards, the proposed attention-based sampling scheme assigns different weights for unlabeled breast cancer images by evaluating their uncertainty with the basic segmentation model. Finally, the active learning strategy selects unlabeled images with the highest weights for manual labeling, thus boosting the performance of the basic segmentation model via retraining with new labeled samples. Testing on four datasets, experimental results show that the proposed framework could greatly improve segmentation accuracy by about 15% compared with an existing method, while largely decreasing the cost of data annotation.

Keywords: breast cancer; image segmentation; active learning; deep learning

Citation: Fu, X.; Cao, H.; Hu, H.; Lian, B.; Wang, Y.; Huang, Q.; Wu, Y. Attention-Based Active Learning Framework for Segmentation of Breast Cancer in Mammograms. *Appl. Sci.* **2023**, *13*, 852. https://doi.org/10.3390/app13020852

Academic Editor: Shaohua Wan

Received: 5 December 2022
Revised: 30 December 2022
Accepted: 6 January 2023
Published: 7 January 2023

1. Introduction

Breast cancer is one of the most serious malignant tumors that threatens the health of women. It is reported that about 12.5 percent of women are affected by breast cancer worldwide [1]. In China, the incidence of breast cancer is increasing by 0.5% per year [2], which has become one of the most dangerous and deadly diseases for women's health. Early-stage breast cancer screening can aid in detecting disease early, thus largely increasing the chances of recovery [3]. These screenings are conducted by experienced doctors, who check the existence of malignant lesions in mammograms for further diagnosis and evaluation. However, such operations are not only annoying and time-consuming, but also carry the risk of wrong or missed detections with manual examinations [4], which is reported to be as high as 30% in breast cancer screening [5]. Since the advent of artificial intelligence (AI), researchers have adopted computer-aided detection (CAD) technologies to considerably reduce the amount of work involved in breast cancer screening. In the early stages of this application, CAD systems with traditional image processing algorithms were applied to aid in breast cancer diagnosis (masses, microcalcification etc.) [6]. With the rapid development of deep learning, the accuracy of medical image segmentation, regarded as a subfield of image segmentation, has been significantly improved. For example, Wang et al. [7] proposed a breast tumor semantic segmentation method based on a convolutional neural network (CNN), which provided accurate results with the extracted biomarkers in radiology imaging. Later, Hann et al. [8] utilized a multiple UNet network to generate multiple segmentations for fusion, which were further thresholded to generate the final segmentation result.

However, these impressive achievements of deep learning in breast cancer screening were built on a large amount of labeled data, and accuracy would greatly drop without it [9]. Due to privacy issues, it is difficult to obtain sufficient breast cancer images for labeling. Moreover, labeling generally requires medical experts with professional knowledge and experience, making it time-consuming and costly to acquire enough labeled data. Achieving desirable segmentation results with fewer samples has thus recently become a research focus.

Benefiting from its ability to achieve high performance with few labeled samples, active learning has become a feasible approach for less-thoroughly labeled breast cancer data [10]. Several successful applications have been implemented to show the remarkable power of active learning in medical image analysis. For example, Vishwesh et al. [11] combined an active learning framework with deep learning for medical image segmentation, where new active learning strategies guided the segmentation model to learn the diversity of uncertain and unlabeled data, thus greatly achieving convergence in accuracy with less labelling. Later, Li et al. [12] proposed a novel active learning framework for histopathology image analysis, where two groups of unlabeled data were selected in each training iteration, one annotated by experts and the other selected from high-confidence unlabeled samples to assign pseudo-labels. Both manual labeling and pseudo-label generation Were able to largely alleviate the problem of scarce labeled samples.

Based on the advantages of using deep learning and active learning for automatic screening tasks, we propose an attention-based active breast cancer segmentation model which is capable of achieving desirable segmentation results without a high quantity of labeled images. The proposed model consists of a basic segmentation model, an attention-based sampling scheme and an active learning based labeling strategy. Specifically, a multi-scale fusion and enhancement module based on UNet is first adopted for segmentation. Afterwards, a novel attention mechanism is used to evaluate the similarity between the unlabeled and segmented samples, thus offering weights as criteria to measure the uncertainty or informativeness of unlabeled samples with respect to the trained and basic segmentation model. Finally, an active learning strategy is used to sort unlabeled samples with weights, thus determining selections that need to be further manually labelled. We observed that these selected samples often contained appearance or shape features which were unacknowledged by the basic segmentation model subjected to the current training dataset. With iterations of retraining with the most informative unlabeled samples, the proposed model stably approached the upper bound of segmentation performance for high accuracy.

The contributions of this paper are summarized as follows:

- We propose an attention-based active breast cancer segmentation framework which effectively improved the accuracy of segmentation with few training samples, thus alleviating the high cost of labeling breast cancer images.
- A novel attention-based sampling scheme is proposed which measures the most informative unlabeled samples via calculating similarity weights.
- We adopt an active learning strategy for the global optimization of accuracy performance, which iteratively selects appropriate unlabeled samples to first manually label and then retrain the model to boost performance.

The rest of this paper is organized as follows. Section 2 reviews related work on the segmentation of breast cancer images. Section 3 presents an overview of the proposed method. Details of the basic segmentation model, attention-based sampling scheme and active-learning-based labeling strategy are also discussed in Section 3. Section 4 presents and discusses the experimental results. Finally, Section 5 concludes the paper.

2. Related Work

In this section, we give a brief literature review, including prior works on traditional segmentation methods for breast cancer images, deep-learning-based segmentation methods for breast cancer images, as well as active learning methods.

2.1. Traditional Segmentation Methods for Breast Cancer Images

Traditional segmentation methods usually apply image processing technologies for segmentation. However, they generally suffer from the drawbacks of low accuracy and are sensitive to the quality of sampled images.

For example, Cheng et al. [13] proposed a near-automatic ultrasound image segmentation algorithm which builds a solid foundation of computer-aided diagnosis for breast cancer. Later, Eziddin et al. [14] proposed the segmentation of mammograms using an iterative fusion process of information obtained from multiple knowledge sources, including context information, image processing algorithms, prior knowledge and so on. Gnonnou et al. [15] proposed a structural method to separate breast margins at pixel-level, thus accurately extracting tumor regions. Later, Kaushal et al. [16] proposed an automated segmentation technique followed by self-driven post-processing operations to detect cancerous cells effectively. Recently, Jing et al. [17] proposed a simple but effective segmentation method with the concept of global thresholding, which successfully segmented tumor regions in breast histopathology images. During the process, partial contrast stretching and median filtering are specially designed to improve image quality for segmentation.

2.2. Deep-Learning-Based Segmentation Methods for Breast Cancer Images

Inspired by the remarkable performance of deep learning methods in image classification and segmentation tasks [18,19], researchers have proposed several works on the segmentation of breast cancer images with various kinds of networks.

For example, Su et al. [20] proposed a fast scanning deep convolutional neural network (FCNN) to achieve pixel-wise region segmentation, successfully eliminating the redundant computation of the original CNN without sacrificing performance. Later, Simin et al. [21] proposed the combination of deep learning with traditional features for medical image classification, where a CNN model is first used to extract image features, and then support vector machines are used for feature learning and classification. Then, Roy et al. [22] used a dropout strategy to generate different Monte Carlo segmentations, where they computed the dissimilarity of these segmentations to measure the structural uncertainty of the image. In such way, they could confidently choose the best matching segmentations from candidates for output.

Despite their strengths, the shortage of sufficient training data affects the performance of deep learning methods, and researchers have focused on active learning to improve the effectiveness of deep learning. For example, Shen et al. [23] proposed a novel deep active learning model for the image segmentation of breast cancer on immunohistochemistry images. They not only achieved significant performance improvements in the segmentation of breast cancer images, but the system also showed promise for implementation as a real-world application.

Recently, difference comparisons between multiple candidate segmentation maps has become an effective method for sampling in active learning. For example, Wang et al. [24] believe that easy samples tend to obtain similar segmentations in K models, where they use K different models to segment images and measure the similarity of outputs, thus building connections between different models for further comparisons. Recently, Zhang et al. [25] generated two segmentation maps before and after processing of their proposed attention module, thus calculating the similarity coefficient of maps to guide sampling of their active learning framework.

2.3. Active Learning Methods

To reduce the cost of labeling, active learning selects the most valuable and informative samples from unlabeled samples for the labeling task, which relieves the dependence of deep learning models on large training datasets. Due to its ideal function in achieving desirable performance with few labelled samples, researchers in the medical image analysis community have proposed several methods for CAD.

For example, Ayerdi et al. [26] proposed an interactive image segmentation system using active learning, which allows rapid segmentation without the requirement of manual intervention. Later, Sharma et al. [27] adopted active learning to perform biomedical image segmentation with limited labeled data, where they combined UNet and an active learning query strategy to select additional samples for annotation, thus capturing the most uncertain and representative samples. Then, Li et al. [28] proposed a deep active learning framework which combines an attention-gated fully convolutional network (ag-FCN) and a distributional difference-based active learning algorithm (dd-AL) to iteratively annotate samples. Later, Lai et al. [29] proposed a semi-supervised active learning framework with region-based selection criteria which iteratively selects regions for annotation queries to rapidly expand the diversity and number of marker sets.

Most recently, Gaillochet et al. [30] proposed a test-time augmentation method for active learning in medical image segmentation, which exploits the power of uncertain information provided by data transformation. Bai et al. [31] proposed a difference-based active learning (DEAL) method for bleed segmentation, which successfully bridged the gap between class activation maps (CAMs) and ground truth with few annotations.

3. Method

To deal with the high labeling cost of breast cancer images, we propose an attention-based deep active learning framework for segmentation in mammograms. First, the overall structure is given, offering a global view of how the proposed framework works. Then, we present the basic breast cancer segmentation model with multi-scale feature fusion and enhancement. Afterwards, we describe an attention-based sampling scheme to assign weights for unlabeled samples under uncertainty. Finally, we describe an active-learning-based labelling strategy to choose the unlabeled samples for manually labeling and retraining, thus reducing the cost of manually labeling a large quantity of unlabeled samples.

3.1. Overall Structure

The existing segmentation models based on deep learning generally require a large number of labeled images for training at substantial cost. Thus, it is crucial to achieve as high an accuracy as possible for segmentation with few labeled samples. To achieve this goal, we propose the overall framework as shown in Figure 1, which consists of three steps, i.e., the basic segmentation model, attention-based sampling and active-learning-based labelling.

During Step A, a labeled set of breast cancer images are first used to train the basic segmentation model, which obtains distinguished feature maps via multi-scale feature fusion and enhancement. In Step B, uncertainty sampling is first adopted to classify good and bad segmentation results for breast cancer in mammograms. Then, a novel attention mechanism is built to calculate weights for samples of the unlabeled breast cancer set. During Step C, unlabeled samples with higher weights, implying that they are more informative for learning, are selected to be manually labeled by professional medical experts. All these samples can then be further used to retrain the breast cancer segmentation model, thus boosting performance in an iterative manner.

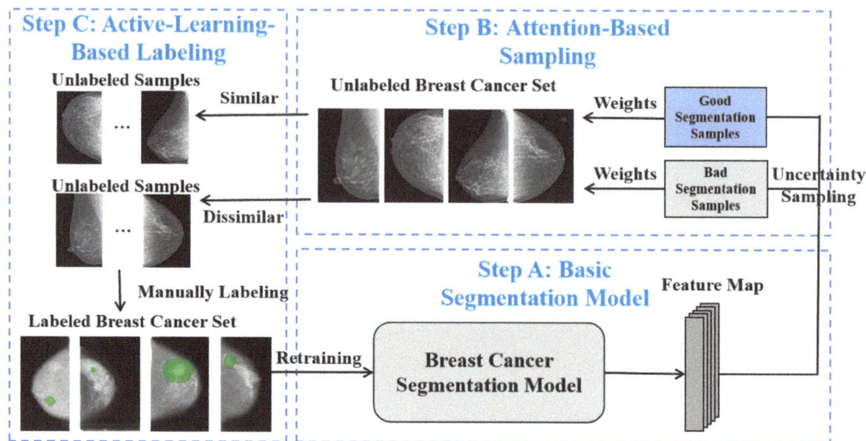

Figure 1. Framework of the proposed framework, which consists of (**A**) Basic Segmentation Model, (**B**) Attention-based Sampling, and (**C**) Active-Learning-based Labeling.

On the basis of structure design, we propose a loss function to involve the traditional IoU loss $Loss_I$ and binary cross-entropy loss $Loss_B$, i.e., $Loss = Loss_I + Loss_B$. Specifically, IoU loss $Loss_I$ can be calculated via

$$Loss_I = 1 - \frac{\sum_{i \in I} p_i \widehat{p}_i}{\sum_{i \in I} p_i + \widehat{p}_i - p_i \widehat{p}_i}, \tag{1}$$

where I refers to the input breast cancer image, i represents each pixel in the image, and $\widehat{p}_i$ and p_i represent the predicted and true labeled values for each pixel, respectively. Moreover, the binary cross-entropy loss $Loss_B$ can be calculated via

$$Loss_B = -\sum_{i \in I} p_i \log(\widehat{p}_i) + (1 - p_i) \log(1 - \widehat{p}_i), \tag{2}$$

Note that $Loss_I$ and $Loss_B$ are calculated in each iteration stage to achieve convergence of the training process.

3.2. Basic Segmentation Model for Breast Cancer

Although skip connections in UNet could avoid the loss of detailed information caused by continuous down-sampling, they cannot capture multi-scale information with strong restrictions on locality. To obtain multi-scale feature maps, we cascade the feature maps at multiple layers with different receptive fields, where skip connections are used across different layers.

On the basis of the UNet segmentation model, the proposed basic model further involves the strength of the promotion feature module (PFM) [32] for the features of each output layer, which fuses multi-scale feature maps to enhance their representation capability. Note that the PFM works as a feature fusion and enhanced block in our former work, which not only fuses the features from multiple scales, but also selectively forgets useless information and enhances informative information, thus constructing more effective feature representation.

Specifically, an input image I in the labeled breast cancer set is sent to the proposed basic segmentation model for feature extraction:

$$F^i = Seg_t^i(I), \tag{3}$$

where F_i refers to feature maps corresponding to the ith output layer of the UNet model, and the function $Seg^t()$ refers to the segmentation model during the tth iteration of retraining.

Then, feature maps of multiple layers are sent to different PFMs for fusion and enhancement:

$$F_E^i = G_e^i(G_f^i(F^i)), where\ i = 1, 2, ..., n \tag{4}$$

where F_E^i represents the output feature map after processing of the ith PFM, functions G_e and G_f represent enhance and forget operations in PFM, and n is the total number of layers in the segmentation model.

In the later fusion step, the original feature map F^i corresponding to the i-th layer is fused with its enhanced version F_E^i via skip connections:

$$F^{i-1} = F_E^i \oplus X_{UP}(F^i), where\ i = 1, 2, ..., n \tag{5}$$

where F^{i-1} refers to the output feature map of the $i - 1$th layer, and the function $X_{UP}()$ is an up-sampling operation.

Finally, the generated feature map F_P^1 combines the high-layer semantic features with shallow features, thus enhancing representation capability for segmentation via multi-scale feature extraction fusion. We set $n = 5$ for all experiments in this paper.

3.3. Attention-Based Sampling Scheme

To obtain more segmentation related knowledge with as few labeled samples as possible, it is essential to obtain more distinguished feature representation from the training stage. We thus judge the informativeness for one specific sample based on its segmentation result with the following equation:

$$Info_i = \begin{cases} 1, if\ U_i \geq \alpha \\ 0, otherwise \end{cases} \tag{6}$$

where $Info_i$ implies whether the ith sample is useful for learning knowledge or not, α is a pre-set parameter based on segmentation performance of experiments, and U_i calculates similarity coefficient with the following equation:

$$U_i = \frac{S_{i,q} \cap S_{i,g}}{S_{i,q} \cup S_{i,g}}, \tag{7}$$

where S_q and S_g respectively represent the predicted and ground-truth feature maps of the ith breast cancer image, which is used to represent the uncertainty and guides the selection of unlabeled breast cancer images. We consider the segmentation result of the i-th breast cancer image as good only if $Info_i = 1$; otherwise, it is considered bad.

With such criteria for judging informativeness, we choose samples from an unlabeled breast cancer set, which are either dissimilar to good segmentation samples or similar to bad segmentation samples, thus greatly improving the learning capability of the segmentation model for the features of difficult samples. Essentially, a soft attention model is generally formed as a dimension of interpretability into internal representations by selectively focusing on specific information. The core procedure of soft attention model [33] can calculate weights based on similarity between an input signal and pre-trained weights. Therefore, we propose a novel attention mechanism which assigns weights based on similarity calculations between unlabeled and labeled samples. In other words, the proposed attention mechanism assigns smaller weights if the unlabeled samples are more similar to the good segmentation samples. On the contrary, it would give higher weights to unlabeled samples with greater similarity to bad samples.

Defining the input unlabeled breast cancer image as query Q and the set of labeled segmentation samples as W, a multi-layer perceptron(MLP) is utilized to calculate the similarity or correlation between Q and one of the pre-trained samples W_i as $sim(Q, W_i) = MLP(Q, W_i)$.

Afterwards, we adopt the Softmax function to perform normalization on the calculated similarity and emphasize the informative parts based on their inherent ability:

$$\alpha_i = softmax(sim(Q, W_i)) = \frac{e^{sim(Q,W_i)}}{\sum_{j=1}^{L} e^{sim(Q,W_i)}};$$

(8)

where L refers to the number of samples in the labeled segmentation results.

3.4. Active Labeling Strategy

Essentially, we believe that unlabeled samples with larger attention weight could contribute to the classification capability of the segmentation model, thus boosting the segmentation performance of the model by using these samples for retraining. Therefore, we propose a labeling strategy for breast cancer samples based on an active learning method, as described in Algorithm 1.

Specifically, we sort the unlabeled samples by weights and choose the unlabeled samples with higher weights for manual labeling, represented as Step 8 in the algorithm. The specific processes of sample calculation, selection and manual labeling and retraining are shown in Figure 2. We use both similarity and dissimilarity weights to select unlabeled images. Then, these samples are first roughly labeled by automatic labeling software and then manually adjusted by experts. Afterwards, they are added as labeled samples into the set of labeled breast cancer images. Meanwhile, these samples are deleted from the unlabeled image set, which can be represented as

$$\begin{cases} U_{t+1} = U_t - I_t \\ L_{t+1} = L_t + I_t \end{cases}$$

(9)

where t refers to the iteration time of training, and $L_t \cap U_t = \varnothing$ to ensure the consistent processing of different iterations.

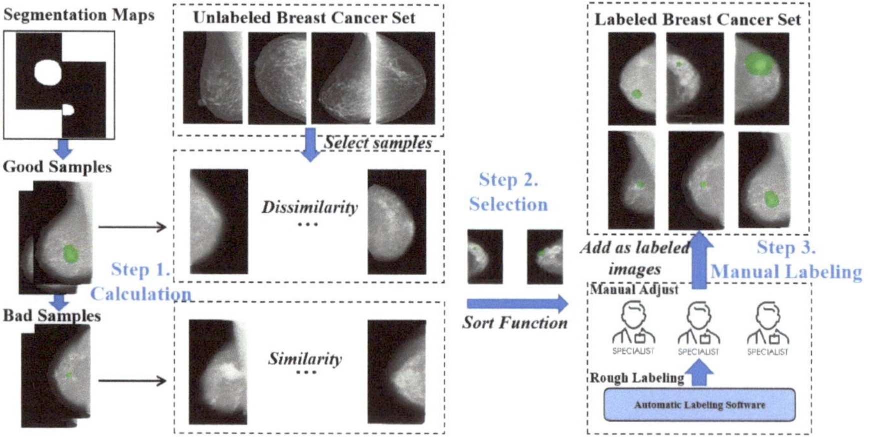

Figure 2. Key steps, i.e., Calculation, Selection, and Manual Labeling, in the proposed active-learning-based labeling strategy.

Finally, the newly constructed labeled set L_{t+1} is used to retrain the basic segmentation model. With all these steps, an active learning iteration process is completed, where the performance of the breast cancer segmentation model could be improved step by step.

Algorithm 1 Labeling strategy of breast cancer samples based on active learning.

Require: Unlabeled sample set U, labeled sample set L
Ensure: Labeled dataset L_t
  1: **While:**
  2: **if** $L \neq \varnothing$ **then**
  3:     Train the breast cancer segmentation model
  4:     Use the updated segmentation model to infer the labeled breast cancer images in the labeled sample set L, and output the feature maps F_P
  5:     Calculate the similarity of the i-th image
  6:     Calculate the attention weights
  7: **end if**
  8: Sort the breast cancer images in the unlabeled sample set according to attention weights

  9: Manually label the selected images which are assigned with higher weights
10: Update U_{t+1} and L_{t+1}
11: Retrain the segmentation model for breast cancer images
12: **EndWhile**

---

4. Experimental Results

This section first introduces the datasets and measurements. Then, it describes the ablation experiments constructed to verify the effectiveness of the proposed modules. Afterwards, we demonstrate the performance of the segmentation models on four datasets to verify the effectiveness of the proposed framework. We also qualitatively compare the proposed active learning framework with existing methods. Finally, we offer the implementation details for readers' convenience.

4.1. Datasets and Measurements

We collected breast cancer images from a cooperating hospital, which are not released due to privacy reasons. The dataset includes 1462 labeled breast cancer images, with resolution 6781×3676 pixels . Since images were acquired by different scanners, we divided all our samples into four parts based on the type of scanner, i.e., Breast-A, Breast-B, Breast-C and Breast-D.

Various measurements were used to verify the effectiveness of the breast cancer image segmentation results, i.e., the mean Dice similarity coefficient (mDice), the mean intersection ratio (mIoU), and the mean absolute error (MAE). Where TP, FP and FN indicate true positive, false positive and false negative samples, mDice can be calculated as:

$$mDice = \frac{2TP}{FP + 2TP + FN}. \tag{10}$$

Note that a higher mDice implies a greater similarity between two samples.

IoU is defined as the area of the intersection divided by the area of the union of the predicted bounding box, which can be evaluated by

$$IoU = \frac{\mathrm{area}\left(B_p \cap B_{gt}\right)}{\mathrm{area}\left(B_p \cup B_{gt}\right)}, \tag{11}$$

where B_p and B_{gt} are the predicted and ground-truth segmentation results, respectively.

By calculating the Euclidean distance between the predicted and the ground-truth results, MAE can be defined as:

$$MAE = \frac{1}{n} \sum_{i=1}^{n} |Y_i - \hat{Y}_i|, \tag{12}$$

where n refers to the total number of samples, and Y_i and $\hat{Y}_i$ are the predicted and ground-truth labels, respectively. Note that a lower MAE implies a better segmentation result.

4.2. Ablation Experiments

To evaluate the effectiveness of the attention-based sampling mechanism and the active labeling strategy, we designed several ablation experiments as shown in Table 1. Note that PFMs represent multiple promotion feature modules used in the basic segmentation model, Att refers to the proposed attention-based sampling scheme, and Act is the proposed active labeling strategy.

After adding PFMs, Att and Act, the segmentation performance on breast cancer images gradually improved on our four breast cancer image datasets, proving the effectiveness of these three modules. Specifically, we found that UNet+PFMs could achieve more precise boundaries of polyp regions and performed accurate segmentation when compared with the basic network (i.e., UNet). However, the shallow usage of boundary information without multi-scale refinement for boundary regions leads to uncompact performance towards larger and more regularized-shape polyp regions. In contrast, due to the usage of PFMs to extract a more distinguishing feature map by fusing multi-scale information, boundaries achieved by the proposed method were much more obvious with clear contour lines, thus providing better segmentation performance.

The ablation experiment on Att proved that the attention-based sampling scheme improved segmentation performance on all datasets. The attention-based design helped in effectively selecting more valuable samples for the further manual labeling process. It is beneficial to focus on the most informative unlabeled samples, which brings the feature information required by the current model for performance improvement with the fewest updating iterations.

Act, representing the active labeling strategy, enlarges size of the labeled breast cancer set, thereby improving the effect of tumor segmentation. Due to the guidance of optimized selection on unlabeled samples, we observed that informative samples for the current trained model were added to the labeled samples set for further retraining.

Table 1. Ablation experiments with different network structure designs on Breast-A, Breast-B, Breast-C and Breast-D datasets, where PFMs, Att and Act represent multiple promotion feature modules used in the basic segmentation model, the proposed attention-based sampling scheme and the proposed active labeling strategy, respectively.

Dataset	Method	mDice	IoU	MAE
Breast-A	UNet	0.356	0.266	0.018
	UNet+PFMs	0.412	0.332	0.013
	UNet+PFMs+Att	0.431	0.361	0.013
	UNet+PFMs+Att+Act	**0.462**	**0.394**	**0.009**
Breast-B	UNet	0.403	0.29	0.013
	UNet+PFMs	0.492	0.371	0.009
	UNet+PFMs+Att	0.515	0.382	0.112
	UNet+PFMs+Att+Act	**0.543**	**0.401**	**0.006**
Breast-C	UNet	0.553	0.429	0.033
	UNet+PFMs	0.677	0.512	0.026
	UNet+PFMs+Att	0.693	0.518	0.029
	UNet+PFMs+Att+Act	**0.725**	**0.533**	**0.023**
Breast-D	UNet	0.369	0.261	0.040
	UNet+PFMs	0.451	0.382	0.311
	UNet+PFMs+Att	0.478	0.396	0.335
	UNet+PFMs+Att+Act	**0.512**	**0.422**	**0.027**

4.3. Comparative Experiments

In this subsection, we describe our comparative experiments and present heatmap visualizations, segmentation results and the effectiveness of active learning.

Figure 3 shows the generated heatmap of breast cancer segmentation achieved by the proposed framework, the comparative method and the ground truth on four datasets. We used UNet in comparisons of heatmap visualization and segmentation results. The last layer of the network can generate a heatmap for each input breast cancer image, which can be used to generate segmentation results. Comparison of heatmaps shows that the proposed framework could more accurately identify the breast cancer region, and thus obtained a better performance in the segmentation task. Even in the case of a blurred image boundary, the heatmap implies that further segmentation results would maintain high accuracy by focusing on the dominant parts of cancer regions.

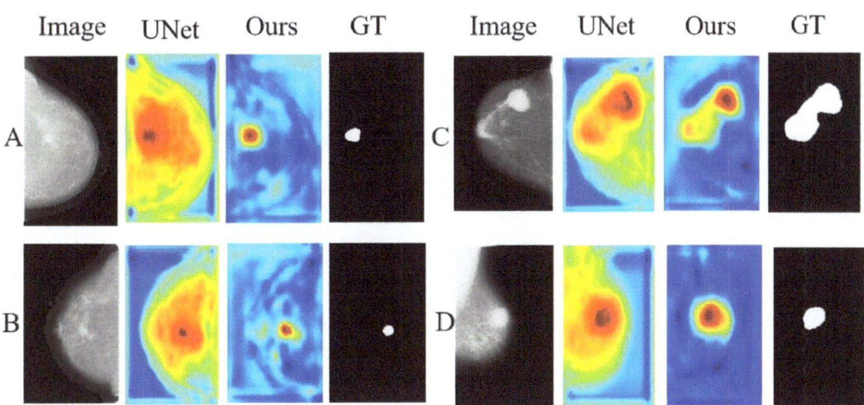

Figure 3. Heatmaps of breast cancer segmentation achieved by the proposed framework, the comparative method and the ground truth results. (**A–D**) refer to Breast-A, Breast-B, Breast-C and Breast-D datasets, respectively.

Figure 4 shows the qualitative comparison results of the breast cancer segmentation. Compared with the comparative segmentation model, i.e., UNet, the proposed framework achieved better segmentation results that were similar to the ground-truth results. Moreover, the proposed active learning strategy could refine the distinguished feature information of breast cancer using unlabeled samples, achieving more accurate pixel-level classification results. In addition, the proposed framework generates and refines the boundary region through an effective iterative update strategy, thus achieving global optimization progressively.

To verify that the active learning strategy can effectively reduce the cost of labeling, we conducted comparative experiments by selecting random sampling and CoreSet [34] as the comparison methods. Figure 5 shows plots of mean Dice for each iteration during the active learning. It is worth noting that the proposed active learning strategy not only had a higher mean Dice value, but also converged in fewer iterations. This proves that the proposed strategy selected more informative samples in each iteration, thus reducing the cost of labeling samples. Without measuring the uncertainty of sample labeling, other comparison methods might suffer from unstable convergence with increasing iterations because they adopt samples without helpful information, or even containing noisy information. Although all methods converged eventually, the compared methods tended to have lower mean Dice values due to the influence of noisy samples.

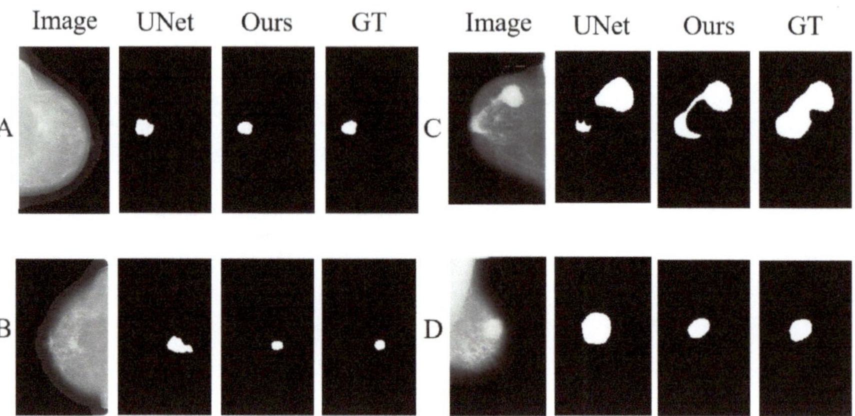

Figure 4. Qualitative comparisons between the segmentation results achieved by the proposed framework, the comparative method, and the ground truth. (**A–D**) refer to Breast-A, Breast-B, Breast-C and Breast-D datasets, respectively.

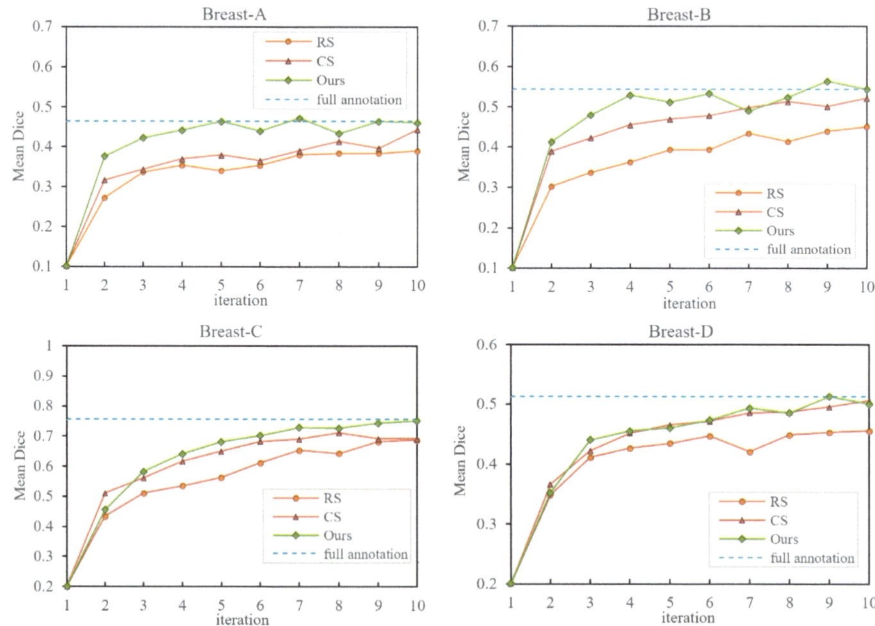

Figure 5. Plots of mean Dice coefficients, comparing different sampling strategies. It's noted that RS and CS refer to random selecting sampling and CoreSet sampling [34], respectively.

4.4. Implementation Details

Due to the scarcity of training samples, we used various data-enhancement methods to expand the training samples so that they met the requirements of model training. The size of input breast images was first adjusted to 352×352 pixels in the training and inference process. Then, we used image flipping to expand the number of training samples, in both horizontal and vertical directions. Finally, we not only randomly adjusted the contrast, brightness and sharpness of breast cancer images, but also randomly dilated and eroded image labels. All these operations were designed for data enhancement.

All experiments were carried out under the Linux Ubuntu operating system with a single Titan V GPU. We use Adaptive Moment Estimation (Adam) as the optimizer, while the initial learning rate was set to 0.0001 and the learning rate was adjusted using learning rate decay. In the active learning strategy experiment , we used 100 unlabeled breast cancer images for initialization, and set the number of iterations to 10 for training of the basic segmentation model. Afterwards, we added the selected unlabeled samples to the labeled dataset, and trained 25 epochs for retraining in each iteration.

5. Conclusions

Due to the high cost of labeling training samples, herein we propose an attention-based active learning framework for the segmentation of breast cancer in mammograms. Specifically, we propose an attention sampling scheme to assign weights for unlabeled samples by evaluating their uncertainty. We also propose an active labeling strategy to select valuable unlabeled samples for manual labeling, thus enlarging the scale of the training set and improving the performance of the segmentation model. Testing on four datasets, experimental results showed that the proposed framework could greatly improve segmentation accuracy. The active learning scheme and attention strategy we adopted can be easily applied to other models and effectively reduce the data size required for model training.

In the future, we will try to introduce semi-supervised learning and unsupervised learning in active learning to further improve the generalization ability of the segmentation model on different datasets. Moreover, we will design specific algorithms to solve problems in breast cancer segmentation such as microcalcification and architectural distortion, thus improving the segmentation accuracy.

Author Contributions: Conceptualization, X.F. and Y.W. (Yirui Wu); methodology, X.F., H.H. and H.C.; software, X.F. and H.C.; validation, B.L., Y.W. (Yirui Wu) and Q.H.; formal analysis, X.F.; investigation, Y.W. (Yansong Wang); resources, X.F.; data curation, X.F. and Y.W. (Yirui Wu); writing—original draft preparation, X.F., H.C. and Y.W. (Yirui Wu); writing—review and editing, X.F., H.C. and Y.W. (Yirui Wu);visualization, B.L., Y.W. (Yansong Wang) and Q.H.; supervision, Y.W. (Yirui Wu) and Q.H.; project administration, Y.W. (Yirui Wu) and Q.H.; funding acquisition, X.F. and Y.W. (Yirui Wu). All authors have read and agreed to the published version of the manuscript.

Funding: This work was supported in part by a grant from General Scientific Research Project of Zhejiang Education Department (Y202147224), National Key R&D Program of China under Grant No. 2021YFB3900601, the Fundamental Research Funds for the Central Universities under Grant B220202074, the Fundamental Research Funds for the Central Universities, JLU, Joint Foundation of the Ministry of Education (No. 8091B022123).

Informed Consent Statement: Informed consent was obtained from all subjects involved in the study.

Data Availability Statement: Data available on request due to restrictions eg privacy or ethical.

Conflicts of Interest: The authors declare no conflict of interest.

References

1. Elmoufidi, A. Deep Multiple Instance Learning for Automatic Breast Cancer Assessment Using Digital Mammography. *IEEE Trans. Instrum. Meas.* **2022**, *71*, 1–13. [CrossRef]
2. Loizidou, K.; Skouroumouni, G.; Nikolaou, C.; Pitris, C. Automatic Breast Mass Segmentation and Classification Using Subtraction of Temporally Sequential Digital Mammograms. *IEEE J. Transl. Eng. Health Med.* **2022**, *10*, 1–11. [CrossRef] [PubMed]
3. Seely, J.; Alhassan, T. Screening for breast cancer in 2018—What should we be doing today? *Curr. Oncol.* **2018**, *25*, 115–124. [CrossRef] [PubMed]
4. Huang, Q.; Miao, Z.; Zhou, S.; Chang, C.; Li, X. Dense Prediction and Local Fusion of Superpixels: A Framework for Breast Anatomy Segmentation in Ultrasound Image With Scarce Data. *IEEE Trans. Instrum. Meas.* **2021**, *70*, 1–8. [CrossRef]
5. Kim, H.E.; Kim, H.H.; Han, B.K.; Kim, K.H.; Han, K.; Nam, H.; Lee, E.H.; Kim, E.K. Changes in cancer detection and false-positive recall in mammography using artificial intelligence: A retrospective, multireader study. *Lancet Digit. Health* **2020**, *2*, e138–e148. [CrossRef]
6. Chen, C.; Wang, Y.; Niu, J.; Liu, X.; Li, Q.; Gong, X. Domain Knowledge Powered Deep Learning for Breast Cancer Diagnosis Based on Contrast-Enhanced Ultrasound Videos. *IEEE Trans. Med. Imaging* **2021**, *40*, 2439–2451. [CrossRef] [PubMed]

7. Wang, Y.; Jin, Z.; Tokuda, Y.; Naoi, Y.; Tomiyama, N.; Suzuki, K. Development of Deep-learning Segmentation for Breast Cancer in MR Images based on Neural Network Convolution. In Proceedings of the 2019 8th International Conference on Computing and Pattern Recognition, Beijing, China, 23–25 October 2019; pp. 187–191.

8. Hann, E.; Biasiolli, L.; Zhang, Q.; Popescu, I.A.; Werys, K.; Lukaschuk, E.; Carapella, V.; Paiva, J.M.; Aung, N.; Rayner, J.J.; et al. Quality control-driven image segmentation towards reliable automatic image analysis in large-scale cardiovascular magnetic resonance aortic cine imaging. In *International Conference on Medical Image Computing and Computer-Assisted Intervention*; Springer: Cham, Switzerland, 2019; pp. 750–758.

9. Chen, J.; Jiao, J.; He, S.; Han, G.; Qin, J. Few-Shot Breast Cancer Metastases Classification via Unsupervised Cell Ranking. *IEEE ACM Trans. Comput. Biol. Bioinform.* **2021**, *18*, 1914–1923. [CrossRef] [PubMed]

10. Belharbi, S.; Ayed, I.B.; McCaffrey, L.; Granger, E. Deep Active Learning for Joint Classification & Segmentation with Weak Annotator. In Proceedings of the IEEE/CVF Winter Conference on Applications of Computer Vision (WACV), Waikoloa, HI, USA, 3–8 January 2021; pp. 3337–3346.

11. Nath, V.; Yang, D.; Landman, B.A.; Xu, D.; Roth, H.R. Diminishing Uncertainty within the Training Pool: Active Learning for Medical Image Segmentation. *IEEE Trans. Med. Imaging* **2021**, *40*, 2534–2547. [CrossRef] [PubMed]

12. Li, W.; Li, J.; Wang, Z.; Polson, J.; Sisk, A.E.; Sajed, D.P.; Speier, W.; Arnold, C.W. PathAL: An Active Learning Framework for Histopathology Image Analysis. *IEEE Trans. Med. Imaging* **2022**, *41*, 1176–1187. [CrossRef] [PubMed]

13. Cheng, J.Z.; Chen, K.W.; Chou, Y.H.; Chen, C.M. Cell-based image partition and edge grouping: A nearly automatic ultrasound image segmentation algorithm for breast cancer computer aided diagnosis. In Proceedings of the Medical Imaging 2008: Computer-Aided Diagnosis, San Diego, CA, USA, 19–21 February 2008; Volume 6915, pp. 743–754.

14. Eziddin, W.; Montagner, J.; Solaiman, B. An iterative possibilistic image segmentation system: Application to breast cancer detection. In Proceedings of the 2010 13th International Conference on Information Fusion, Edinburgh, UK, 26–29 July 2010; pp. 1–8.

15. Gnonnou, C.; Smaoui, N. Segmentation and 3D reconstruction of MRI images for breast cancer detection. In Proceedings of the International Image Processing, Applications and Systems Conference, Sfax, Tunisia, 5–7 November 2014; pp. 1–6.

16. Kaushal, C.; Singla, A. Automated segmentation technique with self-driven post-processing for histopathological breast cancer images. *CAAI Trans. Intell. Technol.* **2020**, *5*, 294–300. [CrossRef]

17. Jing, T.Y.; Mustafa, N.; Yazid, H.; Rahman, K.S.A. Segmentation of Tumour Regions for Tubule Formation Assessment on Breast Cancer Histopathology Images. In *Proceedings of the 11th International Conference on Robotics, Vision, Signal Processing and Power Applications*; Springer: Berlin/Heidelberg, Germany, 2022; pp. 170–176.

18. Shi, G.; Wu, Y.; Liu, J.; Wan, S.; Wang, W.; Lu, T. Incremental Few-Shot Semantic Segmentation via Embedding Adaptive-Update and Hyper-class Representation. In Proceedings of the ACM International Conference on Multimedia, Tokyo, Japan, 13–16 December 2022; pp. 5547–5556.

19. Wu, Y.; Guo, H.; Chakraborty, C.; Khosravi, M.; Berretti, S.; Wan, S. Edge Computing Driven Low-Light Image Dynamic Enhancement for Object Detection. *IEEE Trans. Netw. Sci. Eng.* **2022**. [CrossRef]

20. Su, H.; Liu, F.; Xie, Y.; Xing, F.; Meyyappan, S.; Yang, L. Region segmentation in histopathological breast cancer images using deep convolutional neural network. In Proceedings of the 2015 IEEE 12th International Symposium on Biomedical Imaging (ISBI), Brooklyn, NY, USA, 16–19 April 2015; pp. 55–58.

21. He, S.; Ruan, J.; Long, Y.; Wang, J.; Wu, C.; Ye, G.; Zhou, J.; Yue, J.; Zhang, Y. Combining deep learning with traditional features for classification and segmentation of pathological images of breast cancer. In Proceedings of the 2018 11th International Symposium on Computational Intelligence and Design (ISCID), Hangzhou, China, 8–9 December 2018; Volume 1, pp. 3–6.

22. Roy, A.G.; Conjeti, S.; Navab, N.; Wachinger, C. Inherent brain segmentation quality control from fully convnet monte carlo sampling. In Proceedings of the International Conference on Medical Image Computing and Computer-Assisted Intervention, Granada, Spain, 16–20 September 2018; Springer: Berlin/Heidelberg, Germany, 2018; pp. 664–672.

23. Shen, H.; Tian, K.; Dong, P.; Zhang, J.; Yan, K.; Che, S.; Yao, J.; Luo, P.; Han, X. Deep active learning for breast cancer segmentation on immunohistochemistry images. In Proceedings of the International Conference on Medical Image Computing and Computer-Assisted Intervention, Lima, Peru, 4–8 October 2020; Springer: Berlin/Heidelberg, Germany, 2020; pp. 509–518.

24. Wang, J.; Chen, Z.; Wang, L.; Zhou, Q. An Active Learning with Two-step Query for Medical Image Segmentation. In Proceedings of the 2019 International Conference on Medical Imaging Physics and Engineering (ICMIPE), Shenzhen, China, 22–24 November 2019; pp. 1–5.

25. Zhang, Z.; Li, J.; Tian, C.; Zhong, Z.; Jiao, Z.; Gao, X. Quality-driven deep active learning method for 3D brain MRI segmentation. *Neurocomputing* **2021**, *446*, 106–117. [CrossRef]

26. Ayerdi, B.; Graña, M. Random forest active learning for retinal image segmentation. In Proceedings of the 9th International Conference on Computer Recognition Systems CORES, Wroclaw, Poland, 25–27 May 2015; Springer: Berlin/Heidelberg, Germany, 2016, pp. 213–221.

27. Sharma, D.; Shanis, Z.; Reddy, C.K.; Gerber, S.; Enquobahrie, A. Active learning technique for multimodal brain tumor segmentation using limited labeled images. In *Domain Adaptation and Representation Transfer and Medical Image Learning with Less Labels and Imperfect Data*; Springer: Berlin/Heidelberg, Germany, 2019; pp. 148–156.

28. Li, H.; Yin, Z. Attention, suggestion and annotation: A deep active learning framework for biomedical image segmentation. In Proceedings of the International Conference on Medical Image Computing and Computer-Assisted Intervention, Lima, Peru, 4–8 October 2020; Springer: Berlin/Heidelberg, Germany, 2020; pp. 3–13.
29. Lai, Z.; Wang, C.; Oliveira, L.C.; Dugger, B.N.; Cheung, S.C.; Chuah, C.N. Joint Semi-supervised and Active Learning for Segmentation of Gigapixel Pathology Images with Cost-Effective Labeling. In Proceedings of the IEEE/CVF International Conference on Computer Vision, Montreal, BC, Canada, 11–17 October 2021; pp. 591–600.
30. Gaillochet, M.; Desrosiers, C.; Lombaert, H. TAAL: Test-Time Augmentation for Active Learning in Medical Image Segmentation. In Proceedings of the MICCAI Workshop on Data Augmentation, Labelling, and Imperfections, Singapore, 22 September 2022; Springer: Berlin/Heidelberg, Germany, 2022; pp. 43–53.
31. Bai, F.; Xing, X.; Shen, Y.; Ma, H.; Meng, M.Q.H. Discrepancy-Based Active Learning for Weakly Supervised Bleeding Segmentation in Wireless Capsule Endoscopy Images. In Proceedings of the International Conference on Medical Image Computing and Computer-Assisted Intervention, Singapore, 18–22 September 2022; Springer: Berlin/Heidelberg, Germany, 2022; pp. 24–34.
32. Wu, B.; Wu, Y.; Wan, S. An Image Enhancement Method for Few-shot Classification. In Proceedings of the IEEE International Conference on Embedded and Ubiquitous Computing, Shenyang, China, 20–22 October 2021; pp. 1–7.
33. Vaswani, A.; Shazeer, N.; Parmar, N.; Uszkoreit, J.; Jones, L.; Gomez, A.N.; Kaiser, L.; Polosukhin, I. Attention is All you Need. In Proceedings of the NIPS, Long Beach, CA, USA, 4–9 December 2017; pp. 5998–6008.
34. Sener, O.; Savarese, S. Active Learning for Convolutional Neural Networks: A Core-Set Approach. In Proceedings of the International Conference on Learning Representations, Vancouver, BC, Canada, 30 April–3 May 2018.

Article

Rainfall Similarity Search Based on Deep Learning by Using Precipitation Images

Yufeng Yu *, Xingu He, Yuelong Zhu and Dingsheng Wan

College of Computer and Information, Hohai University, Nanjing 210098, China
* Correspondence: yfyu@hhu.edu.cn

Abstract: Precipitation images play an important role in meteorological forecasting and flood forecasting, but how to characterize precipitation images and conduct rainfall similarity analysis is challenging and meaningful work. This paper proposes a rainfall similarity research method based on deep learning by using precipitation images. The algorithm first extracts regional precipitation, precipitation distribution, and precipitation center of the precipitation images and defines the similarity measures, respectively. Additionally, an ensemble weighting method of Normalized Discounted Cumulative Gain-Improved Particle Swarm Optimization (NDCG-IPSO) is proposed to weigh and fuse the three extracted features as the similarity measure of the precipitation image. During the experiment on similarity search for daily precipitation images in the Jialing River basin, the NDCG@10 of the search results reached 0.964, surpassing other methods. This indicates that the method proposed in this paper can better characterize the spatiotemporal characteristics of the precipitation image, thereby discovering similar rainfall processes and providing new ideas for hydrological forecasting.

Keywords: precipitation image; feature extraction; similarity analysis; multivariate feature fusion; Improved Particle Swarm Optimization

1. Introduction

In recent years, flash floods caused by extreme rainfall have led to extensive social and economic losses [1]. Due to the influence of precipitation intensity, precipitation distribution, and other factors, there are many uncertainties in the time, location, grade, and process of floods, which pose great obstacles to early flood warning and prevention. Therefore, extracting spatiotemporal features of rainfall-runoff processes, identifying and classifying them, so as to discover similar rainfall-flood patterns from historical rainfall events to provide guidance and technical support for hydrological forecasting and water resource utilization, has become an urgent task in the application field of hydrology and water resources [2,3].

The rainfall-flood similarity analysis uses fuzzy mathematics, data mining, and machine learning methods to identify the similar (closest) sequence pairs to the current real-time rainfall-flood sequence from the historical hydrological time series patterns by defining the similarity measure [4]. The most direct application of rainfall-flood similarity analysis is to determine whether a current rainfall-flood process is similar or equivalent to a process in a historical period [5]. In this sense, research on similarity analysis methods has significant potential for rainfall-runoff process forecasting, environmental evolution analysis, and hydrological regularity discovery [6,7].

Rainfall similarity analysis is an important part of rainfall-flood similarity analysis and flood risk assessment [8]. It can not only discover the rules of similar rainfall-flood patterns in history but also provide new ideas and technical support for rainfall-flood forecasting. Zhang [9] established a similarity analysis model for precipitation stations using the K-means clustering algorithm based on the Davies–Bouldin index. Then the single precipitation type histogram similarity model was adapted to analyze the clustering

Citation: Yu, Y.; He, X.; Zhu, Y.; Wan, D. Rainfall Similarity Search Based on Deep Learning by Using Precipitation Images. *Appl. Sci.* **2023**, *13*, 4883. https://doi.org/10.3390/app13084883

Academic Editor: Yu-Dong Zhang

Received: 22 February 2023
Revised: 8 April 2023
Accepted: 10 April 2023
Published: 13 April 2023

results and obtain similar stations. Xiao [5] proposed a rainfall event similarity analysis model for rainfall forecasting, which evaluated the similarity between two rainfalls from multiple perspectives, such as the quantity similarity, pattern similarity, earth mover's distance, and rainfall spatial distribution similarity; the experimental results showed the similar rainfall analysis method is effective and applicable. Ohno [10] developed a new forecasting technique for predicting whether water levels will exceed a 'flood' threshold or not by using deep learning methods based on weather forecast precipitation images, which provided a new idea and reference for extending the flood forecast period.

Traditional similarity analysis of rainfall mainly uses time series text data, which cannot effectively represent the spatial distribution of rainfall. Moreover, the existing research on rainfall similarity lacks comprehensive measurement methods for multivariate characteristics of rainfall and is greatly insufficient in terms of interpretation. In the past decade, with the development of information technology, the hydrological departments have accumulated a large amount of spatiotemporal data, and data types have been expanded from traditional time series to semi-structured and unstructured. As is shown in Figure 1, different colors represent different precipitation grade and provide the spatiotemporal distribution information of precipitation in the Jialing River Basin. However, as time and spatial scales accumulate, it becomes more difficult to discover useful knowledge from these increasingly big data. Therefore, utilizing the latest machine learning and artificial intelligence algorithms to carry out feature extraction and fast similarity analysis on the accumulated big data of precipitation images to provide technical support for the identification of similar rainfall-flood processes and flood control is becoming a meaningful and hot research issue [11,12].

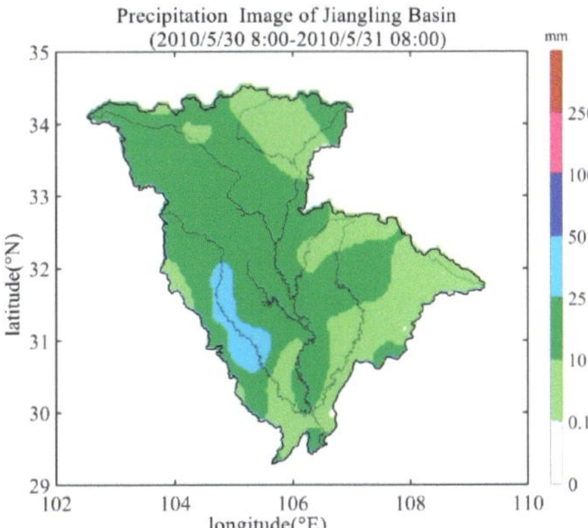

Figure 1. Example of precipitation image.

This paper proposes a rainfall similarity research method based on deep learning by using precipitation images. The novelty of this article lies in that the regional precipitation, precipitation distribution, and precipitation center are extracted as the characters of the precipitation image, and then appropriate distance measures for each feature are defined to better characterize the similarity between images. After that, an ensemble weighting method of normalized depreciation cumulative gain-improved particle swarm optimization (NDCG-IPSO) is proposed to weight and fuse the distance measures of three extracted features as the similarity measure for daily precipitation image similarity search.

The remaining part of this paper is organized as follows: Section 2 presents the related work to this area of research. Section 3 presents the brief of NDCG-IPSO. Several experiments with the proposed method using real-world precipitation images are reported in Section 4. Finally, Section 5 gives conclusions and suggestions for further research.

2. Related Studies

2.1. Image Feature Extraction

Feature extraction (FE) is an important and necessary step in many processes related to image retrieval [13], image encryption [14], and pattern recognition [15], which was used to extract the most distinct and useful information presented in an image dataset, to form a low-dimensional feature space to represent and describe the images for the next searching, browsing, or retrieving. Generally, color, shape, and texture are common characteristics extracted for image retrieval [16].

As shown in Figure 1, the precipitation image has similar shape and texture features. Therefore, only the color features are extracted to characterize the different precipitation images. In this paper, the global color histogram is used to extract the color features of the precipitation image, and the regional precipitation of the basin is calculated according to the practical significance of each color. Moreover, the image is divided into m*n grids, and the block color histogram is used to extract more detailed information, such as spatial distribution and rainfall center, for a better description of the precipitation image.

2.2. Similarity Search

Similarity search (also known as the nearest neighbor search) is the problem of searching the data items that are nearest to a query item under some distance measure from a search (reference) dataset, which is the foundation for data mining tasks such as clustering and classification and has been applied to time series prediction in image retrieval. Generally, a similarity search relies on a distance measurement that quantifies how close two elements are in the feature space. The closer the elements are, the higher the similarity between them. Currently, there are many distance measurements, such as Minkowski distance [17], Dynamic Time Warping Distance (DTW) [18], and editing distance [19].

Minkowski distance is also known as Lp distance. If two time series $Q = \{q_1, q_2, \ldots, q_n\}$ and $C = \{c_1, c_2, \ldots, c_n\}$, then the Minkowski distance $D(Q, C)$ between Q and C is calculated as the following equation.

$$Lp = D(Q,C) = \left(\textstyle\sum_{i=1}^{n} |q_i - c_i|^p \right)^{\frac{1}{p}} \tag{1}$$

The value of p is a positive integer. Minkowski distance is defined differently according to different values of p. If $p = 1$, L_1 is called the Manhattan distance; if $p = 2$, L_2 is called the Euclidean distance, which is the most widely used measure in time series similarity research; If $p = \infty$, L_∞ is the Chebyshev distance.

2.3. Deep Learning

Deep learning refers to a class of machine learning techniques which attempts to mimic the human brain, which is organized with a deep architecture and processes information through multiple stages of transformation and representation [20]. Deep learning methods allow a system to learn complex functions that directly map raw sensory input data to the output without relying on human-crafted features using domain knowledge [21]. Over the past several years, a rich family of deep learning techniques has been proposed and extensively applied to a variety of applications, including speech recognition, object recognition, and natural language processing, among others [22,23].

Particle Swarm Optimization (PSO) is a stochastic swarm-based deep learning optimization algorithm inspired by simulating a simplified social system of a flock of birds that fly towards their unknown destination (fitness function) in search of the locations of food resources [24]. The PSO algorithm is initialized with random particles (birds) with a specific position and velocity for the purpose of computing the objective function of an optimization problem. The best personal and global fitness positions are computed over each iteration of running the PSO algorithm. The position and velocity of each bird are updated according to the calculated fitness functions until the optimal solution is obtained. Assuming that in D-dimensional space, the location of the particle and the flight speed is expressed as the vector $x_i = [x_{i1}, x_{i2}, \ldots, x_{iD}]$ and $v_i = [v_{i1}, v_{i2}, \ldots, v_{iD}]$, respectively. The position of the particle is updated as follows:

$$x_i^{t+1} = x_i^t + v_i^{t+1} \tag{2}$$

The speed of the ith particle is updated as follows from iteration number to iteration number +1:

$$v_i^{t+1} = wv_i^t + c_1 r_1 \left(p_{ibest}^t - x_i^t\right) + c_2 r_2 \left(p_{gbest}^t - x_i^t\right) \tag{3}$$

where w is the inertia weight, which represents the retention degree of the particle to the last velocity; c_1 is the individual learning factor, which represents the learning ability of the particle to the individual optimal solution. c_2 is the social learning factor, which represents the learning ability of the particle to the current optimal solution of the population. p_{ibest}^t is the individual optimal solution found by p_i at time t; p_{gbest}^t is the population optimal solution found by the whole population particle at time t. In addition, the algorithm also has a parameter to represent the population size, that is, the number of particles, and a parameter to represent the maximum number of iterations of the algorithm.

3. Rainfall Similarity Search Based on NDCG-IPSO

The performance of the image search system crucially depends on the feature representation and similarity measurement. Therefore, the process of precipitation image similarity search mainly consists of two steps. Firstly, three features, namely the regional precipitation, the precipitation distribution, and the precipitation center, are extracted from the historical precipitation images and stored in the historical database. Then, the above three features are extracted from the precipitation image to be queried, and the images with higher similarity values are retrieved from the historical images as the similarity query results. The process flow of precipitation image matching is shown in Figure 2.

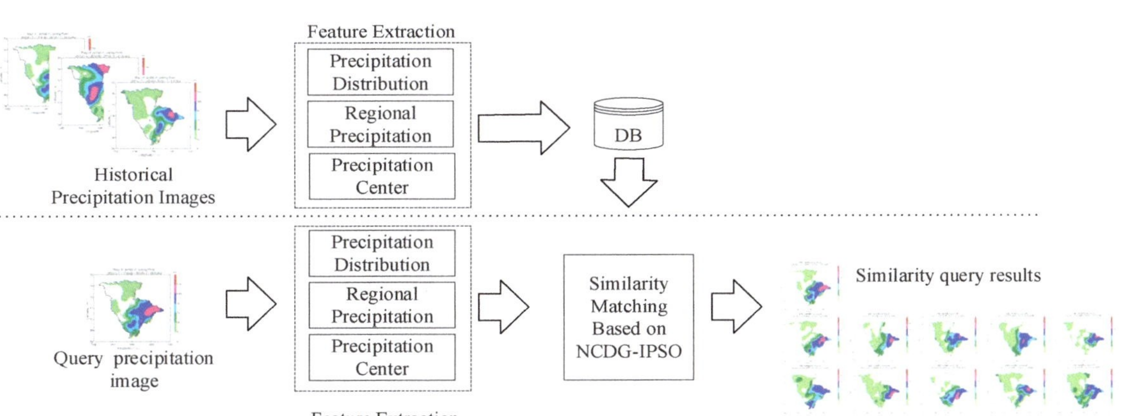

Figure 2. The process flow of precipitation image searching.

3.1. Feature Extraction

3.1.1. Regional Precipitation

Regional precipitation usually represents the total amount of precipitation within a given area at a specific time. In the precipitation image, each color corresponds to a range of precipitation amount. Hence, the regional precipitation feature can be obtained by weighting the color histogram, which can record the frequency of each color in the precipitation image.

Let the color histogram corresponding to the precipitation image contain K colors, namely $C_1, C_2, \ldots, C_K$, the occurrence number of each color in the image is $num(C_i)$ $(1 \le i \le K)$, and the precipitation amount corresponding to each color is $pm(C_i)$, then the regional precipitation within a given area can be calculated as follows:

$$P = \sum_{i=1}^{K} num(C_i) pm(C_i) \tag{4}$$

Moreover, let P_1 and P_2 be the regional precipitation feature of the two precipitation images. The defined Manhattan distance D_p to measure the similarity between the regional precipitation features of the two images can be calculated as follows:

$$D_P = |P_1 - P_2| \tag{5}$$

3.1.2. Precipitation Distribution

Regional precipitation can roughly represent the total amount of regional precipitation within a given area, but it is difficult to reflect the spatial distribution characteristics of precipitation. Therefore, the block-based color histogram is used to divide the watershed image into rectangular blocks with m rows and n columns after truncating redundant annotation. That is, the precipitation image is divided into $m*n$ small grids. Additionally, the regional precipitation for each small grid can be calculated according to formula (4), respectively. Let $P_{(i,j)}$ be the regional precipitation for the grid located at the ith row and the jth column. The precipitation distribution matrix, denoted as R, is defined to characterize the spatial distribution feature for the precipitation image, which can be calculated as follows:

$$R = \begin{bmatrix} P_{(1,1)}, P_{(1,2)}, \ldots, P_{(1,n)} \\ P_{(2,1)}, P_{(2,2)}, \ldots, P_{(2,n)} \\ \cdots \\ P_{(m,1)}, P_{(m,2)}, \ldots, P_{(m,n)} \end{bmatrix} \tag{6}$$

Figure 3 shows the precipitation distribution matrix after blocking operation for the precipitation image.

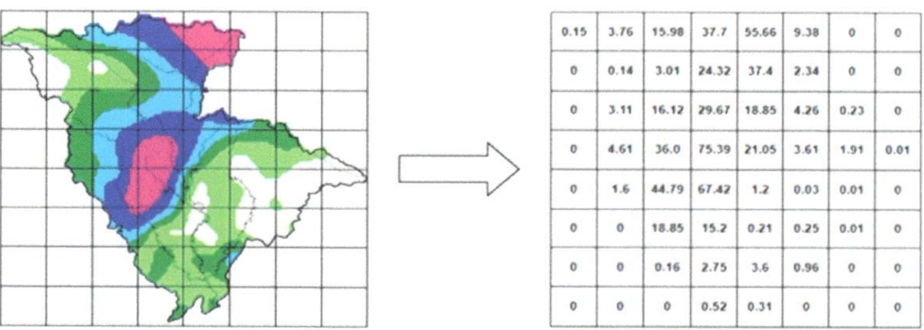

0.15	3.76	15.98	37.7	55.66	9.38	0	0
0	0.14	3.01	24.32	37.4	2.34	0	0
0	3.11	16.12	29.67	18.85	4.26	0.23	0
0	4.61	36.0	75.39	21.05	3.61	1.91	0.01
0	1.6	44.79	67.42	1.2	0.03	0.01	0
0	0	18.85	15.2	0.21	0.25	0.01	0
0	0	0.16	2.75	3.6	0.96	0	0
0	0	0	0.52	0.31	0	0	0

Figure 3. The precipitation distribution matrix.

Let $R_{(a,b)}$ be one of the blocks around $R_{(i,j)}$ in image B, where $a \in \{i-1, i, i+1\}, b \in \{j-1, j, j+1\}$. Let $D_{AB(i,j)}$ be the distance between elements in the ith row and the jth column of precipitation distribution matrix R_A and R_B, which can be calculated using the method shown in Figure 4. If the distance between $R_{A(i,j)}$ and $R_{B(i,j)}$, denoted as D_1, is smaller than that between $R_{A(i,j)}$ and $R_{B(a,b)}$ ($a \in \{i-1, i, i+1\}, b \in \{j-1, j, j+1\}$), denoted as D_2, $D_{AB(i,j)}$ can be represented by D_1, else $D_{AB(i,j)}$ is represented by the mean of D_1 and D_2.

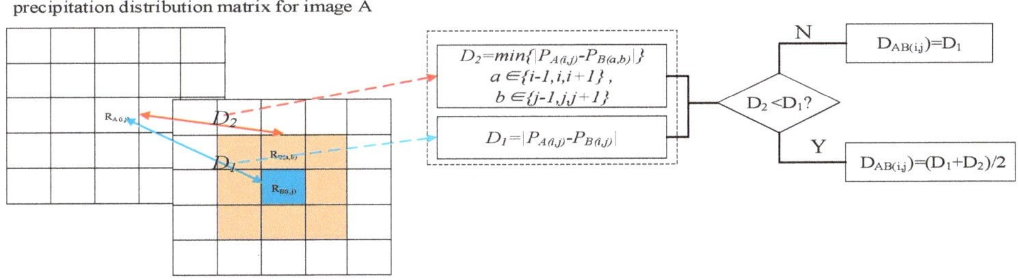

Figure 4. The distance between $R_{A(i,j)}$ and $R_{B(i,j)}$ of two precipitation distribution matrixes.

Thus, the distance of the precipitation distributions of two precipitation images, defined as D_R, can be calculated as follows:

$$D_R = \sum_{i=1}^{m} \sum_{j=1}^{n} D_{AB(i,j)} \tag{7}$$

3.1.3. Precipitation Center

Flood processes are largely influenced by the precipitation center. When the precipitation center is located upstream of the watershed, the long distance to the watershed section leads to a long lag time for the flood peak and presents a short and plump flood process. Meanwhile, when the precipitation center is located downstream, the short distance to the watershed section will make it a short lag time for the flood peak and form a sharp and thin flood hydrograph. Hence, it is important to take the precipitation center as a major feature in the precipitation image similarity research. Combined with the actual precipitation situation and image similarity retrieval requirements, take the block with the maximum precipitation in the precipitation image after being divided into blocks as the precipitation center.

Let $P_{(i,j)}$ be the maximum precipitation of the blocks in the precipitation image, then the precipitation center $C_{(i,j)}$ could be the block at the ith row and the jth column. Let $C_{(i_1,j_1)}$ and $C_{(i_2,j_2)}$ be the precipitation centers of two precipitation images. The Euclidean distance D_c is defined as the difference between the two precipitation centers, which can be calculated as follows:

$$D_C = \sqrt{(i_1 - i_2)^2 + (j_1 - j_2)^2} \tag{8}$$

3.2. Image Similarity Search Based on NDCG-IPSO

The precipitation image has different features. Hence, it may lose other feature information and reduce the accuracy of similarity retrieval if it applies a single feature to represent and measure the similarity of the precipitation image. Therefore, how to fuse the distances of the multiple features to represent the comprehensive distance between two precipitation images is becoming a necessary and urgent task. Given two precipitation images A and B, let D_P, D_R, and D_C represent the regional precipitation distance, precipitation distribution distance, and precipitation center distance between two images, respectively, the fusion distance D between two images A and B can be calculated as follows:

$$D = \gamma_1 D_P + \gamma_2 D_R + \gamma_3 D_C \tag{9}$$

where γ_1, γ_2, and γ_3 are undetermined coefficient weights for the distance of regional precipitation, precipitation distribution, and precipitation center.

There are three kinds of methods namely the subjective weight method, objective weight method, and subjective-objective comprehensive weight method, to determine the undetermined coefficient. Subjective weighting relies on expert's experiential knowledge, leading to subjectivity and variability. Objective weighting depends on the problem domain and sample data, but its results are poorly interpretable with low persuasiveness. Integration weighting methods can combine subjective and objective features, compensating for the shortcomings of both approaches [25].

The NDCG-IPSO is a new subjective–objective comprehensive weight method proposed to improve the efficiency of precipitation image similarity searches, which uses IPSO to adjust the weight of multiple indicators to make the evaluation results close to the evaluation results by experts based on subjective experience, and then applies the NDCG as indicators to evaluate the image search results weighted by multiple features. The process of the NDCG-IPSO is shown in Figure 5. The method combines the advantages of the objective weighting method and the subjective weighting method and makes the weighting result meet the requirements in the way of fitting and approximation.

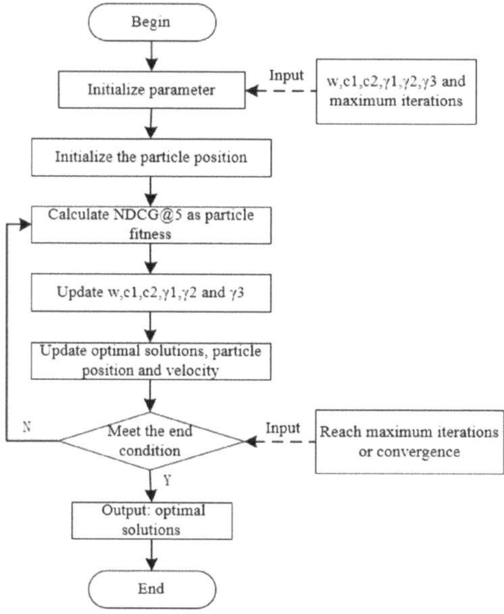

Figure 5. Flow chart of NDCG-IPSO.

3.2.1. Evaluation Metrics

NDCG is used as the metric to evaluate the image search results weighted by multiple features, representing the normalized value of the discounted cumulative gain [26]. Suppose a batch similarity search task is for E_1, E_2, ... , E_n. The search result for E_1 is e_{i1}, e_{i2}, ... , e_{ik}. Each e_{ij} in the results is another entity that the search system considers to be similar to the entity E_i, which has a real score of similarity degree with E_i. The cumulative gain of the K term before the search result of entity E_i is defined as $CG_i@K$, which can be calculated as follows:

$$CG_i@K = \sum_{j=1}^{K} rel_{ij} \qquad (10)$$

DCG discounts the gain of the lower-ranked items to have a significant influence on the gain for the top-ranked items in the search result list. The cumulative loss gain of K term before entity E_i search result $DCG@K$ is calculated as follows:

$$DCG_i@K = \sum_{j=1}^{K} \frac{rel_{ij}}{\log_2(j+1)} \qquad (11)$$

The normalized correlation coefficient is the *DCG@K* value of the ideal search result, denoted as *IDCG@K*, and the calculation formula of *NDCG@K* is as follows:

$$NDCG_i@K = \frac{DCG_i@K}{IDCG_i@K} \tag{12}$$

3.2.2. Parameter Optimization

The IPSO is proposed to change the inertia weight w, and learning factors c_1 and c_2 in the original PSO algorithm to adjust the coefficient weights γ_1, γ_2, and γ_3 in Formula (9) to find the optimal weights for precipitation image retrieval. IPSO changes the above three parameters adaptively with the increase in the number of iterations to avoid the algorithm falling into the partial optimal solution. The major improvements in IPSO include:

- Inertia weight w;

Parameter w represents the degree of retention of the particle to the speed of the last iteration, which can adjust the global and local search capabilities of the algorithm. The larger w takes at the early stage of the iteration, the stronger the particle global search ability is. The smaller w takes at the later stage of the iteration, the stronger the particle local search ability is. Therefore, IPSO takes the strategy of decreasing the number of iterations k linearly to balance the global and local search capabilities of particles, where the relationship between w and the number of iterations k is represented as follows:

$$w(k) = w_{max} - (w_{max} - w_{min}) * \frac{k_{max} - k}{k_{max}} \tag{13}$$

where w_{max} is the initial maximum inertia weight, w_{min} is the minimum inertia weight when iterating to the maximum algebra, k_{max} is the maximum number of iterations, and $w(k)$ is the inertia weight value when iterating for k times.

- Learning ratio c_1 and c_2;

Parameters c_1 and c_2 represent the ability of a particle to learn from the individual and the group optimal solution, respectively, and usually take the same value between 0 and 4 based on experience. Moreover, if c_1 takes a large value and c_2 takes a small value in the early stages of the iteration it can enhance the global search ability of the particles. Meanwhile, if c_1 takes a small value and c_2 takes a large value in the later stages of the iteration, it can improve the local search ability of particles. Therefore, IPSO improves parameters c_1 and c_2 with the symmetric linear strategy [27] to optimize the learning ability of individual optimal solutions and group optimal solutions for particles. The improvements in c_1 and c_2 are presented as follows:

$$c_{1.begin} = c_{2.end} = c_{mid} + \Delta c \tag{14}$$

$$c_{2.begin} = c_{1.end} = c_{mid} - \Delta c \tag{15}$$

$$c_1(k) = c_{1.begin} - \frac{k}{k_{max}}\left(c_{1.begin} - c_{1.end}\right) = c_{mid} + \Delta c\left(1 - 2\frac{k}{k_{max}}\right) \tag{16}$$

$$c_2(k) = c_{2.begin} - \frac{k}{k_{max}}\left(c_{2.end} - c_{2.begin}\right) = c_{mid} - \Delta c\left(1 - 2\frac{k}{k_{max}}\right) \tag{17}$$

where $c_{1.begin}$ and $c_{1.end}$ represent the initial and termination values of c_1, so do $c_{2.begin}$ and $c_{2.end}$. Δc represents the maximum variable length of c_1 and c_2; c_{mid} is the middle value of c_1 and c_2.

4. Experiment and Result Analysis

4.1. Study Area and Data Preprocessing

The Jialing River is the largest branch of the Yangtze River, which is about 1120 km long and covers an area of approximately 160,000 square kilometers within a geographical range of 29°17′30″ N–34°28′11″ N and 102°35′36″ E–109°01′08″ E (Figure 6). It belongs to the humid monsoon climate region, with an average annual precipitation of 931 mm. In normal years, the floods caused by rainfall in the basin are mostly concentrated in July and August with the characteristics of rapid fluctuation, short duration, fast flow rate, and high flood peak, which pose a great threat to the life and property of the residents along the river.

Figure 6. Location of the Jialing Basin.

To verify the NDCG-IPSO method for image similarity search, daily precipitation images from Jialing Basin from 1 January 2010, to 12 December 2019, were used for training and validation. Therefore, the experiment chose 30 precipitation images with different rainfall grades (6 images of light rain, 9 images of moderate rain, and 15 images of heavy rain) as the query samples and 10 matching samples for each query sample from the historical precipitation image. Additionally, each matching sample was assigned a similarity score from 0 (totally dissimilar) to 2 (very similar) according to expert experience to measure how similar the query sample and the matched sample were. In the experiment, the query samples were divided into the training sample set and test sample set according to the ratio of 2:1.

4.2. Results Analysis

The NDCG-IPSO was used to conduct image similarity search experiments on the daily precipitation images of the Jialing Basin. γ_1, γ_2, and γ_3 were initialized randomly, and their sum was guaranteed to be 1. The inertia weight w was set to 0.9, which linearly decreases to 0.4 as the number of iterations increases according to the Formula (13). Individual learning factors c_1 and social learning factors c_2 were set at 2.5 and 1.25. The particle number and the iteration number of the IPSO were set to 30 and 80 to obtain the optimal parameters of multi-feature distances for the precipitation image, which was shown in Table 1.

Table 1. Optimal weights of the three feature distances for precipitation image.

Feature Distance	γ_1	γ_2	γ_3
coefficient	0.46	0.12	0.42

Therefore, we adopted $\gamma_1 = 0.46$, $\gamma_2 = 0.12$, and $\gamma_3 = 0.42$ to calculate the comprehensive distance between precipitation images in subsequent experiments. The experimental results using indicators *NDCG@5* and *NDCG@10* were shown in Table 2.

Table 2. Optimal weights of the three feature distances for precipitation image.

Images	Accuracy	NDCG@5	NDCG@10
Training sample		0.984	0.978
Test sample		0.978	0.964

As seen in Table 2, the NDCG-IPSO can obtain higher index values on both training and test samples. Particularly, the average accuracy of *NDCG@5* and *NDCG@10* of the method on the test samples were 0.978 and 0.964, respectively, which were very close to 1. The experimental results prove the following two points: On the one hand, the three features extracted in this paper can well represent the spatial and temporal characteristics of the precipitation image and meet the needs of the precipitation image analysis, which can be used as indicators for image similarity of daily precipitation images. On the other hand, the NDCG-IPSO has a good effect on fusing feature distances defined in this paper into the comprehensive distance and thus quickly retrieves similar images from precipitation images.

Figure 7 shows the similarity search results of the precipitation image based on the NDCG-IPSO. For the precipitation image to be queried in the first line shown in Figure 7, lines 2 and 3 display the top 10 images that are very similar in terms of regional precipitation, precipitation distribution, and precipitation center, which can prove the effectiveness of this method and provide technical support for the analysis of similar hydrological processes.

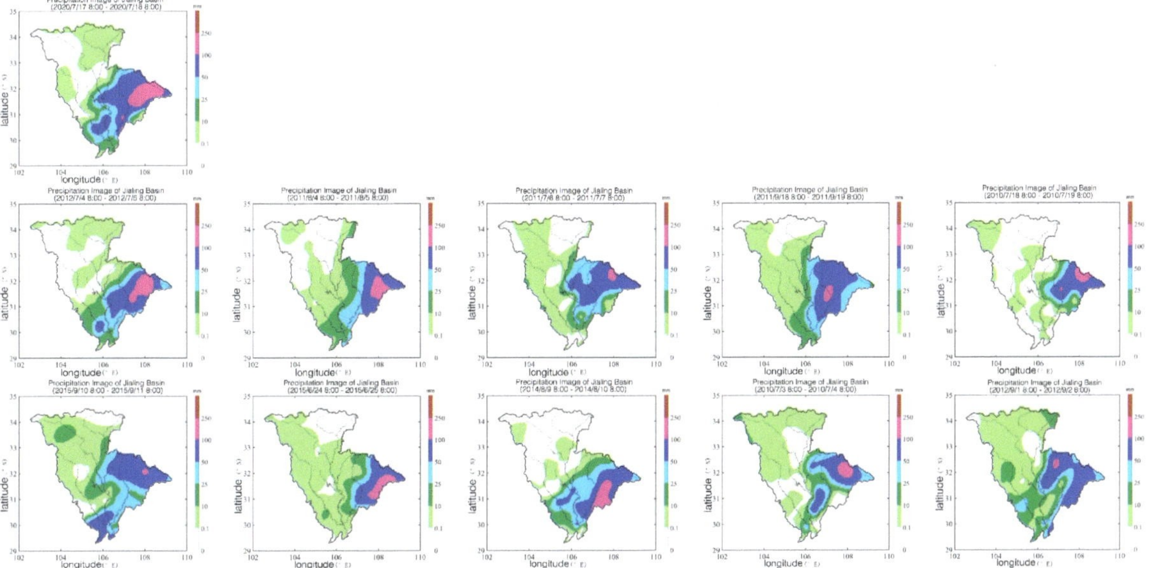

Figure 7. Image search results of NDCG-IPSO.

4.3. Comparative Analysis

To further verify the applicability and robustness of NDCG-IPSO, this experiment compared the image search results of our method with color histogram (CH) [28,29], BORDA [30], principal component analysis (PCA) [31], and the NDCG-PSO under the same conditions. Table 3 shows the image search results of the five methods on the same dataset.

Table 3. Comparative analysis of image search results based on different methods.

Method Accuracy	NDCG@5	NDCG@10
NDCG-IPSO	0.978	0.964
Global CH [28]	0.721	0.662
Block CH [29]	0.834	0.803
BORDA [30]	0.836	0.796
PCA [31]	0.702	0.623
NDCG-PSO	0.959	0.923

A color histogram is widely used in many image retrieval systems, which search similar images by extracting color histograms in images and calculating the distance between the histograms. Considering that precipitation images mainly adopt color features to represent different precipitation information, the color histogram is perhaps the most suitable method for precipitation image similarity searches. The global CH [28] and the block CH [29] with 3 × 3 blocks are used to search similar precipitation images. The results shown in Table 3 indicate that though global CH it can well characterize the color features of precipitation images, it ignores the spatial information of color features and results in low accuracy of similarity retrieval. The block CH considers part of the spatial information, and its searching accuracy is better than that of the global CH. However, block CH does not consider the physical meaning of the different colors on the precipitation image, which leads to worse searching accuracy than that of NDCG-IPSO.

The PCA and BORDA are two commonly used multi-index comprehensive evaluation methods, which are also widely used in the mining of multivariate hydrological similarity. The PCA conducts principal component analysis on all three feature distances and takes the feature with cumulative variance contribution rates greater than 85% as the principal component, and then weighs the features according to their variance contribution rates to obtain the search results after sorting. The BORDA sorts the feature distance once and synthesizes the similarity search results of those feature distances by BORDA to obtain the final query result. Although PCA, BORDA and NDCG-IPSO extract precipitation spatial distribution, precipitation center, and regional precipitation and comprehensively consider three distances to search similar images, NDCG-IPSO uses machine learning algorithms to optimize ensemble weighing method parameters and thus achieves better results than that of PCA and BORDA.

The only difference between NDCG-PSO and NDCG-IPSO is that the latter adopts the inertia weight and two learning factors in the PSO algorithm with an increase in the number of iterations. This adaptive adjustment improves the search performance and accuracy of the NDCG-IPSO. Figure 8 shows the fitness function value, namely *NDCG@5*, varying with the number of iterations in the NDCG-PSO and NDCG-IPSO. It can be seen from Figure 8 that NDCG-IPSO obtains the optimal particle fitness of 0.984 after 18 iterations; while NDCG-PSO gets stuck in a local optimum of 0.959 after 27 iterations. This indicates that IPSO can improve the search accuracy and optimization speed of image similar search and thus provides support for improving similar precipitation image retrieval.

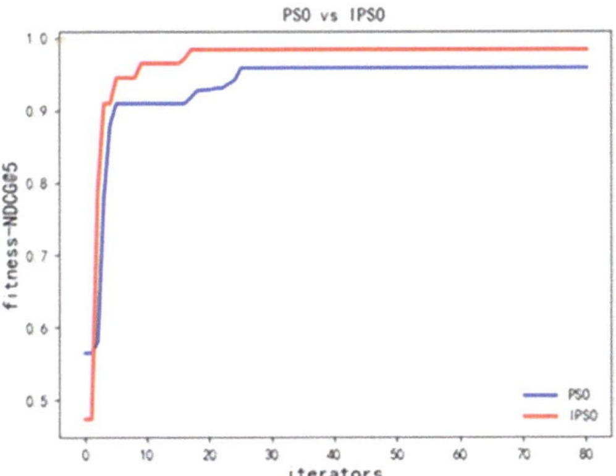

Figure 8. Comparison of experimental results between PSO algorithm and IPSO algorithm.

5. Conclusions

This paper proposes a rainfall similarity research method based on deep learning by using precipitation images. Firstly, the regional precipitation, precipitation distribution, and precipitation center are extracted from the precipitation images, and the similarity measurement for each feature is calculated separately.

Additionally, an ensemble weighting method of normalized depreciation cumulative gain-improved particle swarm optimization (NDCG-IPSO) is proposed to weigh and fuse the three extracted features as the similarity measure of the precipitation image. Finally, the comparing experiment of our method with CH, PCA, BORDA, and NDCG-PSO on the daily precipitation images in the Jialing River Basin illustrates that the methods proposed in this paper can better characterize the spatiotemporal characteristics of the precipitation image and discover similar rainfall processes, which will provide a new idea for hydrological forecasting.

Although some achievements have been made, many problems must still be solved. One problem is that it only considers the daily precipitation images similarity searching. However, a rainfall process may be composed of multiple single-day precipitation images.

Hence, how to conduct the rainfall process similarity search based on the similarity measurement method of NDCG-PSO and thus build a rainfall-flood similarity pattern repository to provide guidance and technical support for hydrological forecasting and water resource utilization is the direction of our future work.

Author Contributions: Conceptualization, Y.Y.; Methodology, Y.Y., X.H., Y.Z. and D.W.; Software, X.H.; Validation, Y.Z.; Formal analysis, X.H.; Writing—original draft, X.H.; Writing—review & editing, Y.Y. and D.W.; Visualization, X.H.; Supervision, Y.Y., Y.Z. and D.W.; Project administration, Y.Y., Y.Z. and D.W. All authors have read and agreed to the published version of the manuscript.

Funding: This research was funded by the National Key R&D Program of China (No. 2021YFB3900605 and 2018YFC1508100).

Institutional Review Board Statement: Not applicable.

Informed Consent Statement: Not applicable.

Data Availability Statement: Not applicable.

Acknowledgments: The authors would like to thank the Bureau of Hydrology, Changjiang water Resources Commission, especially Chen Yubin and Zhang Xiao, for their great help in data provision, algorithm design, and model training of this paper.

Conflicts of Interest: The authors declare no conflict of interest.

References

1. Li, B.; Liang, Z.; Bao, Z.; Wang, J.; Hu, Y. Changes in streamflow and sediment for a planned large reservoir in the middle Yellow River. *Land Degrad. Dev.* **2019**, *30*, 878–893. [CrossRef]
2. Stenta, H.R.; Riccardi, G.A.; Basile, P.A. Grid size effects analysis and hydrological similarity of surface runoff in flatland basins. *Hydrol. Sci. J.* **2017**, *62*, 1736–1754. [CrossRef]
3. Liang, Z.; Xiao, Z.; Wang, J.; Sun, L.; Li, B.; Hu, Y.; Wu, Y. An improved chaos similarity model for hydrological forecasting. *J. Hydrol.* **2019**, *577*, 123953. [CrossRef]
4. Dilmi, D.; Barthès, L.; Mallet, C.; Aymeric, C. Modified DTW for a quantitative estimation of the similarity between rainfall time series. *EGU Gen. Assem.* **2017**, *19*, EGU2017-16005.
5. Xiao, Z.; Liang, Z.; Li, B.; Hou, B. New flood early warning and forecasting method based on similarity theory. *J. Hydrol. Eng.* **2019**, *24*, 04019023. [CrossRef]
6. Barthel, R.; Haaf, E.; Giese, M.; Nygren, M.; Heudorfer, B.; Stahl, K. Similarity-based approaches in hydrogeology: Proposal of a new concept for data-scarce groundwater resource characterization and prediction. *Hydrogeol. J.* **2021**, *29*, 1693–1709. [CrossRef]
7. Wang, H.; Xing, C.; Yu, F. Study of the hydrological time series similarity search based on Daubechies wavelet transform. In *Unifying Electrical Engineering and Electronics Engineering, Proceedings of the 2012 International Conference on Electrical and Electronics Engineering, London, UK, 4–6 July 2012*; Springer: New York, NY, USA, 2013; pp. 2051–2057.
8. Yang, J.; Wan, D.; Yu, Y. Similarity Search Method of Hydrological Time Series based on Fragment Alignment Distance and Dynamic Time Warping. In Proceedings of the IEEE 2022 5th International Conference on Advanced Electronic Materials, Computers and Software Engineering (AEMCSE), Wuhan, China, 22–24 April 2022; pp. 214–220.
9. Zhang, L.; Zhu, Y.; Li, S.; Gao, X. Study on Similarity Model of Precipitation Series Based on Precipitation Type Histogram. *J. China Hydrol.* **2013**, *33*, 10–16.
10. Ohno, G.; Kazunori, I. Flood Forecast Based on Deep Learning Using Distribution MAP of Precipitation. In Proceedings of the 22nd IAHR APD Congress, Sapporo, Japan, 14–17 September 2020.
11. Wang, X.; Liu, Y.; Chen, Y.; Liu, Y. An adaptive density-based time series clustering algorithm: A case study on rainfall patterns. *ISPRS Int. J. Geo-Inf.* **2016**, *5*, 205. [CrossRef]
12. Gang, J.; Zhao, W. RETRACTED ARTICLE: Remote sensing image-based rainfall changes in plain areas and IoT motion image detection. *Arab. J. Geosci.* **2021**, *14*, 1–17. [CrossRef]
13. Pradhan, J.; Kumar, S.; Pal, A.K.; Banka, H. Texture and colour region separation based image retrieval using probability annular histogram and weighted similarity matching scheme. *IET Image Process.* **2020**, *14*, 1303–1315. [CrossRef]
14. Wu, Y.; Zhang, L.; Berretti, S.; Wan, S. Medical Image Encryption by Content-Aware DNA Computing for Secure Healthcare. *IEEE Trans. Ind. Inform.* **2023**, *19*, 2089–2098. [CrossRef]
15. Divakar, R.; Singh, B.; Bajpai, A.; Kumar, A. Image pattern recognition by edge detection using discrete wavelet transforms. *J. Decis. Anal. Intell. Comput.* **2022**, *2*, 26–35.
16. Alsmadi, M.K. Content-based image retrieval using color, shape and texture descriptors and features. *Arab. J. Sci. Eng.* **2020**, *45*, 3317–3330. [CrossRef]
17. Kumbure, M.M.; Luukka, P. A generalized fuzzy k-nearest neighbor regression model based on Minkowski distance. *Granul. Comput.* **2022**, *7*, 657–671. [CrossRef]
18. Gassouma, M.S.; Benhamed, A.; El Montasser, G. Investigating similarities between Islamic and conventional banks in GCC countries: A dynamic time warping approach. *Int. J. Islam. Middle East. Financ. Manag.* **2023**, *16*, 103–129. [CrossRef]
19. Ristad, E.S.; Yianilos, P.N. Learning String-Edit Distance. *IEEE Trans. Pattern Anal. Mach. Intell.* **1997**, *20*, 522–532. [CrossRef]
20. Benti, N.E.; Chaka, M.D.; Semie, A.G. Forecasting Renewable Energy Generation with Machine learning and Deep Learning: Current Advances and Future Prospects. *Preprints.org* **2023**, 2023030451. [CrossRef]
21. Zhang, L.; Zhang, B. Deep learning for remote sensing data: A technical tutorial on the state of the art. *IEEE Geosci. Remote Sens. Mag.* **2016**, *4*, 22–40. [CrossRef]
22. Frana, R.P.; Monteiro, A.; Arthur, R.; Lano, Y. An overview of deep learning in big data, image, and signal processing in the modern digital age. *Trends Deep. Learn. Methodol.* **2021**, 63–87. [CrossRef]
23. Wu, Y.; Guo, H.; Chakraborty, C.; Khosravi, M.; Berretti, S.; Wan, S. Edge Computing Driven Low-Light Image Dynamic Enhancement for Object Detection. *IEEE Trans. Netw. Sci. Eng.* **2022**, 3151502. [CrossRef]
24. Kennedy, J.; Eberhart, R. Particle swarm optimization. In Proceedings of the ICNN'95-International Conference on Neural Networks, Perth, WA, Australia, 27 November–1 December 1995; Volume 4, pp. 1942–1948.
25. Paramanik, A.R.; Sarkar, S.; Sarkar, B. OSWMI: An Objective-Subjective Weighted method for Minimizing Inconsistency in multi-criteria decision making. *Comput. Ind. Eng.* **2022**, *169*, 108138. [CrossRef]
26. Furui, K.; Ohue, M. Compound virtual screening by learning-to-rank with gradient boosting decision tree and enrichment-based cumulative gain. In Proceedings of the 2022 IEEE Conference on Computational Intelligence in Bioinformatics and Computational Biology (CIBCB), Ottawa, ON, Canada, 15–17 August 2022; pp. 1–7.
27. Feng, K.; Li, X.; Qian, X.; Wu, L.; Zheng, H.; Chen, M.; Li, M.; Liu, B. Atmospheric Optical Turbulence Profile Model Fitting Based on Improved Particle Swarm Algorithm. *Laser Optoelectron. Prog.* **2022**, *59*, 73–84.
28. Liu, C.; Sui, X.; Kuang, X.; Liu, Y.; Gu, G.; Chen, Q. Optimized Contrast Enhancement for Infrared Images Based on Global and Local Histogram Specification. *Remote Sens.* **2019**, *11*, 849. [CrossRef]

29. Li, K.; Sun, X.; Gao, B.; Zhou, J. Weighted Histogram Block Detection Algorithm for Digital Trunking Terminal. In Proceedings of the International Conference on Intelligent Automation and Soft Computing, Chicago, IL, USA, 28–30 May 2021; Springer: Cham, Switzerland, 2021; pp. 166–172.
30. Muhammad, M.; Oscar, V. Pairwise consensus and the Borda rule. *Math. Soc. Sci.* **2022**, *116*, 17–21.
31. Boudou, A.; Viguier-Pla, S. Principal components analysis and cyclostationarity. *J. Multivar. Anal.* **2022**, *189*, 104875. [CrossRef]

Article

Faster RCNN Target Detection Algorithm Integrating CBAM and FPN

Wenshun Sheng *, Xiongfeng Yu, Jiayan Lin and Xin Chen

Pujiang Institute, Nanjing Tech University, Nanjing 211200, China
* Correspondence: sws@njpji.edu.cn

Abstract: In the process of image shooting, due to the influence of angle, distance, complex scenes, illumination intensity, and other factors, small targets and occluded targets will inevitably appear in the image. These targets have few effective pixels, few features, and no obvious features, which makes it difficult to extract their effective features and easily leads to false detection, missed detection, and repeated detection, thus affecting the performance of target detection models. To solve this problem, an improved faster region convolutional neural network (RCNN) algorithm integrating the convolutional block attention module (CBAM) and feature pyramid network (FPN) (CF-RCNN) is proposed to improve the detection and recognition accuracy of small-sized, occluded, or truncated objects in complex scenes. Firstly, it incorporates the CBAM attention mechanism in the feature extraction network in combination with the information filtered by spatial and channel attention modules, focusing on local efficient information of the feature image, which improves the detection ability in the face of obscured or truncated objects. Secondly, it introduces the FPN feature pyramid structure, and links high-level and bottom-level feature data to obtain high-resolution and strong semantic data to enhance the detection effect for small-sized objects. Finally, it optimizes non-maximum suppression (NMS) to compensate for the shortcomings of conventional NMS that mistakenly eliminates overlapping detection frames. The experimental results show that the mean average precision (MAP) of target detection of the improved algorithm on PASCAL VOC2012 public datasets is improved to 76.2%, which is 13.9 percentage points higher than those of the commonly used Faster RCNN and other algorithms. It is better than the commonly used small-sample target detection algorithm.

Keywords: CF-RCNN; CBAM; FPN; target detection; attention mechanism

Citation: Sheng, W.; Yu, X.; Lin, J.; Chen, X. Faster RCNN Target Detection Algorithm Integrating CBAM and FPN. *Appl. Sci.* **2023**, *13*, 6913. https://doi.org/10.3390/app13126913

Academic Editors: Yirui Wu and Shaohua Wan

Received: 28 January 2023
Revised: 30 May 2023
Accepted: 5 June 2023
Published: 7 June 2023

1. Introduction

The traditional machine learning method aims to train the classifier by manually dividing the area and filtering the data based on the existing specific data model. This method has a high time cost, and the obtained classifier does not perform well on datasets that are quite different from the specific data model, exposing the shortcomings of low robustness and weak generalization ability. An effective classifier not only needs to be able to continuously detect changes in the shape and state of the target, but also needs to be able to accurately predict and deal with system robustness problems caused by special changes.

The fusion of convolutional neural networks [1] in traditional machine learning models can better solve these problems. Because the convolutional neural network can independently learn and summarize the data characteristics of the target object, the selection and feature extraction of feature regions can be performed independently on the datasets of different scenarios, and it is applicable even in complex environments. After fusing the convolutional neural network, the target detection algorithm does not need manual intervention, which not only saves labor cost and strengthens the generalization ability, but also improves the system's robustness. Compared with the traditional machine learning manual annotation method, it shows better performance in practical applications. Therefore, applying the convolutional neural network to assist target detection can enhance the ability of the model to extract image features and improve the accuracy of model object classification.

For the VGG16 network model [2] in object recognition applications, due to the shallow layer, the summary of object features is insufficient and the expression ability is weak. In order to solve such problems, a CF-RCNN algorithm integrating the CBAM (Convolutional Block Attention Module) attention mechanism [3] and FPN (Feature Pyramid Network) feature pyramid structure [4] is proposed. The algorithm uses the VS-ResNet network with a stronger expression ability and deeper layers to replace the VGG16 network in the feature extraction module of the traditional Faster RCNN algorithm, which strengthens the ability of the target detection model to analyze image features; VS-ResNet changes the ResNet-50 network [5] to the group convolution [6] mode, reducing the number of super parameters and the complexity of the algorithm model. The original residual structure is changed to an inverse residual structure [7], so that the Re*LU* activation function [8] can better save function information. The risk priority number (RPN) is optimized to reduce the error rate of candidate frame filtering in the case of dense objects. Based on the traditional Faster RCNN algorithm, CF-RCNN fuses the FPN feature pyramid structure to improve the detection capability for small targets; the CBAM attention mechanism is introduced to focus on efficient information, to improve the accuracy of truncated or occluded object detection.

2. Related Works

In recent years, many scholars have performed active exploration in the field of small-scale target detection and truncated target detection and have achieved good results.

Salau et al. [9] presented a modified GrabCut algorithm for localizing vehicle plate numbers. The approach extends the use of the traditional GrabCut algorithm with the addition of a feature extraction method that uses geometric information to give accurate foreground extraction.

Yang et al. [10] proposed a multi-scale feature attention fusion network named parallel feature fusion with CBAM (PFF-CB) for occlusion pedestrian detection. Feature information of different scales can be integrated effectively into the PFF-CB module. The PFF-CB module uses a convolutional block attention module (CBAM) to enhance the important feature information in space and channels.

The authors in [11] proposed the DF-SSD algorithm, which is based on a dense convolutional network (DenseNet) for detection and uses feature fusion. It uses the algorithm framework of the Single Shot MultiBox Detector (SSD) [12] for reference and introduces DenseNet-S-32-1 to replace the original VGG16 network. In addition, DF-SSD also uses a multi-scale feature fusion mechanism. Their proposed residual prediction module is designed to enhance feature propagation along the feature extraction network, i.e., to use a small convolution filter to predict the target class and the offset of the bounding box position.

The authors in [13] proposed an improved double-head RCNN small target detection algorithm. A transformer and a deformable convolution module are introduced into ResNet-50, and a feature pyramid network structure CARAFE-FPN based on content perception feature recombination is proposed. In that regional recommendation network, the anchor generation scale is reset according to the distribution characteristics of the small target scale, so that the detection performance of the small target is further improved.

In [14], the authors proposed a BiFPN structure in the EfficientDet network, which is composed of a weighted bidirectional feature pyramid network, and added cross-scale connection to enhance the representation ability of features, and added corresponding weight to each input to better perform small target detection task.

In [15], the authors performed deep optimization based on the YOLOv5 architecture and proposed a multi-scale multi-attention target detection algorithm YOLO-StrVB is based on the STR (Swin Transformer) network structure. The method mainly aims at small target detection under a complex background, reconstructs the framework to build a multi-scale network, increases the target detection layer, and improves the target feature extraction capability under different scales. Then, a bidirectional feature pyramid network is added for multi-scale feature fusion, and a jump connection is introduced. On this basis, the baseline backbone network end integrates STR architecture.

Fu et al. [16] proposed a deconvolutional single shot detector (DSSD) algorithm, in which ResNet with stronger learning ability is used as the backbone network, and a deconvolution layer is introduced to further reduce the missing rate of small targets.

Singh et al. [17] started from the perspective of training, considered the level of data, and selectively back-propagated the gradient of target instances with different sizes according to different changes in image scale, to realize scale normalization on an image pyramid (SNIP). While capturing the change in the object as much as possible, the model was efficiently trained with objects with appropriate proportions, and the detection performance was significantly improved.

Although the above algorithms improve the detection performance of the model to a certain extent, they do not fully consider the effect of local features on truncated target detection, and fail to optimize the light weight of the feature extraction module. The CF-RCNN algorithm proposed in this paper is based on group convolution, which significantly reduces the weight of the feature extraction network. The CBAM attention mechanism is introduced to fully consider local features. The FPN structure is integrated to improve the detection ability of small-sized targets and truncated targets. The CF-RCNN algorithm is superior to many of the most advanced methods. The specific advantages and disadvantages are compared in Table 1.

Table 1. Comparison of advantages and disadvantages of CF-RCNN and advanced algorithms.

Detection Method	Advantage	Disadvantage
GrabCut algorithm	The proposed algorithm is not country-specific and can be used to detect LPs in complex environments	This method is not extended to motorcycle license localization
DF-SSD algorithm	The algorithm has a fast processing speed and a compact model	This algorithm is not applicable to large objects and objects without specific relationships
Scalable and Efficient Object Detection	This method uses fewer parameters and has a wide range of resource limits	The compound scaling method mentioned is complex and difficult to operate
Multi-Head Attention Detection	This method is suitable for small target detection in complex environments	There are obvious false detection and missing detection phenomena
SNIP	This method is effective in target recognition and detection under extreme scale changes	This method does not have universal generalization and the mAP value is low
CF-RCNN	This method has a strong detection ability for small-sized blocked or truncated targets, and has strong stability in complex environments	It is easy to cause the problem of multi-frame prediction of a single target

The following are the primary contributions of this work:

- The ResNet-50 network model is optimized, and a VS-ResNet network model with stronger expression ability is proposed, which improves the classification accuracy in the target recognition process.
- The improved model of non-maximum suppression is applied to solve the problem of eliminating prediction frame errors in the process of classification and regression.
- A CF-RCNN algorithm is proposed and implemented for small-sized targets and truncated target detection.

3. Model Design and Implementation

3.1. Faster RCNN Model Design

Faster RCNN is a classic network proposed by Girshick in 2015 [18]. It consists of four modules, namely the feature extraction network module (Conv layers) [19], Region Proposal Network (RPN) [20,21] module, ROI Pooling [22] module, and Classification and Regression [23] module. The frame diagram of Faster RCNN is shown in Figure 1.

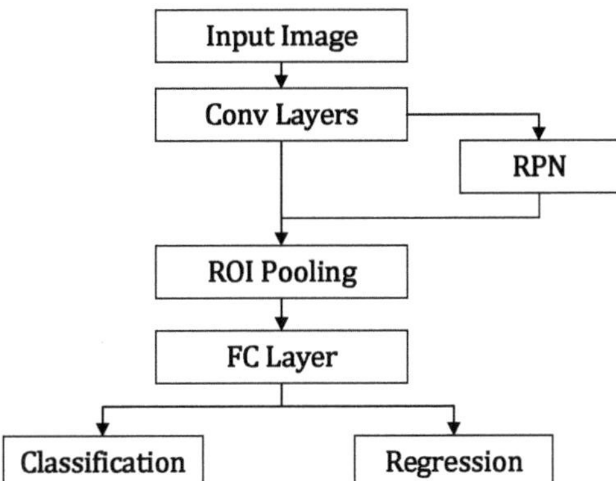

Figure 1. Schematic diagram of Fast RCNN algorithm framework.

After the original image is input, the shared convolution layer computes it. The results can be shared with the RPN. RPN makes region suggestions (about 300) on the feature map after convolutional neural network (CNN) convolution, extracts feature maps according to the region suggestions generated by RPN, performs ROI pooling on the extracted features, then classifies and regresses the processed data through the full connection layer.

The feature extraction network module [19] is an important part of the Fast RCNN algorithm. The flow of the feature extraction module is shown in Figure 2.

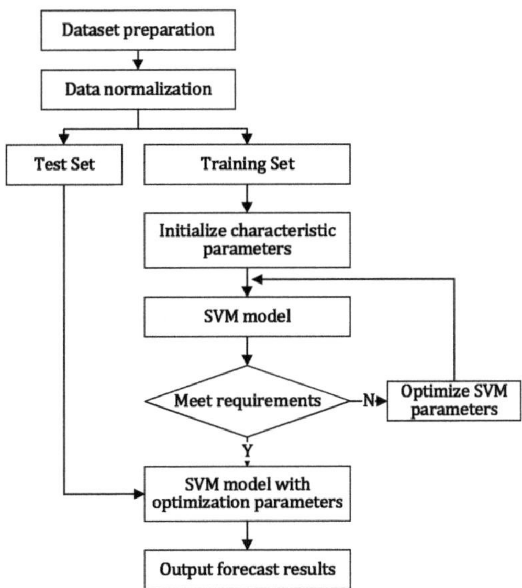

Figure 2. Feature extraction flow chart.

After normalization, the prepared dataset is divided into the training set and the test set. When the training set is trained by the Support Vector Machine (SVM) [24] model, better parameters are obtained by iteration, and finally, the SVM model meeting the requirements is obtained. The test dataset is input into the optimized SVM model, and the output result is finally obtained.

VGG16 is used as the feature extraction network in Faster RCNN. To break through the bottleneck of region selection in the predecessor target detection model, the new Fast RCNN algorithm model innovatively proposes the RPN algorithm, replacing the selective search (SS) [25] algorithm model used in Fast RCNN and RCNN, that is, the Faster RCNN algorithm can be understood as RPN + Fast RCNN. The SS algorithm is a simple image processing without a training function. RPN is a fully convolutional network, and its first several convolutional layers are the same as the first five layers of Faster R-CNN, so RPN can share the calculation results of convolutional layers to reduce the time consumption of region suggestions. The RPN module is used to replace the candidate region selection mode of the original SS algorithm, thus greatly reducing the time cost. RPN selects candidate regions according to image color, shape, and size, and usually selects 2000 candidate boxes that may contain identification objects.

When the detected image is input to the feature extraction network module, the feature image is generated after the module convolution operation, and it is input to the RPN module to select candidate regions. This module uses the center point of the sliding window as the anchor. All the pixels in the feature picture correspond to k anchor points, generating 128-, 256-, 512-pixel areas and 1:1, 1:2, and 2:1 scale windows, respectively. The combined result is a total of 9 windows, as shown in Figure 3.

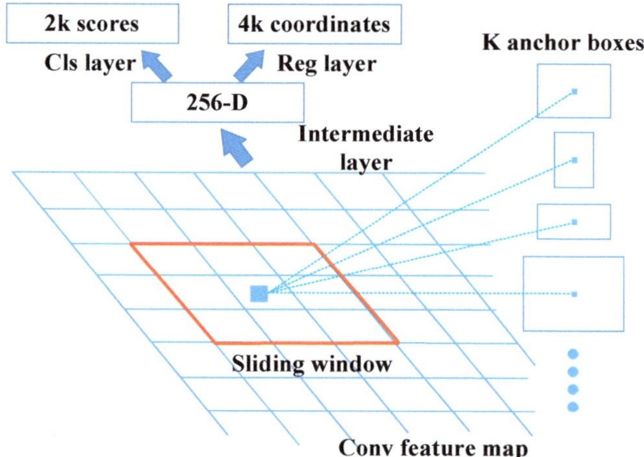

Figure 3. RPN structure diagram.

The generated candidate area and convolution layer feature map are input to the ROI Pooling module. ROI Pooling pools feature maps of different sizes into a uniform size, which facilitates their input to the next layer. Following the ROI Pooling module, the target classification and position regression module processes the data output from the ROI Pooling layer to obtain the object category of the candidate area and the modified image block diagram, and its processing formula can be expressed by Formula (1).

$$\begin{aligned}
\hat{G}_x &= P_w d_x(P) + P_x \\
\hat{G}_y &= P_h d_y(P) + P_y \\
\hat{G}_w &= P_w \exp(d_w(P)) \\
\hat{G}_h &= P_h \exp(d_h(P))
\end{aligned} \tag{1}$$

where P_x, P_y, P_w, and P_h are the horizontal and vertical coordinates and the width and height values of the candidate box center pixel, respectively; d_x, d_y, d_w, and d_h are the regression parameters of $N + 1$ categories of candidate boxes, with a total of $4 \times (N + 1)$ nodes; exp is an exponential function with the natural number e as the base.

To solve the problem of weak expression ability of the VGG16 network due to its small number of layers, the VS-ResNet network is used to replace the VGG16 network. The VS-ResNet network is based on the ResNet-50 network improvement. In the original Faster RCNN, the CBAM attention mechanism and FPN feature pyramid structure are integrated to enhance the detection capability for small truncated or occluded objects. The specific improvement process is as follows.

3.2. Design of Group Convolution and Inverse Residual Structure

The structure of the packet convolution network refers to the multi-dimensional convolution [26] combination principle of Inception, which changes the single path convolution kernel convolution operation of the characteristic graph into a multi-channel convolution kernel convolution stack, reducing the parameters in the network and effectively reducing the complexity of the algorithm model, but its accuracy will not be greatly affected. VS-ResNet refers to the ResNeXt network [27] and uses 32 groups of 8-dimensional convolutional cores.

When a deep convolution network is used for convolution, the convolution parameter of the partial convolution kernel is 0, which results in partial parameter redundancy. The experiment proves that when the dimension is too low, too much image feature information of the ReLU activation function is lost. The ReLU function is shown in Formula (2).

$$\mathrm{ReLU}(x) = \max(x,0) = \begin{cases} x, x \geq 0 \\ 0, x < 0 \end{cases} \tag{2}$$

To solve this problem, the VS-ResNet network uses a reciprocal residual structure, with convolution kernel sizes of 1×1, 3×3, and 1×1, as shown in Figure 4.

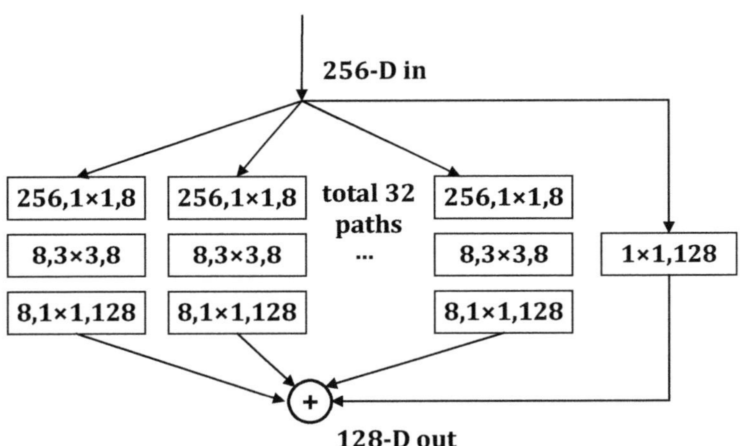

Figure 4. Convolution block structure diagram.

The common residual structure is first based on the output basis of the previous layer, using 1×1 convolution-kernel-size dimension reduction. Convolution kernels of 3×3 size are passed again to extract image features; lastly, dimension elevation of convolution size 1×1 is used. Unlike the ordinary residual structure, the inverted residual structure has exactly the opposite order of dimension increase and dimension reduction, that is, 1×1-size convolution is first used to increase the dimension. Additionally, you have a 3×3 convolution. Finally, a 1×1 convolution kernel is used to reduce the dimension to the original feature map size. When the Re*LU* function is activated linearly in the $x > 0$ region, it may cause the function value to be too large after activation, thus affecting the stability of the model. The Re*LU* function does not limit output, allowing very high values on the positive side. However, Re*LU*6 limits the value of the positive side to 6, which can eliminate most of the value, so VS-RCNN uses the RE*LU*6 function instead of the Re*LU* function, as shown in Formula (3).

$$\mathrm{Re}LU6(x) = \min[\max(x, 0), 6] \tag{3}$$

3.3. Reference to Auxiliary Classifier

With the deepening of the network, the convergence of the feature extraction network becomes more difficult. It is necessary to retain a certain degree of the front-end network propagation gradient to alleviate the gradient disappearance. VS-ResNet adds an auxiliary classifier [28] after convolution layer 2 and convolution layer 4 of the ResNet-50 network, respectively to retain the low-dimensional output information of convolution layer 2 and convolution layer 4. In the final classification task, combined with the actual application, a fixed utility value is set and shallow feature reuse is used for auxiliary classification. The utility value of the auxiliary classifier in the VS-ResNet network is set to 0.1. The auxiliary classifier consists of an average pooling layer, a convolution layer, and two full connection layers, as shown in Figure 5. After the image is extracted by convolution layer 2 in ResNet-50, a feature map is generated and input to the first auxiliary classifier. Firstly, the number of network parameters is reduced by an average pooling downsampling layer, which uses a pool size of 14 × 14 with a step distance of 6. Then, the convolution layer is entered, which uses 128 convolution kernels of 1 × 1 with a sliding step of 1. The results are then flattened and fed into the fully connected layer that follows this layer.

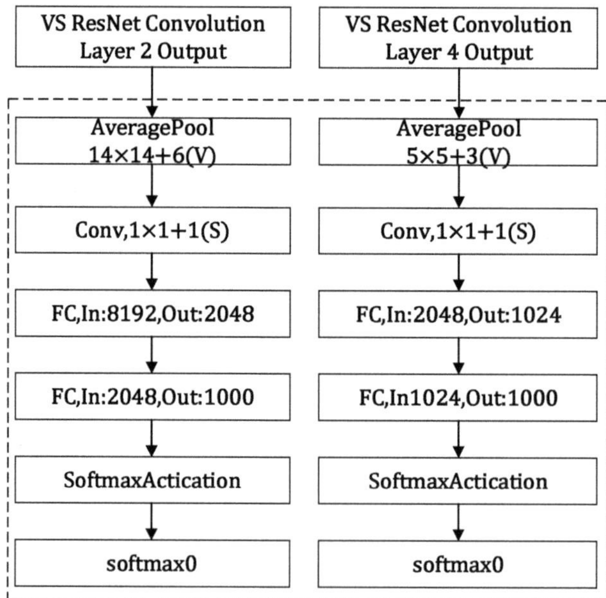

Figure 5. Auxiliary classifier diagram.

Dropout [29] at 50% is performed in two fully connected layers to prevent overfitting. Some parameters of the second auxiliary classifier are different from those of the first auxiliary classifier. The size of the pool is 5 × 5 instead of 14 × 14, and the step distance is changed from 6 to 3, which is based on the position of the auxiliary classifier in the convolutional neural network. The input of the two fully connected layers is 2048 and

1024 neurons, respectively. The output results of the two auxiliary classifiers are multiplied by the utility ratio set in VS-ResNet and then added to the final classification result. The addition of the auxiliary classifiers increases the gradient of network backpropagation and alleviates the phenomenon of gradient disappearance.

3.4. Detailed Structure of VS-ResNet Network

For the feature extraction network module, VS-ResNet changes the residual block structure from a funnel model to a bottleneck model, so that the activation function information can be better preserved, as shown in Table 2. Referring to the structural parameters of ResNet-50, the number of convolution channels of the first residual structure is changed from the original [64,64,256] to [256,256], and the last three residual structures are also modified successively from [128,128,512] to [512,512,256]. [256,256,1024] is [1024,1024,512] and [512,512,2048] is [2048,2048,1024]. The group convolution method is referenced in the inverted residual structure, and the number of groups is set to 32. By referring to the model of the Swin Transformer algorithm [30], the original hierarchical structure [3,4,6,3] is modified to [3,3,9,3], and auxiliary classifiers are added after the second and fourth layers to accelerate the convergence speed of the network to a certain extent and alleviate the phenomenon of gradient disappearance.

Table 2. Improved ResNet-50 network architecture.

Network Layer	Output	52 Layer
Convolution layer 1	112×112	$7 \times 7, 64$, Step-length 2
		3×3, Max pooling, Step-length 2
Convolution layer 2	56×56	$\begin{bmatrix} 1 \times 1, 256 \\ 3 \times 3, 256, \mathrm{C} = 32 \\ 1 \times 1, 128 \end{bmatrix} \times 3$
	The first auxiliary classifier	
Convolution layer 3	28×28	$\begin{bmatrix} 1 \times 1, 512 \\ 3 \times 3, 512, \mathrm{C} = 32 \\ 1 \times 1, 256 \end{bmatrix} \times 3$
Convolution layer 4	14×14	$\begin{bmatrix} 1 \times 1, 1024 \\ 3 \times 3, 1024, \mathrm{C} = 32 \\ 1 \times 1, 512 \end{bmatrix} \times 9$
	The second auxiliary classifier	
Convolution layer 5	7×7	$\begin{bmatrix} 1 \times 1, 2048 \\ 3 \times 3, 2048, \mathrm{C} = 32 \\ 1 \times 1, 1024 \end{bmatrix} \times 3$
	1×1	Convolutional pooling layer, 1000 neurons, Softmax classifier

3.5. Incorporate the CBAM Attention Mechanism

The attention mechanism can selectively ignore some inefficient information in the image, focus on the efficient information, reduce the resource consumption of the inefficient part, improve the network utilization, and enhance the ability of object detection. Therefore, the CBAM attention mechanism is integrated into the feature extraction network, and the channel attention mechanism and spatial attention mechanism are connected [31] to form a simple but effective attention module, whose structure is shown in Figure 6.

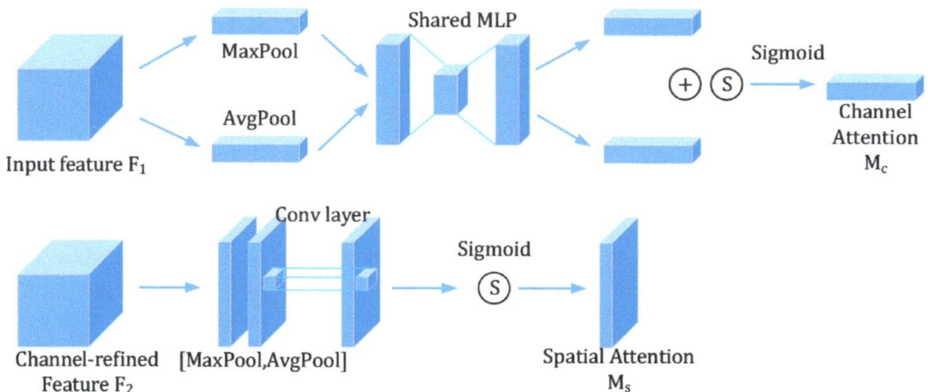

Figure 6. Schematic diagram of CBAM attention mechanism.

In the channel attention module, the global average pooling and maximum pooling of the same input feature space are performed to obtain the spatial information of the feature map, and then the obtained feature space information is input into the multi-layer awareness mechanism module of the next layer for dimension reduction and dimension-increase processing. The weight of the two shared convolution layers in the multi-layer awareness network is shared. Then, the characteristics of the perceptual network output are added and then processed with the sigmoid activation function to obtain channel attention. The calculation formula is shown in Formula (4).

$$Mc(F) = \varepsilon[MLP(F_{avg}^c) + MLP(F_{max}^c)] \tag{4}$$

Among them, Mc is the channel attention module calculation factor, ε is the sigmoid activation function, MLP is the multi-layer perceptron, and F represents the feature vector.

Spatial attention features are complementary to channel attention and reflect the importance of input values in spatial dimensions. The calculation formula is shown in Formula (5). First, global average pooling and global maximum pooling of one channel dimension are performed on the feature map, then the two features are spliced, and finally the dimension is reduced to one by 7×7 convolution post-channel processing using the sigmoid function [32] to generate spatial attention feature maps.

$$Ms(F) = \varepsilon\left\{conv_{7\times7}[unit(F_{avg}^s, F_{max}^s)]\right\} \tag{5}$$

Among them, Ms is the space attention module calculation factor, ε is the sigmoid activation function, MLP is the multi-layer perceptron, F represents the feature vector, *unit* is the channel combination, and *conv* is the convolution operation.

In order to facilitate the use of pre-trained models during the experiment, CBAM is not embedded in all convolutional residual blocks, but only acts after different convolutional layers.

3.6. Introduce FPN Feature Pyramid Structure

In order to alleviate the unsatisfactory detection ability of the Faster RCNN algorithm for small-sized targets, the FPN feature pyramid network model is introduced into the Faster RCNN target detection algorithm. The FPN network model is divided into two network routes. One of the network routes produces multi-scale features from bottom to top, connecting the high-level features with high semantics and low resolution and the low-level features with high resolution and low semantics [33]. Another network route is from top to bottom; after some layer changes, the rich semantic information contained in the upper layer is transferred layer by layer to the low-layer features for fusion [34].

Compared with the SSD algorithm, FPN also uses multi-level features and multi-scale anchor frames. SSD predicts the low-level data alone, where it is difficult to ensure strong semantic features, and the detection effect is not ideal for small targets.

Figure 7 is a schematic diagram of the FPN feature pyramid structure. In the figure, the bottom-up feature map with computed order on the left is {C2, C3, C4, C5}, and the top-down feature pyramid structure on the right is {P2,P3,P4,P5}. The CF-RCNN algorithm uses the above VS-ResNet as the backbone extraction network of the FPN feature pyramid structure. The image part on the left side of Figure 7 is a downsampling model. During the feature extraction operation of this model, the value of the step size is set as {4,8,16,32}. In the upsampling model of the right image, the upper feature map is convolved with a convolution kernel of 1×1 size, the step size is set to 1, and the number of channels is 256, to adjust the dimension to be consistent so that it can be fused with the lower feature. Then, after 3×3 size convolution, the aliasing situation in the 2-fold upsampling process is eliminated to obtain the feature map. {P2, P3, P4, P5} share the weight of RPN and Fast RCNN, and use a different anchor size of $\{32^2, 64^2, 128^2, 256^2\}$ and the anchor ratio of {1:2, 1:1, 2:1} on the {P2, P3, P4, P5} feature map to select candidate boxes.

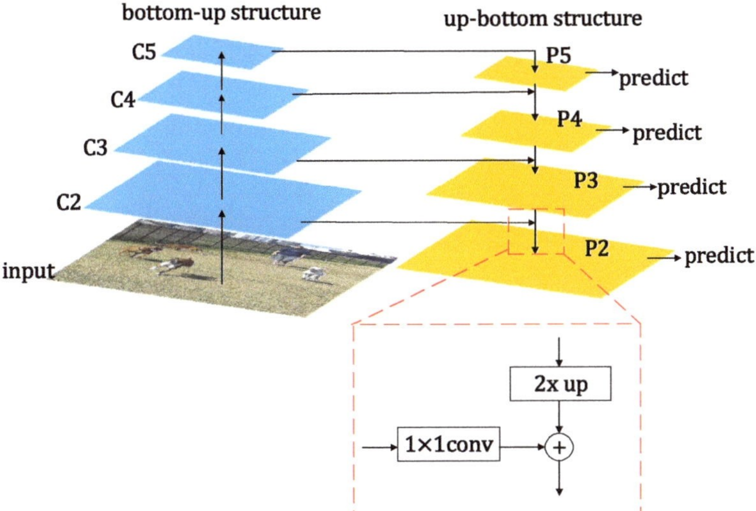

Figure 7. FPN Feature Pyramid Structure.

3.7. Improvement of Non-Maximum Suppression NMS Algorithm

Non-Maximum Suppression (NMS) is an important part of the target detection algorithm, and its main function is to eliminate redundant candidate boxes generated in the RPN network. The elimination criterion is shown in Formula (6).

$$S_i = \begin{cases} S_i, IOU(M, b_i) < N_t \\ 0, IOU(M, b_i) \geq N_t \end{cases} \tag{6}$$

where S_i is the score of the i th candidate box; M indicates the candidate box with the largest score; b_i is the candidate box to be scored; IOU is the combined ratio of b_i; and N_t is the threshold value of IOU set in NMS.

When the ratio between the candidate box to be scored and the candidate box with the largest score exceeds the threshold, the NMS deletes the candidate box to be scored. As a result, NMS mistakenly deletes overlapping candidate boxes in partially crowded and truncated complex scenarios. To solve this problem, CF-RCNN uses the Soft-NMS algorithm. The Soft-NMS score function is shown in Formula (7).

$$S_i = \begin{cases} S_i, IOU(M, b_i) < N_t \\ S_i(1 - IOU(M, b_i)), IOU(M, b_i) \geq N_t \end{cases} \tag{7}$$

If the combined ratio is less than the threshold, the score remains unchanged. If it is greater than or equal to the threshold, S_i is the difference between S_i multiplied by 1 and the combined ratio. Compared with NMS, Soft-NMS does not directly overlap candidate boxes, but resets S_i to a smaller score.

3.8. Overall Structure of CF-RCNN

Combined with the above improvement measures, the detailed structure of the CF-RCNN algorithm is shown in Figure 8.

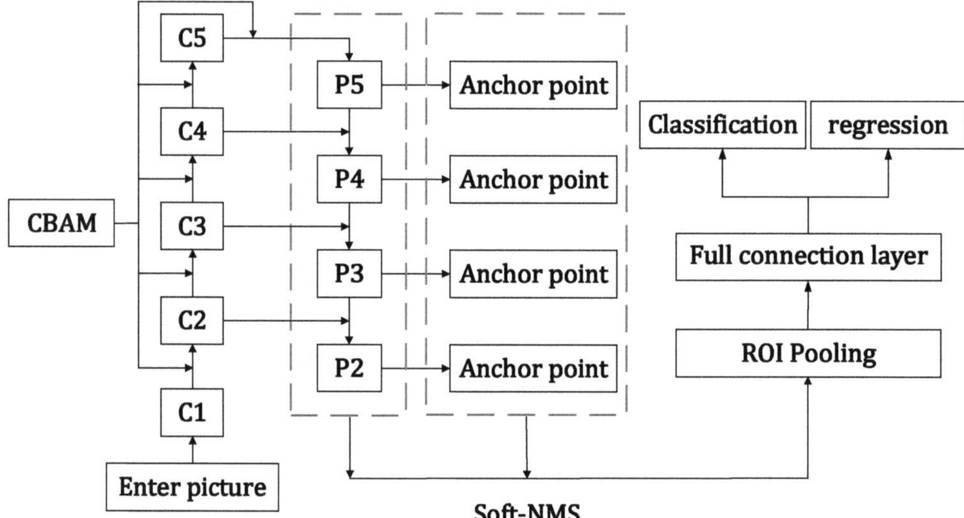

Figure 8. Overall structure of CF-RCNN.

Firstly, CF-RCNN takes the image stream as input, extracts image features based on the inverted residual ResNet-50 model, and obtains intermediate feature vector graphics. Secondly, the CBAM attention mechanism is used to calculate the attention weight in the two dimensions of space and channel, and adjusts the parameters of the intermediate feature map. Third, based on the FPN feature pyramid structure, the feature map is multi-scaled. Fourth, on each feature map, the RPN network uses a single scale to select regions that may contain objects and balances the positive and negative sample ratios through Soft-NMS. Finally, the classification and position regression operations are performed on the objects in the selected area.

4. Experimental Results and Analysis

The experiment is based on the deep learning network framework PyTorch 1.81, the programming language is Python 3.8, the CPU configuration is a 10-core Intel(R) Xeon(R) Gold 5218R CPU @ 2.10 GHz, and the memory is 64 GB. The GPU is configured as an RTX 3090 + CUDA 1.1, with 24 GB of video memory. The operating system uses Ubuntu 18.04 version.

4.1. Public Dataset Preparation

In order to verify the effectiveness of the improvement measures in CF-RCNN for improving the performance of object detection, object recognition and object detection experiments are performed on the CIFAR-10 dataset [35] and the PASCAL VOC2012 dataset [36], respectively. The CIFAR-10 dataset has a total of 60,000 images, which are divided into 10 types of objects to be recognized. Each type of object to be recognized has a capacity of 6000 images, and the ratio of the training set and test set is 5:1. The PASCAL VOC2012 dataset contains 20 detection categories and a total of 17,125 images, of which 5718 images are used in the training set and 5824 images are used in the validation set.

4.2. Object Recognition Experiment Results Analysis

The experiment compares the change in the error rate of the VGG16, ResNet-50, and VS-ResNet networks on the CIFAR-10 dataset with the increase in the number of iterations. The number of iterations is set to 35, the training batch is 16, and the learning rate is set to 0.0001. The experimental results are shown in Table 3 and Figure 9.

Table 3. Comparison of the lowest error rates of different convolutional networks.

Network Model	Number of Plies	Minimum Error Rate
VGG16	16	0.1106
ResNet-50	50	0.0941
VS-ResNet-50	52	0.0809

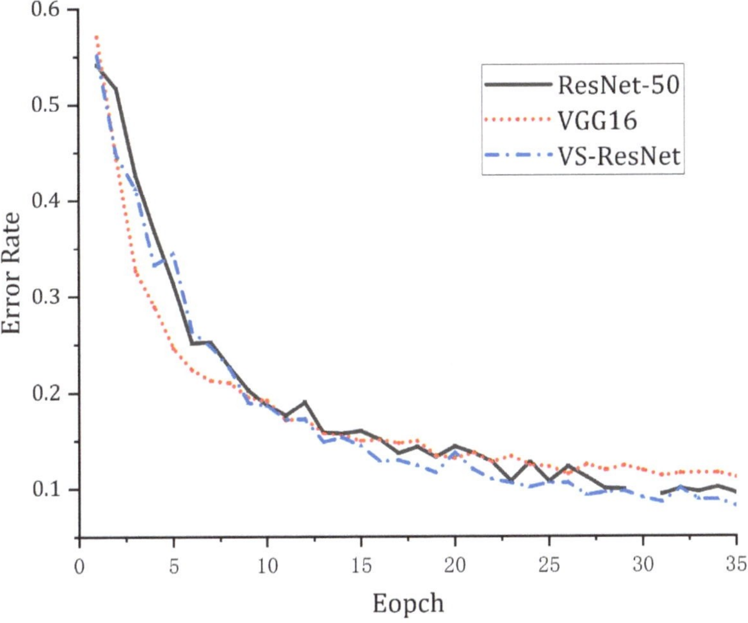

Figure 9. Comparison of error rates of different convolutional networks.

The data in Table 3 show that the ResNet-50 network shows better performance in the object classification task than the traditional VGG16 network, with an increase of 1.65 percentage points in the accuracy rate. Compared with the initial ResNet-50, the VS-ResNet network increases by 1.32 percentage points, and it increases by 2.97 percentage points compared with the VGG16 model. In Figure 9, the abscissa is the number of iterations, and the ordinate is the test error rate. It can be seen from Figure 9 that the VGG16 model has a certain advantage in the convergence speed, and the advantage is obvious in the first

10 iterations. After 15 iterations, the descending gradient of the model becomes smaller and gradually tends to saturation. The VS-ResNet model alleviates the problem of the slow convergence speed of ResNet-50 to a certain extent. The convergence improvement effect of the first 5 iterations is obvious, and the classification error rate after 15 iterations has obvious advantages over ResNet-50 and VGG16. To sum up, the VS-ResNet network accelerates the convergence speed of ResNet-50 and reduces the classification error rate.

4.3. Analysis of Target Detection Experiment Results

In order to verify the impact of different optimization strategies on the performance of the algorithm, the target detection experiment is based on the traditional Faster RCNN algorithm, combined with the above optimization measures.

4.3.1. AP_{50} Detection Accuracy Comparison

The experiment sets the initial learning rate to 0.005, the momentum parameter to 0.9, and the weight decay coefficient to 0.0005. AP_{50} refers to the detection accuracy when the IOU threshold is 0.5 [37,38]. The batch size is 16 and the number of iterations is 20 k. The experimental results are shown in Table 4.

Table 4. Impact of different optimization strategies.

VGG16	VS-ResNet	CBAM	FPN	Soft-NMS	Detection Accuracy AP_{50} (%)	Average Processing Time (ms)
√					62.3	275
	√				65.6	283
	√	√			69.9	292
	√	√	√		74.5	311
	√	√	√	√	76.2	316
√		√	√	√	72.4	302

Note: √ indicates that the optimization strategy uses this module.

4.3.2. AP Detection Accuracy Comparison

Comparing the data in Table 4, it can be seen that compared with Faster RCNN, CF-RCNN's detection accuracy AP_{50} increases by 13.9 percentage points to 76.2%, and the average processing time is slightly lower than that of Faster RCNN, but it can still meet the real-time requirements. In order to verify the impact of the improvement measures on the detection ability of small-sized objects, objects smaller than 32×32 pixels are selected as the experimental dataset to compare the detection accuracy. The AP value is the average AP accuracy achieved for 10 different IOU thresholds of 0.50:0.05:0.95. The results are shown in Table 5.

Table 5. Effect of FPN structure on fast RCNN performance.

Algorithm	AP (%)
Faster RCNN	5.9
Faster RCNN + FPN	19.8
Faster RCNN + FPN + Soft-NMS	21.3

It can be seen from Table 5 that after FPN and Soft-NMS are introduced into the Faster RCNN model, the AP value for small-sized object detection is increased by 15.4 percentage points, which alleviates the weak detection performance of the target detection algorithm based on small-sized objects.

4.3.3. Truncated Objects Detection Accuracy Comparison

To verify whether the improved algorithm optimizes the detection ability of the original algorithm for truncated targets, the picture is manually edited, some detected objects are intercepted, and CF-RCNN and Faster RCNN are used to detect them. The test results are shown in Figure 10. The experimental results show that the accuracy of CF-RCNN is higher than that of Fast RCNN, and the accuracy of the image frame and the possibility of classification are higher.

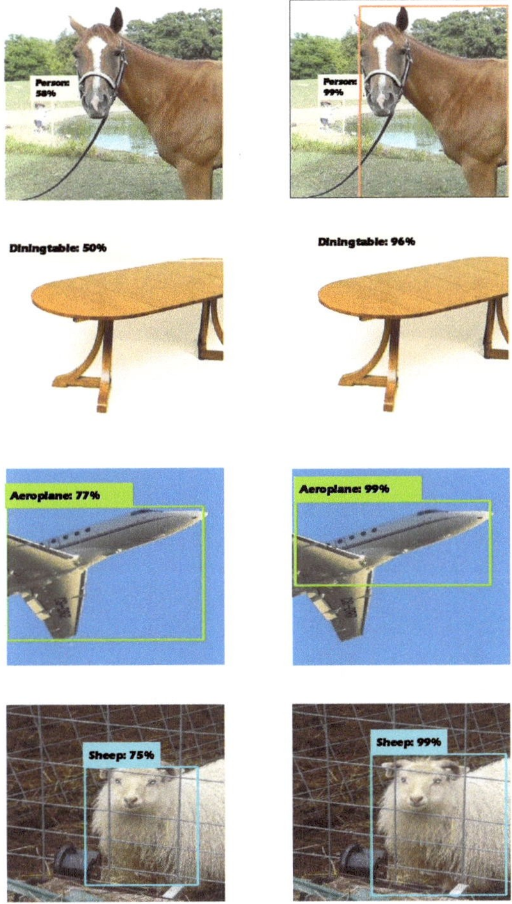

Figure 10. Comparison of truncated target detection results.

In the first row, CF-RCNN identifies a horse with only half a body, but Faster RCNN does not do this. The recognition rate of the pictures in other corresponding picture groups is also improved. The detection results of the third row of aircraft show that CF-RNN can not only effectively identify small truncated targets, but also identify large-size targets, and the recognition of the aircraft body is more complete.

4.3.4. Recall Rate Comparison

The data in Table 6 are the recall rate changes on the PASCAL VOC2012 dataset before and after algorithm optimization. Average Recall (AR) is the average recall rate, which refers to taking the maximum recall rate for different IOUs and then calculating the average [39]. The experimental data show that the recall rate of the improved algorithm on PASCAL VOC 2012 changes from the original 53.9% to 62.8%, which increases by 9.3 percentage points.

Table 6. Comparison of recall rate before and after algorithm optimization.

Algorithm	AR (%)
Faster RCNN	53.9
CF-RCNN	62.8

4.3.5. Algorithm Efficiency Comparison

Different classical algorithms and CF-RCNN algorithms are used for comparative experiments, and the AP_{50} values and average processing time changes of several algorithms are compared. The experiments are based on the same dataset to ensure that the experimental environment of several algorithms is consistent. As shown in Table 7, the classical algorithms of the experiment include YOLOv5 [40] and SSD algorithms.

Table 7. Performance comparison between classical target detection algorithm and CF-RCNN algorithm.

Algorithm	AP_{50} (%)	Average Processing Time (ms)
YOLOv5	72.9	84
SSD	60.8	91
CF-RCNN	76.2	111

4.3.6. Adaptability Comparison

Compared with the other two classical target detection algorithms, CF-RCNN has a longer processing time, but meets the basic real-time requirements, and the detection accuracy is higher than the other two algorithms.

In order to test the adaptability of the CF-RCNN algorithm to the truncation and occlusion of objects, the images with object occlusion or truncation are screened out on the PASCAL VOC 2012 dataset, and Faster RCNN and CF-RCNN are used for comparative experiments. Figure 11 shows part of the experimental results.

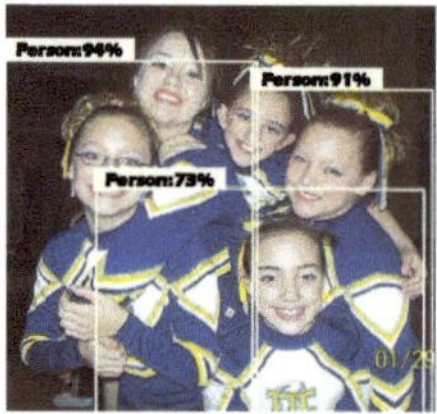

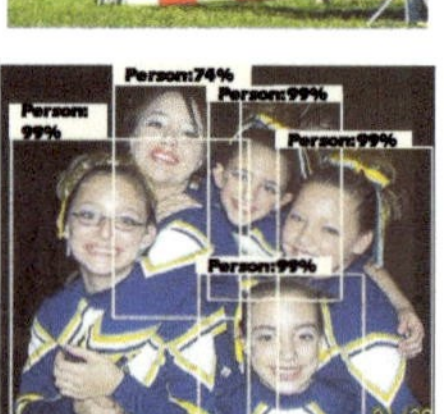

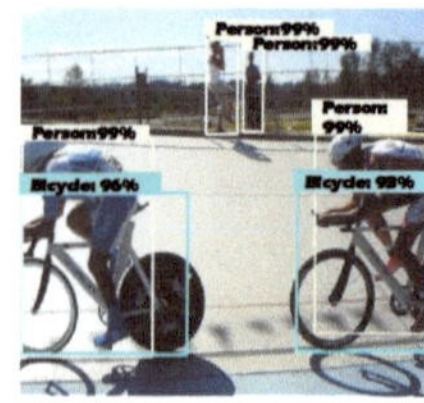

Figure 11. *Cont.*

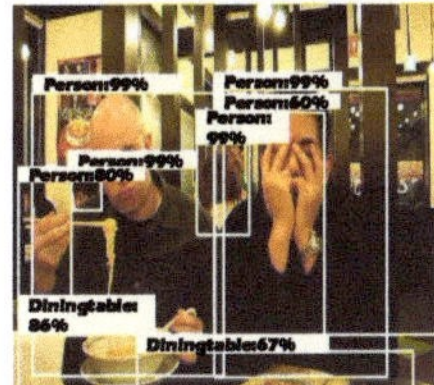

Figure 11. Comparison of detection results before and after algorithm optimization.

From the comparison of the experimental results in Figure 11, it can be seen that the traditional Faster RCNN algorithm can basically meet the detection requirements of truncated or occluded objects in a simple environment. The results of object classification are not significantly different from those of the CF-RCNN algorithm, but they are obviously inferior to the CF-RCNN algorithm in object location selection. In the case of truncation or occlusion of objects with complex scenes, traditional Faster RCNN has most false detection and missed detection phenomena; the CF-RCNN algorithm makes up for such defects well, and better alleviates the unsatisfactory detection results in the case of occlusion or truncation.

5. Conclusions

In recent years, more and more scholars have entered the field of computer vision and artificial intelligence for exploratory research, which has led to considerable development of the technology in this field. The original Faster RCNN algorithm is not ideal for the detection of small-sized objects and occluded or truncated objects. The improved CF-RCNN algorithm solves this problem well and makes up for the defects. Compared with the VGG16 network, the VS-ResNet network not only retains the characteristics of the VGG16 network with fast iterative convergence speed but also improves the recognition accuracy of the network. The introduction of the FPN and CBAM modules has greatly enhanced the detection ability of the algorithm for small-sized occluded or truncated objects. Experiments show that not only can the CF-RCNN algorithm achieve better detection results in simple scenes, but it also has a better stability based on complex environments. However, it is easy to cause the problem of multiple prediction frames for a single object. The implementation process of the algorithm is detailed, but it also increases the time complexity of the algorithm. The next step will explore and study the method of cascading multiple detectors to solve this problem.

Author Contributions: Conceptualization, W.S. and X.Y.; methodology, W.S. and X.Y.; investigation, X.Y. and J.L.; software, J.L. and X.C.; supervision, W.S. and X.Y.; writing—original draft preparation, J.L. and X.C.; writing—review and editing, W.S., X.Y. and X.C.; project administration, W.S., X.Y. and J.L. All authors have read and agreed to the published version of the manuscript.

Funding: This work was sponsored by the Key University Science Research Project of Jiangsu Province (19KJD520005), Qinglan Project of Jiangsu Province of China (Su Teacher's Letter (2021) No. 11), and the Young teacher development fund of Pujiang Institute Nanjing Tech University ((2021) No.73).

Institutional Review Board Statement: No research involving humans or animals.

Informed Consent Statement: Informed consent was obtained from all subjects involved in the study.

Data Availability Statement: Data are available on request due to restrictions, e.g., privacy or ethical.

Acknowledgments: The authors would like to thank the editor and the anonymous reviewer whose constructive comments will help to improve the presentation of this paper.

Conflicts of Interest: The authors declare no conflict of interest.

Abbreviations

RCNN	Region Convolutional Neural Network
CBAM	Convolutional Block Attention Module
FPN	Feature Pyramid Networks
CF-RCNN	Region convolutional neural network integrating convolutional block attention module and feature pyramid networks algorithm
NMS	Non-Maximum Suppression
MAP	Mean Average Precision
VGG16	VGG16 network model
VS-ResNet	VS-ResNet network
ResNet-50	ResNet-50 network
RPN	Risk Priority Number
GrabCut	GrabCut algorithm
PFF-CB	Parallel Feature Fusion with CBAM
DenseNet	Dense Convolutional network
ResNeXt	ResNeXt network
AP	Average Precision
AR	Average Recall
CNN	Convolutional Neural Network

References

1. Zhang, K.; Feng, X.H.; Guo, Y.R.; Su, Y.K.; Zhao, K. Overview of deep convolutional neural networks for image classification. *J. Image Graph.* **2021**, *26*, 2305–2325.
2. Rocha, D.A.; Ferreira, F.M.F.; Peixoto, Z.M.A. Diabetic retinopathy classification using VGG16 neural network. *Res. Biomed. Eng.* **2022**, *38*, 761–772. [CrossRef]
3. Fu, H.X.; Song, G.Q.; Wang, Y.C. Improved YOLOv4 Marine Target Detection Combined with CBAM. *Symmetry* **2021**, *13*, 623. [CrossRef]
4. Li, Y.C.; Zhou, S.L.; Chen, H. Attention-based fusion factor in FPN for object detection. *Appl. Intell.* **2022**, *52*, 15547–15556. [CrossRef]
5. Walia, I.S.; Kumar, D.; Sharma, K.; Hemanth, J.D.; Popescu, D.E. An Integrated Approach for Monitoring Social Distancing and Face Mask Detection Using Stacked ResNet-50 and YOLOv5. *Electronics* **2021**, *10*, 2996. [CrossRef]
6. Wang, A.L.; Wang, W.Y.; Zhou, H.M.; Zhang, J. Network Intrusion Detection Algorithm Combined with Group Convolution Network and Snapshot Ensemble. *Symmetry* **2021**, *13*, 1814. [CrossRef]
7. Qu, J.Y.; Liu, C. A flight delay prediction model based on the lightweight network MobileNetV2. *Signal Process.* **2022**, *38*, 973–982.
8. Shi, G.C.; Wu, Y.R.; Liu, J.; Wan, S.H.; Wang, W.H.; Lu, T. Incremental Few-Shot Semantic Segmentation via Embedding Adaptive-Update and Hyper-class Representation. In Proceedings of the 30th ACM International Conference on Multimedia, Lisboa, Portugal, 10 October 2022; pp. 5547–5556. [CrossRef]
9. Salau, A.O.; Yesufu, T.K.; Qgundare, B.S. Vehicle plate number localization using a modified GrabCut algorithm. *J. King Saud Univ. Comput. Inf. Sci.* **2019**, *33*, 399–407. [CrossRef]
10. Yang, G.Y.; Wang, Z.Y.; Zhuang, S.N.; Wang, H. PFF-CB: Multiscale occlusion pedestrian detection method based on PFF and CBAM. *Comput. Intell. Neurosci.* **2022**, *2022*, 3798060. [CrossRef] [PubMed]
11. Zhai, D.R.; Shang, D.; Wang, S.; Dong, S. DF-SSD: An improved SSD object detection algorithm based on DenseNet and feature fusion. *IEEE Access* **2020**, *8*, 24344–24357. [CrossRef]
12. Jawad, T.; Maisha, B.; Nazmus, S. Targeted face recognition and alarm generation for security surveillance using single shot multibox detector (SSD). *Int. J. Comput. Appl.* **2019**, *177*, 8–13.
13. Wang, D.; Hu, L.; Fang, J.; Xu, Z. Small object detection algorithm based on improved double-head RCNN for UAV aerial images. *J. Beijing Univ. Aeronaut. Astronaut.* **2023**. prepublish. [CrossRef]
14. Pang, N.R.; Le, Q.V. EfficientDet: Scalable and efficient object detection. In Proceedings of the 2020 IEEE Conference on Computer Vision and Pattern Recognition, Seattle, WA, USA, 5 February 2020; pp. 10778–10787.
15. Zhang, Z.; Zhang, S.; Wang, H.; Ran, X. Multi-head attention detection of small targets in remote sensing at multiple scales. *Comput. Eng. Appl.* **2022**. prepublish. [CrossRef]

16. Fu, C.-Y.; Liu, W.; Ranga, A.; Tyagi, A.; Berg, A.C. DSSD: Deconvolutional single shot detector. In Proceedings of the 2017 IEEE Conference on Computer Vision and Pattern Recognition (CVPR 2017), Honolulu, HI, USA, 23 January 2017; pp. 2881–2890.

17. Singh, B.; Davis, L.S. An analysis of scale invariance in object detection snip. In Proceedings of the 2018 IEEE Conference on Computer Vision and Pattern Recognition, Salt Lake City, UT, USA, 18–23 June 2018; pp. 3578–3587.

18. Pazhani, A.A.J.; Vasanthanayaki, C. Object detection in satellite images by faster R-CNN incorporated with enhanced ROI pooling (FrRNet-ERoI) framework. *Earth Sci. Inform.* **2022**, *15*, 553–561. [CrossRef]

19. Salau, A.O.; Jain, S. Feature extraction: A survey of the types, techniques, applications. In Proceedings of the 5th IEEE International Conference on Signal Processing and Communication (ICSC), Noida, India, 7–9 March 2019; pp. 158–164.

20. Seong, J.H.; Lee, S.H.; Kim, W.Y.; Seo, D.H. High-precision RTT-based indoor positioning system using RCDN and RPN. *Sensors* **2021**, *21*, 3701. [CrossRef] [PubMed]

21. Catelani, M.; Ciani, L.; Galar, D.; Patrizi, G. Risk assessment of a wind turbine: A new FMECA-based tool with RPN threshold estimation. *IEEE Access* **2020**, *8*, 20181–20190. [CrossRef]

22. Akiyoshi, H.; Shinya, K.; Ryohei, N. Computerized classification method for histological classification of masses on breast ultrasonographic images using convolutional neural networks with ROI pooling. *Electron. Commun. Jpn.* **2022**, *105*, 3.

23. Szostak, D.; Włodarczyk, A.; Walkowiak, K. Machine learning classification and regression approaches for optical network traffic prediction. *Electronics* **2021**, *10*, 1578. [CrossRef]

24. Anissa, B.; Jamal, K.; Arsalane, Z. Face recognition using SVM based on LDA. *Int. J. Comput. Sci. Issues (IJCSI)* **2013**, *10*, 171–179.

25. Kim, Y. Robust selective search. *ACM SIGIR Forum* **2019**, *52*, 170–171. [CrossRef]

26. Meng, Y.B.; Shi, D.W.; Liu, G.H.; Xu, S.J.; Jin, D. Dense irregular text detection based on multi-dimensional convolution fusion. *Opt. Precis. Eng.* **2021**, *29*, 2210–2221. [CrossRef]

27. Zhu, X.L.; He, Z.L.; Zhao, L.; Dai, Z.C.; Yang, Q.L. A Cascade Attention Based Facial Expression Recognition Network by Fusing Multi-Scale Spatio-Temporal Features. *Sensors* **2022**, *22*, 1350. [CrossRef] [PubMed]

28. Zhu, W.; Ma, L.X.; Zhang, P.; Liu, D.Y. Morphological recognition of rice seedlings based on GoogLeNet and UAV images. *J. South China Agric. Univ.* **2022**, *43*, 99–106.

29. Chen, Q.; Zhang, W.Y.; Zhu, K.; Zhou, D.; Dai, H.; Wu, Q.Q. A novel trilinear deep residual network with self-adaptive Dropout method for short-term load forecasting. *Expert Syst. Appl.* **2021**, *182*, 115272. [CrossRef]

30. Liao, Z.H.; Fan, N.; Xu, K. Swin Transformer Assisted Prior Attention Network for Medical Image Segmentation. *Appl. Sci.* **2022**, *12*, 4735. [CrossRef]

31. Li, L.; Fang, B.H.; Zhu, J. Performance Analysis of the YOLOv4 Algorithm for Pavement Damage Image Detection with Different Embedding Positions of CBAM Modules. *Appl. Sci.* **2022**, *12*, 10180. [CrossRef]

32. Wu, Y.R.; Guo, H.F.; Chakraborty, C.; Khosravi, M.; Berretti, S.; Wan, S.H. Edge Computing Driven Low-Light Image Dynamic Enhancement for Object Detection. *IEEE Trans. Netw. Sci. Eng.* **2022**, 1. [CrossRef]

33. Feng, T.; Liu, J.G.; Fang, X.; Wang, J.; Zhou, L.B. A Double-Branch Surface Detection System for Armatures in Vibration Motors with Miniature Volume Based on ResNet-101 and FPN. *Sensors* **2020**, *20*, 2360. [CrossRef]

34. Liu, Z.Y.; Yuan, L.; Zhu, M.C.; Ma, S.S.; Chen, L.Z.T. YOLOv3 Traffic sign Detection based on SPP and Improved FPN. *Comput. Eng. Appl.* **2021**, *57*, 164–170.

35. Lv, X.Y. CIFAR-10 Image Classification Based on Convolutional Neural Network. *Front. Signal Process.* **2020**, *4*, 100–106. [CrossRef]

36. Wang, X.H.; Ye, Z.X.; Wang, W.J.; Zhang, L. High precision semantic segmentation based on multi-level feature fusion. *J. Xi'an Polytech. Univ.* **2021**, *35*, 43–49.

37. Chen, X.S.; Su, T.; Qi, W.M. Printed circuit board defect detection algorithm based on improved faster RCNN. *J. Jianghan Univ.* **2022**, *50*, 87–96.

38. Wu, S.K.; Yang, J.R.; Wang, X.G.; Li, X.P. IoU-Balanced loss functions for single-stage object detection. *Pattern Recognit. Lett.* **2022**, *156*, 96–103. [CrossRef]

39. Chen, C.; Wang, C.Y.; Liu, B.; He, C.; Cong, L.; Wan, S.H. Edge Intelligence Empowered Vehicle Detection and Image Segmentation for Autonomous Vehicles. *IEEE Trans. Intell. Transp. Syst.* **2023**, 1–12. [CrossRef]

40. Jia, D.Y.; He, Z.H.; Zhang, C.W.; Yin, W.T.; Wu, N.K. Detection of cervical cancer cells in complex situation based on improved YOLOv3 network. *Multimed. Tools Appl.* **2022**, *81*, 8939–8961. [CrossRef]

Article

A Combined Multi-Classification Network Intrusion Detection System Based on Feature Selection and Neural Network Improvement

Yunhui Wang [1,2,†], Zifei Liu [3,†], Weichu Zheng [3], Jinyan Wang [1,2], Hongjian Shi [3,*] and Mingyu Gu [4]

[1] National Key Laboratory of Science and Technology on Avionics System Integration, Shanghai 200233, China; wang.yh@outlook.com (Y.W.); wangjy121@avic.com (J.W.)
[2] China National Aeronautical Radio Electronics Research Institute, Shanghai 200233, China
[3] School of Electronic Information and Electrical Engineering, Shanghai Jiao Tong University, Shanghai 200240, China; liuzifei@sjtu.edu.cn (Z.L.); sjtu_zwc0518@sjtu.edu.cn (W.Z.)
[4] Sino-European School of Technology, Shanghai University, Shanghai 200244, China; 22124558@shu.edu.cn
* Correspondence: shhjwu5@sjtu.edu.cn; Tel.: +86-137-1795-9365
† These authors contributed equally to this work.

Featured Application: Increased parallelism in edge network security issues, Feature loss handling.

Abstract: Feature loss in IoT scenarios is a common problem. This situation poses a greater challenge in terms of real-time and accuracy for the security of intelligent edge computing systems, which also includes network security intrusion detection systems (NIDS). Losing some packet information can easily confuse NIDS and cause an oversight of security systems. We propose a novel network intrusion detection framework based on an improved neural network. The new framework uses 23 subframes and a mixer for multi-classification work, which improves the parallelism of NIDS and is more adaptable to edge networks. We also incorporate the K-Nearest Neighbors (KNN) algorithm and Genetic Algorithm (GA) for feature selection, reducing parameters, communication, and memory overhead. We named the above system as Combinatorial Multi-Classification-NIDS (CM-NIDS). Experiments demonstrate that our framework can be more flexible in terms of the parameters of binary classification, has a fairly high accuracy in multi-classification, and is less affected by feature loss.

Keywords: feature loss; network intrusion detection systems; multi-classification; attack-type identification

Citation: Wang, Y.; Liu, Z.; Zheng, W.; Wang, J.; Shi, H.; Gu, M. A Combined Multi-Classification Network Intrusion Detection System Based on Feature Selection and Neural Network Improvement. *Appl. Sci.* **2023**, *13*, 8307. https://doi.org/10.3390/app13148307

Academic Editor: Christos Bouras

Received: 12 June 2023
Revised: 4 July 2023
Accepted: 14 July 2023
Published: 18 July 2023

1. Introduction

Smart edge computing systems are distributed computing architectures that bring computing and data processing power closer to the network edge from traditional centralized cloud computing environments [1]. The design goal of smart edge computing systems is to place computing resources as close as possible to the data generation source or data usage endpoint to provide a lower latency, reduce network bandwidth pressure, and enhance user experience.

Smart edge computing systems face important challenges. Security and privacy have become more complex and critical in distributed environments [2,3]. Protecting systems from the threat of cyber attacks, data breaches, and malicious behavior is critical. At the same time, ensuring the security and privacy of data during transmission and storage, as well as effective authentication, access control, and vulnerability patching mechanisms, are key measures to ensure the security and privacy of smart edge computing systems.

One security measure in smart edge computing systems is NIDS (network intrusion detection system), which is used to monitor and detect potential intrusions in network traffic. NIDS detects intrusion attacks before they cause harm to the system and uses the alarm and protection system to expel them [4]. After the attack, information about the intrusion attack can be collected and added to the knowledge base as a prevention system, thus enhancing prevention capability [5]. NIDS should provide a better picture of the system and facilitate establishing a system security system.

A large proportion of NIDS is based on machine learning for algorithm design. Machine learning methods have been used to try to improve IDSs to detect malicious communications [6]. In addition, benchmark datasets on malware and cyber attacks have been published to provide IDS research [7–9]. Deep neural networks [10,11] can learn complex causal relationships by training large amounts of data; after training is complete, detection can be performed quickly, regardless of the amount of training data [12].

NIDS consists of several stages: data collection, pre-processing, FS (feature selection), and classification. Recently, attention has been focused on FS and optimization techniques to select basic features. Feature selection has proven to be an effective method to prepare large amounts of dimensional data for machine learning and data mining [13]. The FS phase is critical because NIDS must process large amounts of data, and the direct use of raw feature sets leads to a significant decrease in efficiency [14]. At the same time, edge computing systems in IoT environments may not be able to provide the full functionality and features of certain devices due to the diversity of devices, network connectivity, and resource constraints. By performing feature selection, features that are irrelevant or unimportant for the data analysis task in a particular scenario can be eliminated. This reduces the feature dimensionality and reduces the computational complexity of model training and inference, while also reducing sensitivity to feature incompleteness.

However, the detection rate of NIDS also decreases due to the loss of some features caused by FS. This is unacceptable for IoT environments with high traffic [15]. Also, current NIDSs mainly support binary classification; they only determine whether a packet is anomalous or not. For multi-ecological network systems, the large number of data values leads to systems that often have multiple security measures. Binary classification is challenging in informing and guiding other security deployments, which is very detrimental to systems in the IoT.

In this paper, we propose a new network intrusion detection framework named CM-NIDS, which is divided into two main parts: pre-processing and multi-classification. In the pre-processing part, we filter the dataset and select features based on genetic algorithms (GA) to obtain the feature set. In the multi-classification part, we use K-Nearest Neighbor (KNN) to determine the type of packets and an innovative combinatorial architecture based on a neural network to determine the particular type of data packet.

The main contributions of this work are as follows:

(1) Introduction of CM-NIDS: This work presents a new network intrusion detection framework called Combined Multi-Classification-NIDS (CM-NIDS). This framework is based on an improved neural network and supports other security systems by accurately matching the attack type of the packets. The introduction of CM-NIDS addresses the need for effective network intrusion detection and enhances overall system security and parallelism.

(2) KNN-GA-Based Feature Selection: This work proposes a feature selection approach using a combination of the K-Nearest Neighbors (KNN) algorithm and Genetic Algorithm (GA). This approach significantly reduces the number of parameters the neural network uses, leading to reduced communication and memory overhead. Additionally, it helps mitigate the impact of feature incompleteness, thereby improving the model's robustness and performance. Introducing this feature selection technique contributes to more efficient and effective network intrusion detection.

(3) Data Pre-Processing and Feature Selection: This work emphasizes the importance of data pre-processing methods and specifically highlights the significance of feature selection. The authors provide detailed insights into the pre-processing techniques employed, which contribute to improving the quality and relevance of the input data.

(4) Comparative Evaluation: This work evaluates the proposed CM-NIDS framework alongside existing frameworks. The evaluation demonstrates that CM-NIDS achieves higher attack-type matching rates than other methods.

The remainder of this paper is as follows. Section 2 describes the methodology of our proposed CM-NIDS approach. Section 3 illustrates the results and discussion. Section 4 concludes this research work.

2. Related Works

The Internet of Things (IoT) is a transformative technology that connects everyday objects to the internet, enabling them to send and receive data. It involves a vast network of interconnected devices, ranging from simple sensors and actuators to complex machinery, vehicles, and even smart homes and cities. In recent years, as networks have grown, the data traffic from IoT has created challenges for security maintenance. Waseem et al. [16] discuss the current scale and security challenges facing the IoT, provide a detailed analysis of the attacks emerging at each layer of the IoT, and argue for the need for a comprehensive security framework. Ramalingam et al. [17] think that the rapidly expanding volume of computerized matter poses many research challenges for efficiently storing, processing, and viewing large quantities.

A network intrusion detection system (NIDS) detects behavior that compromises the security of a computer system by collecting information. Its ultimate aim is to identify potential attacks from the flow of messages on the network. Ennaji et al. [18] propose five collaborative learning models for the IDS problem, aiming at classifying network traffic and achieving optimal results. This is based on a stacked integrated learning technique considering some essential machine learning algorithms. However, they only consider 10 features that do not meet the needs of real situations.

Over the past decade, researchers have proposed various ML- [19] and DL-based techniques to improve the efficacy of NIDS in identifying malicious threats. Nevertheless, the significant increase in network traffic and associated attack vectors has created challenges for NIDS systems to identify malicious intrusions. Tosin et al. [19] applied four machine learning models in parallel, K-Nearest Neighbor (K-NN), Parsimonious Bayes (NB), Logistic Regression (LR), and Artificial Neural Network (ANN) with multilevel feature selection methods, and compared multiple metrics to determine which model has the best detection capability in terms of accuracy. Desale et al. [20] used the Genetic Algorithm (GA) to optimize a feature analysis, reducing the number of features selected and the cost of feature maintenance.

The IoT environment is prone to feature loss problems due to network transmission problems, energy limitations, sensor failures, and data collection and storage problems, which can be solved with feature selection. Recent research on NIDS has partly focused on FS (feature selection) techniques and optimization techniques to select the most important features. Jyoti et al. [21] provide an overview of feature selection methods used in NIDS, describe existing feature selection frameworks and classifications for NIDS, and argue that there is an urgent need for new FS to meet the demands of big data. Su et al. [22] consider the feature redundancy of FS and propose a learning automata approach to select optimal and salient features for network traffic intrusion detection. Heather et al. [23] pre-process feature detection, and features were pre-screened based on an existing and admittedly more comprehensive dataset to improve the efficiency of machine learning.

3. Materials and Methods

3.1. Framework

Our work is divided into two main modules, pre-processing and multi-classification. The main purpose of pre-processing is to perform feature selection to reduce the impact of feature loss on NIDS and reduce the workload of NIDS, improve efficiency, and reduce memory usage. Multi-classification is based on neural networks to improve the parallelism and accuracy of NIDS. Their organizational structure is shown in Figure 1.

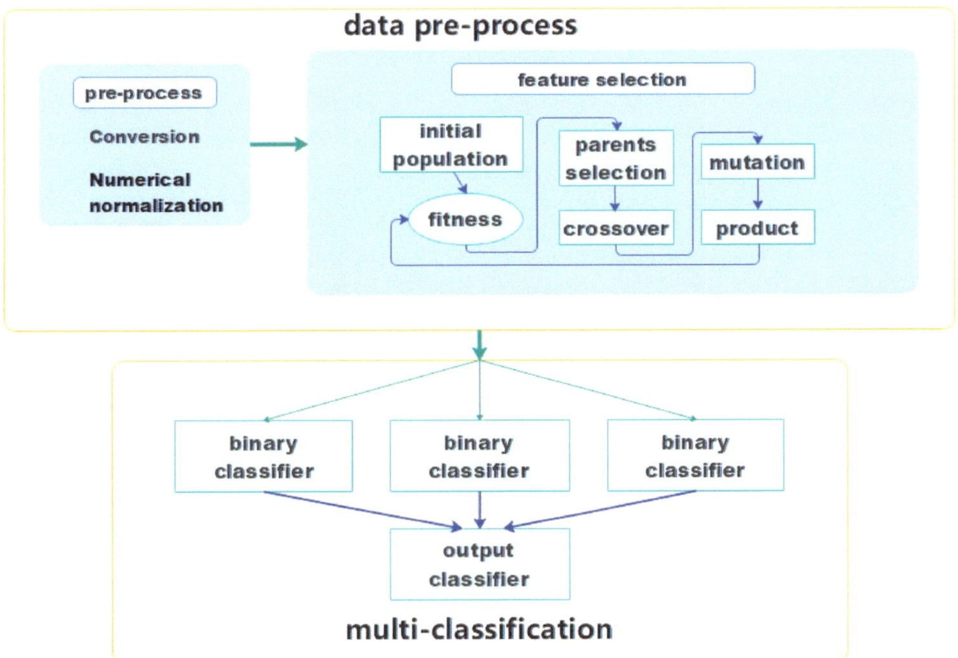

Figure 1. The overall framework of our model. The yellow parts are the two important components of our model, data pre-processing and multi-classification, respectively. The two components of data pre-processing are shown in blue, the process of multi-classification is shown in green boxes, and the arrows indicate the data flow direction.

Pre-processing includes two parts: data pre-processing and feature selection. The output of pre-processing is the filtered dataset and the selected feature set. The work of multi-classification is taking the previous dataset and determining the specific type of attack; its final output is the type of attack for all abnormal packets.

3.2. Pre-Processing

We used the representative KDDCup99 dataset [24], which is a stream-based dataset with additional information from packet-based data or host-based log files. These data are available in a total dataset volume and 10% data volume. They have 41 features with five distinct attack classes. The specific types of attacks they include are shown in Table 1.

Table 1. Specific attack types in KDDCup99, corresponding to five attack classes.

Attack Categories	Specific Attack Types
DoS	Back, land, Neptune, pod, smurf, teardrop
Probe	Ipsweep, nmap, portsweep, satan
R2L	ftp_write, guess_passwd, imap, multihop, phf, spy, warezclient, warezmaster
U2R	buffer overflow, loadmodule, perl, rootkit

For data pre-processing, our input is the initial dataset, and we obtained numerical datasets that can be used directly for FS through pre-processing. The pre-processing of the dataset consists of two steps: the conversion of character-based features to numeric features (the next part we will refer to as the conversion), and numerical normalization. In the conversion part, we performed the following: convert all feature types (including normal) into numeric identifiers.

In the numerical normalization part, we first calculated the mean and mean absolute error for each attribute using the following formula:

$$\overline{x}_k = \frac{1}{n}\sum_{i=1}^{n} x_{ik} \tag{1}$$

$$S_k = \sqrt{\frac{1}{n}\sum_{i=1}^{n}(x_{ik} - \overline{x})^2} \tag{2}$$

where x_k denotes the mean value of the kth attribute, S_k denotes the mean absolute error of the kth attribute, and x_k denotes the kth attribute of the i-th record.

Then, we apply the normalization metric to each data record:

$$Z_{ik} = \frac{x_{ik} - \overline{x}_k}{S_k} \tag{3}$$

where Z_{ik} denotes the kth attribute value of the i-th data record after normalization.

We then normalize each value to the interval [0, 1] as follows:

$$x^* = \frac{x - x_{min}}{x_{max} - x_{min}} \tag{4}$$

where x_{max} is the maximum value of the sample data, x_{min} is the minimum value of the sample data, and x is the previously obtained data. In this way, the data pre-processing gives us a numerical, normalized dataset suitable for FS use.

For feature selection, we used feature selection to obtain the feature set that can be used for classification. FS is the method of selecting unique features based on specific criteria in machine learning. For a dataset, three components do not contribute to feature selection, which we call noise. These noises are irrelevant, redundant, and noisy. We distinguish between these three types of noise by using Table 2 as follows.

Table 2. Different feature components during feature selection.

Feature Type	Description	Accuracy	Training Time
Strongly Relevant	It is necessary for an optimal feature subset.	High	Low
Irrelevant	It cannot help discriminate between supervised or unsupervised data.	Low	High
Redundant	It can be completely replaced with a set of other features.	Low	High
Noisy	It is not relevant to the learning or mining task.	Low	High

In feature selection, feature subsets are selected from the original feature set based on feature redundancy and relevance [25]. We considered GA-KNN as our algorithm. In machine learning, one of the uses of Genetic Algorithms is to obtain the correct number of variables to create predictive models. The use of GA for FS has been more widely discussed and researched [26,27]. Feature selection techniques are used to accurately identify relevant subsets of features, reducing the number of dimensions to improve overall efficiency. Feature selection helps reduce the possibility of overfitting while reducing the effect of feature loss by focusing on certain features.

In this process, the addition of KNN improves the efficiency of GA. This is because it can artificially enhance the composition of the population by selecting new members that are closer to quality individuals as much as possible.

The pseudocode of GA is as Algorithm 1.

Algorithm 1. GA for feature selection

1	Initialize population P from the dataset randomly
2	**for** each iteration **do**
3	Randomly select a subset of the population P as $Psub$
4	Crossing operation:
5	**for** each pair of individual(P^i, P^{i+1}) in $Psub$ **do**
6	Generate a random probability p
7	**if** $Pc > p$ **then**
8	randomly choose position P_a.
9	Exchange the genes
10	$P^i = [\ldots, P_a^{i+1}, P_{a+1}^{i+1}, P_{a+2}^{i+1}, \ldots]$
11	$P^{i+1} = [\ldots, P_a^{i}, P_{a+1}^{i}, P_{a+2}^{i}, \ldots]$
12	**end if**
13	**end for**
14	Mutation operation:
15	**for** each individual P^i in $Psub$ **do**
16	Generate a random probability p
17	**if** $Pm > p$ **then**
18	Random change a gene in Pi with a random device
19	**end if**
20	**end for**
21	Reverse operation:
22	**for** each individual P^i in $Psub$ **do**
23	Randomly choose two positions Pa and Pb where $Pa < Pb$
24	Reverse the genes between Pa and Pb to get $P^{i'}$
25	**if** f $(P^{i'}) >$ f (P^i) **then**
26	$P^i = P^{i'}$
27	**end if**
28	**end for**
29	Add $Psub$ into population P.
30	Calculate the fitness values f $(P^i) = 1/L[P^i]$ for all individuals in P
31	Sorted P by descending order of fitness value.
32	Keep the first several individuals in P according to population size
33	**end for**
34	**return** The best feature set $S^{best} = P^0$

GA simulates natural selection and inheritance. Continuous random selection, crossover, and mutation operations generate individuals better suited to the environment, converging into the fittest individuals. This approach yields quality solutions to problems. The 41 chromosome features are converted into binary code for machine learning algorithms. Each individual's genotype consists of 41 genes. In GA chromosome coding, binary strings (1 s and 0 s) represent selected and non-selected feature subsets. Random binary bit assignment ensures population diversity.

For functional selection, the proposed enhancement of GA in feature selection starts with the training data as the input and then generates the feature chromosomes. Each chromosome will be evaluated. The chromosome with the highest classification accuracy will be selected and considered as the optimized feature chromosome. The next step is to use the optimized feature chromosomes from the training data to train the KNN (K-Nearest Neighbor). The KNN algorithm is a basic classification and regression algorithm—a method in supervised learning. Given a training set M and a test object n, where the object is a vector with an attribute value and unknown category label. The distance (typically Euclidean distance) or similarity (typically cosine similarity) between object m and each object in the training set is calculated. The nearest neighbors are identified, and the top K categories with the highest frequency among them are assigned to the test object z. This classifier helps identify network activity classes as normal or abnormal using the calculated accuracy.

On the other hand, test data are used to evaluate the performance of the classifier to check whether the trained algorithm is able to generalize new information based on the evaluation results encountered from the training data. This process will improve the new generation of chromosomes without sacrificing population diversity.

The selection operation ensures that the best chromosomes representing the best solution will be selected as parents to produce better-performing offspring for the next generation. This study used tournament selection. After selecting the best-performing chromosomes, crossover operations were applied to produce new progeny, resulting in better chromosomes. This is performed by mixing the parents' genes using crossover probabilities (Pc). This study applied a single-point crossover. The mutation is performed with random bit-flip mutations in individual chromosomes, where substrings are randomly selected and flipped using the mutation probability (Pm). Specifically, we randomly selected and reversed a contiguous segment of a selected individual's genes. Here, we used 30 iterations, a crossover rate (Pc) of 0.7, and a variation rate (Pm) of 1/41, i.e., 0.02. The algorithm described above was designed to efficiently select features with the highest accuracy scores in the training data.

3.3. Multi-Classification

For multi-classification, we already obtained the dataset and the feature set. For this section, we obtained the specific type of attack on the set of anomalies. For the sake of description, we consider the normal packet as a particular case of the abnormal packet.

We did not use a neural network directly for the classification of the anomaly dataset. Instead, it was deformed and simplified to some extent. On the one hand, the penalty of error detection is too small for the neural network to provide a good training effect. On the other hand, we simplified the multi-classification step as much as possible; the direct deployment of neural networks is unrealistic due to the parameters and complexity of deploying neural networks at a particular scale [28]. Edge devices usually have a limited computational power and storage resources. Neural networks usually require many computational resources to perform training and inference tasks, especially when the models are large and the number of parameters is large. The computational power of edge devices may need to be improved to meet the high requirements of complex neural networks. Parallel computing can accelerate the training and inference process of neural networks by simultaneously utilizing multiple computing devices or processing units, thus improving overall performance and efficiency. We improved the architecture of the multi-classifier by splitting it into 23 binary classifiers, resulting in the attack set. The framework is shown in Figure 2.

As shown in Figure 2, using neural networks alone for multi-classification means that the packets need to be matched directly for 23 cases, which is demanding regarding the data volume and computing power. We improved the above algorithm by delivering the packets to each of the 23 binary-classification neural networks simultaneously. Each neural network was responsible for determining whether the packet was of one of its corresponding types. We set up a mixer after the 23 neural networks, which aggregated and decided the type of packets.

The results of the sub-models were integrated through the classifier layer. Specifically, each sub-model output a two-dimensional vector representing the probabilities of the corresponding sample belonging to two different classes. These output vectors were processed through the classifier layer, which consisted of a linear layer that mapped the input dimension from 23 to 23.

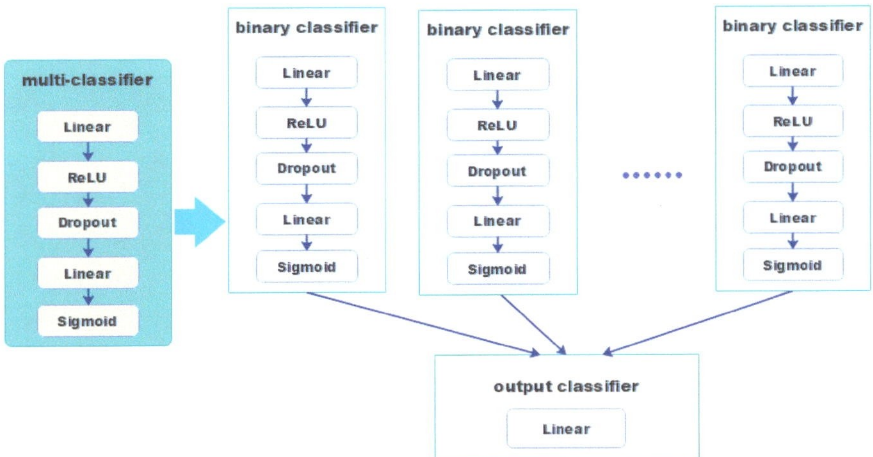

Figure 2. A multi-classification scheme based on binary classification. The left figure shows the general multi-classification model, which uses a multi-classifier directly for attack-type matching based on the extracted feature set and dataset. The figure on the right is a multi-classifier based on 23 binary classifiers. The output classifier will aggregate and filter the results of the binary classifiers.

The pseudocode of CM-NIDS is as Algorithm 2.

The final output vector can be viewed as a prediction for multi-class classification, where each dimension corresponds to a class, and the values indicate the model's confidence or probability for that class.

The integration process is not performed directly within the sub-models but rather through the model's overall structure. The outputs of the sub-models are passed as inputs to the classifier layer, which linearly maps these outputs to generate the final prediction. Through the forward propagation process of the entire network, the input features undergo processing with the feature layer, multiple sub-models, and the classifier layer to obtain the predicted class for the input sample.

Overall, our model leverages multiple sub-models to learn diverse and specialized feature representations, integrates their outputs through a classifier layer for ensemble benefits, allows flexibility in adapting to different tasks and datasets, increases model depth for enhanced expressive capacity, and provides interpretability through a clear delineation of sub-model functionality. These characteristics contribute to an improved performance, broader representation capabilities, and a better understanding and interpretability of the model, advancing the application of neural network models in scientific research.

We obtained specific types of abnormal attacks to facilitate a detailed and reliable reference for other network security systems. Our model can be better adapted to edge networks. First, since the model is split into multiple sub-models, these can be deployed on different nodes or devices of the distributed system, thus enabling parallel computation and processing. Such parallelism can accelerate the inference process of the model and improve the overall computational efficiency and throughput. Second, for edge networks, the structure split into multiple sub-models may be better adapted to the characteristics of the edge environment. Edge networks usually have limited computational resources and bandwidth and must make real-time decisions and reasoning locally. By splitting the model into multiple sub-models, the computation can be distributed to the edge devices, reducing the reliance on the central server and communication overhead. This distributed deployment can improve the real-time responsiveness and reliability of the edge network.

Algorithm 2. NIDS based on neural network with 23 sub-structures

---

Input: Training set $\mathbf{D} = \{(x^{(n)}, y^{(n)})\}^N n = 1$, Validation set $\mathbf{V}$, Learning rate α, Regularization factor λ, Number of network layers $\mathbf{L}$, Number of neurons $\mathbf{M}_l$.

1 Initialize $\mathbf{D}$, $\mathbf{V}$
2 **repeat**
3   Random reordering of the samples in the training set $\mathbf{D}$
4   **for** n = 1 … N **do**
5   Select sample (x(n), y(n)) from the training set $\mathbf{D}$
6   Feedforward calculation of net input $\mathbf{z}^{(l)}$ and activation value $\alpha^{(l)}$ for

        each layer, up to the last layer
        // Feedforward through feature extraction layers

7   **for** l = 1 **to L** $-$ **1 do**
8        $\mathbf{z}^{(l)} = \mathbf{W}^{(l)} * \alpha^{(l-1)} + \mathbf{b}^{(l)}$
9        $\alpha^{(l)} = \sigma(\mathbf{z}^{(l)})$
10   **end for**

        // Feedforward through the 23 sub-structures

11   **for** i = 1 **to 23 do**
12       $\mathbf{z}^{(L,i)} = \mathbf{W}^{(L,i)} * \alpha^{(L-1)} + \mathbf{b}^{(L,i)}$
13       $\alpha^{(L,i)} = \sigma(\mathbf{z}^{(L,i)})$
14   **end for**
15   Backpropagation calculates the error $\delta^{(l)}$ for each layer
16   **for** i = 1 **to 23 do**

        $\delta^{(L,i)} = \partial L(y^{(n)}, \overline{y}^{(n)})/\partial z^{(L,i)}$

17   **end for**
18   **for** l = L $-$ 1 **to 1 do**

        $\delta^{(l)} = (\partial L(y^{(n)}, \overline{y}^{(n)}))/\partial z^{(l)}) * (\mathbf{W}^{(l+1)T} * \delta^{(l+1)})$

19   **end for**
20   **for** l = 1 **to L do**

        $\partial L(y^{(n)}, \overline{y}^{(n)})/\partial \mathbf{W}^{(l)} = \delta^{(l)} * \alpha^{(l-1)T}$
        $\partial L(y^{(n)}, \overline{y}^{(n)})/\partial b^{(l)} = \delta^{(l)}$
        end for

21   Update parameters
22   **for** l = 1 **to L** $-$ **1 do**
23       $\mathbf{W}^{(l)} = \mathbf{W}^{(l)} - \alpha * \partial L(y^{(n)}, \overline{y}^{(n)})/\partial \, \mathbf{W}^{(l)} - \lambda * \mathbf{W}^{(l)}$
24       $\mathbf{b}^{(l)} = \mathbf{b}^{(l)} - \alpha * \partial L(y^{(n)}, \overline{y}^{(n)})/\partial \, \mathbf{b}^{(l)}$
25   **end for**

      //Update parameters for the 23 sub-structures

26   **for** i = 1 **to 23 do**

        $\mathbf{W}^{(L,i)} = \mathbf{W}^{(L,i)} - \alpha * \partial L(y^{(n)}, \overline{y}^{(n)})/\partial \, \mathbf{W}^{(L,i)} - \lambda * \mathbf{W}^{(L,i)}$
        $\mathbf{b}^{(L,i)} = \mathbf{b}^{(L,i)} - \alpha * \partial L(y^{(n)}, \overline{y}^{(n)})/\partial \, \mathbf{b}^{(L,i)}$
        end for

27   **end for**
28 **until** The accuracy of neural networks no longer decreases on the validation

    set $\mathbf{V}$

Output: W, b

---

4. Results

4.1. Settings

The dataset we used is KDDCup99 (http://kdd.ics.uci.edu/databases/kddcup99/kddcup99.html, accessed on 2 November 2022), a stream-based dataset enriched with additional information from packet-based data or host-based log files, so its data is adequate and representative.

This experimental section involves three main parts. First, we explain the environment and settings we used; afterward, we compare our method with others regarding accuracy and speed. Finally, regarding carrying out other experiments, we analyze them in light of the above.

We experimented on a seven-device cluster where each device was a GeForce RTX 2080 Ti GPU. The implementation was based on a PyTorch distributed library. In conjunction with the use of the PyTorch distributed library, we simulated the situation in an edge network, with distributed training through multiple GPU devices.

In terms of evaluation indicators, we considered two main aspects. Firstly, for NIDS, its detection and correctness rates for errors are significant. We used TPR, Precision, and Recall to evaluate this. In statistics, True Positives (TP) are the correctly predicted positive values, which means that the value of the actual class is yes and the value of the predicted class is also yes. True Negatives (TN) are the correctly predicted negative values, which means that the actual class's value is no and the predicted class's value is also no. False Positives and False Negatives occur when the actual class contradicts the predicted class. False Positives (FP) indicate the situation when the real class is no and the predicted class is yes. False Negatives (FN) indicate the case when the real class is yes, but the predicted class is no. In the evaluation index of the binary-classification results, we used the following parameters:

$$\mathrm{TPR} = \frac{\mathrm{TP} + \mathrm{TN}}{\mathrm{TP} + \mathrm{TN} + \mathrm{FP} + \mathrm{FN}} \tag{5}$$

$$\mathrm{recision} = \frac{\mathrm{FP}}{\mathrm{TP} + \mathrm{FP}} \tag{6}$$

$$\mathrm{Recall} = \frac{\mathrm{TP}}{\mathrm{TP} + \mathrm{FN}} \tag{7}$$

Furthermore, for multi-classification networks, correctly identifying the type of attack is fundamental. We call this the type detection rate (TDR):

$$\mathrm{TDR} = \frac{\mathrm{TD}}{\mathrm{TOTAL}} \tag{8}$$

TD is the number of packets for which the type of attack was correctly identified. TOTAL is the total number of packets.

It is also essential for NIDS to work efficiently. We compared efficiency as a metric by using the same dataset and recording the running time of each method.

The backbone of the neural networks we used in our experiments is shown in Figure 3.

4.2. Baseline

We evaluated NIDS primarily in terms of accuracy in dichotomous and multi-classification. Firstly, we compared the differences between our method and other NIDSs regarding False Negatives, accuracy, etc. KNN-NIDS [29] uses KNN as the classification algorithm of the project and uses PCA for dimension reduction. LR-NIDS [30], Bayes-NIDS [31], DT-NIDS [32], and AB-NIDS [33] use a Logistic Regression, Naive Bayesian algorithm, Decision Tree algorithm, Ada Boost algorithm, and Random Forest algorithm, respectively, in the NIDS scenario.

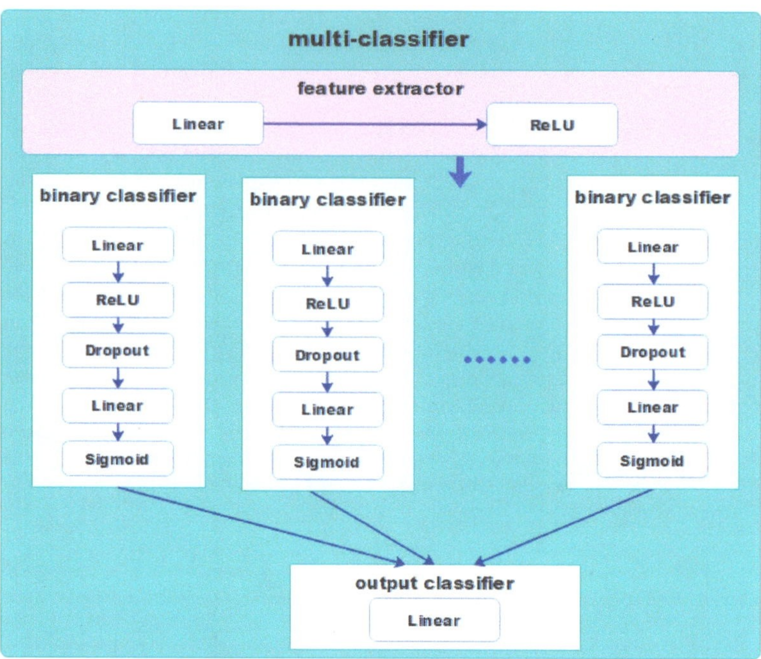

Figure 3. The backbone of the neural network model, used in the experiment of the feature extractor, binary classification, and multi-classification.

In order to ensure good training results and evaluate the classifier's performance, we combined the above considerations. The F1-Score combines two metrics, Precision and Recall, to evaluate the overall performance of the model:

$$F - Sc = \frac{2 \times rcs \times Rcl}{Prcs + Rcl} \tag{9}$$

The higher the value of F-Sc, the better the training effect is proved to be. The results are as Table 3:

Table 3. Comparison of different algorithms in terms of correct packet detection rate.

Methods	TPR	Precision	Recall	F-Sc
CM-NIDS	1.000	1.000	1.000	1.000
KNN-NIDS	0.772	0.157	0.607	0.172
SVM-NIDS	0.788	0.147	0.594	0.153
LR-NIDS	0.619	0.076	0.643	0.135
Bayes-NIDS	0.283	0.588	0.624	0.605
DT-NIDS	0.566	0.082	0.699	0.147
AB-NIDS	0.610	0.076	0.652	0.137

For NIDS, its primary task is to identify whether the packets are anomalous or not, so it has very demanding requirements for the accuracy of binary classification. Our approach (CM-NIDS) has a clear advantage in binary classification. On the TPR metric, CM-NIDS outperforms the optimal algorithm by 21 percentage points, and has a clear advantage on the three remaining commonly used measures of binary classification accuracy (especially on Precision and F-Sc comparisons). This is partly due to our adequate data pre-processing combined with KNN-based feature selection and partly due to our use of a more accurate binary-classification method.

Furthermore, it is fundamental to correctly identify the attack type for multi-classification networks. At the same time, we examined the difference between the model and some classical models in terms of the total tensor used. We were able to see that our model uses a significantly smaller number of tensors. The comparative algorithms we used include APFL [34], Ditto [35], FedBN [36], FedFomo [37], and PerAvg [38]. They are all among more emerging algorithms applicable to NIDS. Our comparison of several algorithms is shown in Table 4.

Table 4. Comparison of different algorithms in terms of TDR and total tensors.

Methods	TDR	Total Tensors
CM-NIDS	0.910	2,100,391
APFL-NIDS	0.500	5,537,952
Ditto-NIDS	0.471	5,537,952
FedBN-NIDS	0.502	2,900,832
FedFomo-NIDS	0.640	6,856,712
PerAvg-NIDS	0.394	2,900,832

In terms of TDR, we compared our approach with other frameworks. It is clear to see that we have an outstanding advantage in terms of attack-type classification. The CM-NIDS algorithm has a TDR value of 0.910, the highest among the provided algorithms. This indicates that the algorithm can classify different data classes more accurately and has a higher detection accuracy in network intrusion detection tasks. Also, the CM-NIDS algorithm uses a total tensor number of 2,100,391, which is less than other algorithms. This may imply that the CM-NIDS algorithm requires less computational and storage resources in the computation process and, thus, is more competitive in terms of efficiency. The smaller number of tensors may imply less memory usage and computational burden and, therefore, it can run more efficiently in the same hardware environment. This is mainly due to our improved approach to multi-classification. It is performed through a combination of binary classification, resulting in a much higher accuracy rate.

4.3. Other Experiments

In terms of parameters and methods, we conducted separate comparative experiments on some of the more critical methods and parameters to demonstrate that our framework is the optimal solution. Our comparison of several algorithms is shown in Table 5.

Table 5. Comparison of different algorithms in terms of accuracy and number of selected features.

Methods	FN	Accuracy
GA + KNN	3	0.991
PCA + KNN	5	0.989
PCA + KNN	10	0.989
PCA + KNN	15	0.990

In the above table, FN refers to the number of selected features, and we compared GA and PCA, KNN, and NN. In terms of run efficiency, KNN was much faster than NN. We have yet to list this set of experiments due to NN's unpredictably long run time. With an equal number of FNs, GA was slightly more accurate than PCA. As the number of FNs increased, the accuracy of PCA + KNN remained essentially constant. However, the FN of GA is the result of the output under the fitness action rather than a pre-set parameter. Therefore, the GA + KNN approach has a more significant advantage in terms of feature selection.

In addition, we deliberately evaluated the robustness of our framework. First, given that the number of parameters is positively correlated with the training effectiveness of the network, we adjusted and retrained the parameters. The results are as Table 6.

Table 6. Comparison of accuracy with the different neural networks with different parameters.

Type	ACR	Type	ACR
True-Set	0.977	APFL-NIDS	0.500
NOM-1	0.909	Ditto-NIDS	0.471
NOM-2	0.885	FedBN-NIDS	0.502
NOM-3	0.790	SVM-NIDS	0.406
KNN-NIDS	0.448		

NOM refers to the number of model parameters. For NOM-1, NOM-2, and NOM-3, the situation of the number of our model parameters is that NOM-2 has the largest number of model parameters, NOM-1 the second largest, and NOM-3 the smallest. It can be noticed that as the number of parameters decreases, the worst case of the number of model parameters remains around 80% and is consistently significantly better than the other frameworks. Even in the case of the NOM-3 dataset, the ACR of our model is about twice that of the other methods.

In a realistic network environment, the number of normal packets will far exceed the abnormal ones. Based on the above premise, we adjusted the ratio of normal to abnormal packets to 10:1, named the True-Set. For the True-Set, binary classification can work more evenly, resulting in better results for multi-classification than the previous training.

To illustrate the rationality of the parameters we set on the neural network, we performed parameter tuning to prove that our solution is optimal. And we examined the performance of our model on other datasets. Figure 4 shows the results of our adjustment of the parameters of Dropout. Here, Dropout = 0.01 is the parameter value we used in experiments.

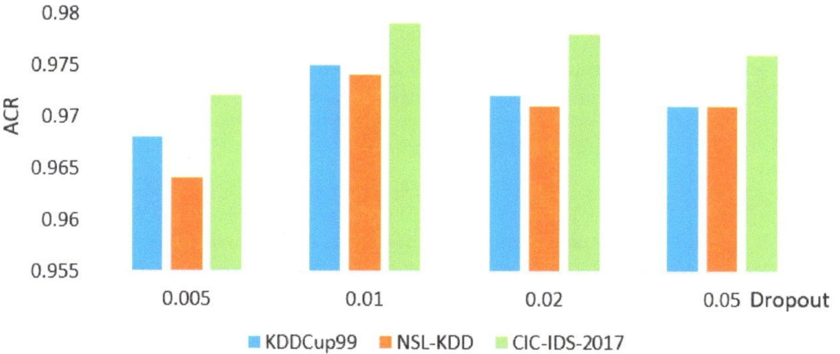

Figure 4. ACR comparison graph for the CM-NIDS algorithm with different Dropout values, where 0.01 was used in our program CM-NIDS.

As shown above, our parameters are the optimal solutions for all cases. At the beginning, there is a significant increase in ACR with the rise of the epoch, but it peaks after the epoch increases to 40. Similarly, the effect of the Dropout parameter on the ACR in the experiment yields the same trend. The main datasets compared are two other classic datasets in the NIDS domain, NSL-KDD (https://www.unb.ca/cic/datasets/nsl.html, accessed on 2 November 2022) and CIC-IDS-2017 (https://www.unb.ca/cic/datasets/ids-2017.html, accessed on 2 November 2022). Overall, the accuracy of our model is stable above 96% under the three datasets, and the multi-classification accuracy of the model is around 97.5% under the optimal parameters, demonstrating the robustness of our model in the multi-classification task. It can be seen that the ACR trends are the same on the three datasets, but the model has a relatively better performance on CIC-IDS-2017, which may be due to the adequate CIC-IDS-2017 dataset.

To illustrate the benefits of using FS, we set up the following experiment and its result is as Table 7:

Table 7. Comparison of NIDS systems with FS and without FS in terms of metrics.

Method	Dataset	With FS?	ACR	Time	Feature No.
CM-NIDS	NOM	YES	0.8144	28.195	3
CM-NIDS	True-Set	YES	0.9737	27.619	3
CM-NIDS	NOM	NO	0.8134	27.313	41
CM-NIDS	True-Set	NO	0.9733	28.766	41

As the time used for the experiments was related to the amount of data, we kept the overall amount of data constant and adjusted the other parameters. As mentioned above, the overall time used remained the same under the different datasets, even though we increased the FS step. At the same time, the number of features was greatly reduced, allowing us to use some of the features for accurate classification without acquiring all of them, in line with the objective reality of the IoT environment and reducing the time required. There was a 92.68% reduction in the number of feature sets after using FS. This also means a smaller memory footprint.

We also performed a comparison of the KNN as well as the two neural network algorithms, the results of which are shown in Table 8.

Table 8. Comparison of accuracy of each algorithm in two cases.

Method	Accuracy	Case
Neural network	100.00%	Binary
Neural network—Single	99.18%	Binary
KNN	99.70%	Binary
Neural network	91.13%	Multi
Neural network—Single	99.15%	Multi
KNN	68.05%	Multi

Neural network—CM refers to the neural network with 23 sub-structures used in our experiments, and Neural network—Single refers to a single neural network. From the above results, it can be seen that the neural network is more accurate than KNN in determining normal/abnormal packets in a multi-classification situation. The binary classification accuracy and multi-classification accuracy of Neural network—CM are both higher than those of Neural network—Single, which proves that our structure is effective.

5. Discussion

We compared our method with others in terms of detection accuracy. In particular, we used a diverse set of accuracy-related metrics. Our binary-classification and multi-classification matching rates outperform other algorithms, especially multi-classification. In addition, we compared the parameters of the neural networks used, the different ways FS is used, the use of FS or not, the different datasets used, and the performance of the models under different datasets. Our framework is very robust.

6. Conclusions

This paper shows work on pre-processing by applying a pre-trained GA-KNN algorithm. First, we use a GA for feature selection to perform dimensionality reduction for attack-type detection, after which we use a KNN algorithm for attack-type matching. This approach can effectively contribute to the attack detection rate of NIDS in terms of multiple classifications, thus informing security systems. Meanwhile, we propose a multi-classification model for anomaly detection based on this and an improved neural network model that provides detailed guidance for other network security systems. We also offer solutions such as modified objective functions for the inherent drawbacks of traditional datasets.

Author Contributions: Conceptualization, Y.W.; methodology, Z.L. and W.Z.; software, Z.L. and W.Z.; validation, Y.W., J.W. and H.S.; formal analysis, Y.W., Z.L. and W.Z.; investigation, Y.W. and J.W.; data curation, H.S.; writing—original draft preparation, Z.L. and W.Z.; writing—review and editing, Y.W., J.W., H.S. and M.G.; visualization, Z.L. and W.Z.; supervision, J.W.; project administration, M.G.; funding acquisition, H.S. All authors have read and agreed to the published version of the manuscript.

Funding: This work was supported in part by the Shanghai Key Laboratory of Scalable Computing and Systems, Internet of Things special subject program, China Institute of IoT (Wuxi), Wuxi IoT Innovation Promotion Center under Grant 2022SP-T13-C, and Industry-university-research Cooperation Funding Project from the Eighth Research Institute in China Aerospace Science and Technology Corporation (Shanghai) under Grant USCAST2022-17.

Institutional Review Board Statement: Not applicable.

Informed Consent Statement: Not applicable.

Data Availability Statement: Data available in a publicly accessible repository that does not issue DOIs. Publicly available datasets were analyzed in this study. This data can be found here: KD-DCup99 (http://kdd.ics.uci.edu/databases/kddcup99/kddcup99.html, accessed on 2 November 2022), NSL-KDD (https://www.unb.ca/cic/datasets/nsl.html, accessed on 2 November 2022), and CIC-IDS-2017 (https://www.unb.ca/cic/datasets/ids-2017.html, accessed on 2 November 2022).

Conflicts of Interest: The authors declare no conflict of interest.

References

1. Panchali, B. Edge Computing- Background and Overview. In Proceedings of the International Conference on Smart Systems and Inventive Technology, Tirunelveli, India, 13–14 December 2018; pp. 580–582.
2. Shi, H.; Wang, H.; Ma, R.; Hua, Y.; Song, T.; Gao, H.; Guan, H. Robust Searching-based Gradient Collaborative Management in Intelligent Transportation System. *ACM Trans. Multimed. Comput. Commun. Appl.* **2022**. [CrossRef]
3. Cai, Z.; Ren, B.; Ma, R.; Guan, H.; Tian, M.; Wang, Y. A Hardware-Assisted Distributed Framework to Enhance Deep Learning Security. *IEEE Trans. Comput. Soc. Syst.* **2023**, 1–9. [CrossRef]
4. Li, D.; Deng, M.; Wang, H. IoT data feature extraction and intrusion detection system for smart cities based on deep migration learning. *Int. J. Inf. Manag.* **2019**, *49*, 533–545. [CrossRef]
5. Sudar, K.M.; Beulah, M.; Deepalakshmi, P.; Nagaraj, P.; Chinnasamy, P. Detection of Distributed Denial of Service Attacks in SDN using Machine learning techniques. In Proceedings of the International Conference on Computer Communication and Informatics, Coimbatore, India, 27–29 January 2021; pp. 1–5.
6. Sharafaldin, I.; Lashkari, A.H.; Ghorbani, A.A. Toward generating a new intrusion detection dataset and intrusion traffic characterization. In Proceedings of the International Converence on Information Systems Security and Privacy, Funchal Madeira, Portugal, 22–24 January 2018; Volume 1, pp. 108–116.
7. Panigrahi, R.; Borah, S. A detailed analysis of CICIDS2017 dataset for designing intrusion detection systems. *Int. J. Eng. Technol.* **2018**, *7*, 479–482.
8. Tavallaee, M.; Bagheri, E.; Lu, W.; Ghorbani, A.A. A detailed analysis of the KDD CUP 99 data set. In Proceedings of the IEEE Symposium on Computational Intelligence for Security and Defense Applications, Ottawa, ON, Canada, 8–10 July 2009; pp. 1–6.
9. Loye, R.A.Y. Training and testing anomaly-based neural network intrusion detection systems. *Int. J. Inf. Secur. Sci.* **2013**, *2*, 57–63.
10. Sato, H.; Kobayashi, R. A machine learning-based NIDS that collects training data from within the organization and updates the discriminator periodically and automatically. In Proceedings of the International Symposium on Computing and Networking Workshops, Matsue, Japan, 23–26 November 2021; pp. 420–423.
11. Zhang, J.; Hua, Y.; Song, T.; Wang, H.; Xue, Z.; Ma, R.; Guan, H. Improving Bayesian Neural Networks by Adversarial Sampling. In Proceedings of the AAAI Conference on Artificial Intelligence, Online, 22 February–1 March 2022; Volume 36, pp. 10110–10117.
12. Alazab, A.; Hobbs, M.; Abawajy, J.; Alazab, M. Using feature selection for intrusion detection system. In Proceedings of the International Symposium on Communications and Information Technologies, Gold Coast, Australia, 2–5 October 2012; pp. 296–301.
13. Guo, H.; Wang, H.; Hua, Y.; Song, T.; Ma, R.; Guan, H. Siren: Byzantine-robust Federated Learning via Proactive Alarming. In Proceedings of the ACM SoCC, Seattle, WA, USA, 1–4 November 2021.
14. Zhang, R.; Chu, X.; Ma, R.; Zhang, M.; Lin, L.; Gao, H.; Guan, H. OSTTD: Offloading of Splittable Tasks with Topological Dependence in Multi-Tier Computing Networks. *IEEE J. Sel. Areas Commun.* **2022**, *41*, 555–568. [CrossRef]
15. Sivasankari, N.; Kamalakkannan, S. Building NIDS for IoT Network using Ensemble Approach. In Proceedings of the International Conference on Communication and Electronics Systems, Coimbatore, India, 22–24 June 2022; pp. 328–333.
16. Iqbal, W.; Abbas, H.; Daneshmand, M.; Rauf, B.; Bangash, Y.A. An In-Depth Analysis of IoT Security Requirements, Challenges, and Their Countermeasures via Software-Defined Security. *IEEE Internet Things J.* **2020**, *7*, 10250–10276. [CrossRef]

17. Ramalingam, P.; MurugaPrabu, S.; Samuel, S.J. Enhanced protection for multimedia content in cloud. In Proceedings of the International Conference on Communication and Signal Processing, Tamilnadu, India, 6–8 April 2017; pp. 1369–1372.
18. Ennaji, S.; Akkad, N.E.; Haddouch, K. A Powerful Ensemble Learning Approach for Improving Network Intrusion Detection System (NIDS). In Proceedings of the International Conference on Intelligent Computing in Data Sciences, Tartu, Estonia, 15–16 November 2021; pp. 1–6.
19. Olayinka, T.C.; Ugwu, C.C.; Okhuoya, O.J.; Olusọla Adetunmbi, A.; Solomon Popoola, O. Combating Network Intrusions using Machine Learning Techniques with Multilevel Feature Selection Method. In Proceedings of the IEEE Nigeria 4th International Conference on Disruptive Technologies for Sustainable Development, Lagos Nigeria, Nigeria, 5–7 April 2022.
20. Desale, K.S.; Ade, R. Genetic algorithm based feature selection approach for effective intrusion detection system. In Proceedings of the International Conference on Computer Communication and Informatics, Coimbatore, India, 8–10 January 2015; pp. 1–6.
21. Verma, J.; Bhandari, A.; Singh, G. Feature Selection Algorithm Characterization for NIDS using Machine and Deep learning. In Proceedings of the IEEE International IoT, Electronics and Mechatronics Conference, Toronto, ON, Canada, 1–4 June 2022; pp. 1–7.
22. Su, Y.; Qi, K.; Di, C.; Ma, Y.; Li, S. Learning automata-based feature selection for network traffic intrusion detection. In Proceedings of the IEEE Third International Conference on Data Science in Cyberspace, Guangzhou, China, 18–21 June 2018; pp. 622–627.
23. Lawrence, H.; Ezeobi, U.; Bloom, G.; Zhuang, Y. Shining New Light on Useful Features for Network Intrusion Detection Algorithms. In Proceedings of the IEEE 19th Annual Consumer Communications & Networking Conference, Las Vegas, NV, USA, 8–11 January 2022.
24. Eldos, T.; Siddiqui, M.K.; Kanan, A. On the KDD'99 Dataset: Statistical analysis for feature selection. *J. Data Min. Knowl. Discov.* **2012**, *3*, 88.
25. Li, J.; Cheng, K.; Wang, S.; Morstatter, F. Feature selection: A data perspective. *Assoc. Comput. Mach.* **2017**, *50*. [CrossRef]
26. Thejaswee, M.; Srilakshmi, P.; Karuna, G.; Anuradha, K. Hybrid IG and GA based Feature Selection Approach for Text Categorization. In Proceedings of the International Conference on Electronics, Communication and Aerospace Technology, Coimbatore, India, 5–7 November 2020; pp. 1606–1613.
27. Ghoualmi, L.; Benkechkache, M.E.A. Feature Selection Based on Machine Learning Algorithms: A weighted Score Feature Importance Approach for Facial Authentication. In Proceedings of the International Informatics and Software Engineering Conference, Ankara, Turkey, 15–16 December 2022; pp. 1–5.
28. Lawrence, T.; Zhang, L. IoTNet: An efficient and accurate convolutional neural network for IoT devices. *Sensors* **2019**, *19*, 5541. [CrossRef] [PubMed]
29. Htun, P.T.; Khaing, K.T. Anomaly intrusion detection system using Random Forests and k-Nearest Neighbor. *Probe* **2013**, *41102*, 2377.
30. Kim, D.S.; Nguyen, H.N.; Park, J.S. Genetic algorithm to improve SVM based network intrusion detection system. In Proceedings of the International Conference on Advanced Information Networking and Applications, Taipei, Taiwan, 28–30 March 2005; Volume 2, pp. 155–158.
31. Sultana, N.; Chilamkurti, N.; Peng, W.; Alhadad, R. Survey on SDN based network intrusion detection system using machine learning approaches. *Peer-Peer Netw. Appl.* **2019**, *12*, 493–501. [CrossRef]
32. Hashemi, M.J.; Keller, E. Enhancing robustness against adversarial examples in Network Intrusion Detection Systems. In Proceedings of the IEEE Conference on Network Function Virtualization and Software Defined Networks, Leganes Madrid, Spain, 10–12 November 2020; pp. 37–43.
33. Tang, T.A.; Mhamdi, L.; McLernon, D.; Zaidi, S.A.R.; Ghogho, M. Deep Recurrent Neural Network for intrusion detection in SDN-based networks. In Proceedings of the IEEE Conference on Network Softwarization and Workshops, Montreal, QC, Canada, 25–29 June 2018; pp. 202–206.
34. Deng, Y.; Kamani, M.M.; Mahdavi, M. Adaptive Personalized Federated Learning. *arXiv* **2020**, arXiv:2003.13461.
35. Li, T.; Hu, S.; Beirami, A.; Smith, V. Ditto: Fair and robust federated learning through personalization. In Proceedings of the International Conference on Machine Learning, Virtual Event, 18–24 July 2021; pp. 6357–6368.
36. Li, X.; Jiang, M.; Zhang, X.; Kamp, M.; Dou, Q. Fedbn: Federated Learning on Non-Iid Features via Local Batch Normalization. *arXiv* **2021**, arXiv:2102.07623.
37. Zhang, M.; Sapra, K.; Fidler, S.; Yeung, S.; Alvarez, J.M. Personalized Federated Learning with First Order Model Optimization. *arXiv* **2020**, arXiv:2012.08565.
38. Fallah, A.; Mokhtari, A.; Ozdaglar, A. Personalized federated learning with theoretical guarantees: A model-agnostic meta-learning approach. In Proceedings of the Advances in Neural Information Processing Systems, Virtual, 6–12 December 2020.

Article

Hash Based DNA Computing Algorithm for Image Encryption

Hongming Li [1], Lilai Zhang [2,3], Hao Cao [2,3] and Yirui Wu [2,3,4,5,*]

[1] Electronic Information Engineering, Ningbo Polytechnic, Ningbo 315000, China
[2] Key Laboratory of Water Big Data Technology of Ministry of Water Resources, Hohai University, Nanjing 211100, China
[3] College of Computer and Information, Hohai Univerisity, Nanjing 211100, China
[4] Key Laboratory of Symbolic Computation and Knowledge Engineering of Ministry of Education, Jilin University, Changchun 130012, China
[5] Nanjing Hezehaichuan Technology Company, Nanjing 211100, China
* Correspondence: wuyirui@hhu.edu.cn

Abstract: Deoxyribonucleic Acid (DNA) computing has demonstrated great potential in data encryption due to its capability of parallel computation, minimal storage requirement, and unbreakable cryptography. Focusing on high-dimensional image data for encryption with DNA computing, we propose a novel hash encoding-based DNA computing algorithm, which consists of a DNA hash encoding module and content-aware encrypting module. Inspired by the significant properties of the hash function, we build a quantity of hash mappings from image pixels to DNA computing bases, properly integrating the advantages of the hash function and DNA computing to boost performance. Considering the correlation relationship of pixels and patches for modeling, a content-aware encrypting module is proposed to reorganize the image data structure, resisting the crack with non-linear and high dimensional complexity originating from the correlation relationship. The experimental results suggest that the proposed method performs better than most comparative methods in key space, histogram analysis, pixel correlation, information entropy, and sensitivity measurements.

Keywords: DNA computing; image encryption; context-aware DNA permutation; hash DNA encoding

1. Introduction

With the breakthrough development of information technology, how to process data with secured and rapid characteristics has become a major concern [1]. Inspired by DNA structure, a novel computing paradigm using a biological molecule to carry genetic information is designed to ensure the feasibility of computing at a molecular level with impressive characterizations of programmability and high-throughput coding, which not only males it possible to replace traditional silicon-based facilities by biological tools [2], but also provides an alternative way to consider computing with more than 0 and 1. Therefore, DNA computing [3], as a popular computing model with considerable potential to meet the security requirement, has nowadays become a hot spot in the data processing domain.

In hardware-based data processing design, DNA logic gates are designed in combination with digital circuits, acting as basic components to form complex DNA circuits. For example, Zhang et al. [4] propose an entropy-driven incision-assisted recovery strategy for reactants in the DNA loop, which can recover reactants in the catalytic loop and improve their recoverability, creating more effective DNA circuits for molecular transformation and synthetic biology. Unlike most artificially catalyzed DNA loops, Yang et al. [5] develop a catalytic DNA logic circuit regulated by DNase, which is controlled by a covalent modification strategy and demonstrates a great potential in more complex computing by building complex cascade circuits.

In algorithm-based data computing design, molecular programming is built on the basis of DNA circuits with powerful modularity. For example, Zhang et al. [6] design a DNA molecular computing platform to analyze miRNA profiles in serum samples, achieving

Citation: Li, H.; Zhang, L.; Cao, H.; Wu, Y. Hash Based DNA Computing Algorithm for Image Encryption. *Appl. Sci.* **2023**, *13*, 8509. https://doi.org/10.3390/app13148509

Academic Editor: Krzysztof Koszela

Received: 3 March 2023
Revised: 21 June 2023
Accepted: 20 July 2023
Published: 23 July 2023

an intelligent diagnosis of cancer. Later, Ma et al. [7] design and implement various types of a DNA computing system, which verifies the feasibility of using DNA computing for intelligent diagnoses in fields of biology and biomedicine.

Despite various designs of clinical applications, DNA computing is popular for building a encryption application, due to its reliable encryption performance and parallel processing capability. Specifically, we could roughly divide encryption methods into two types, i.e., diffusion encryption and permutation encryption. Diffusion encryption refers to the process of spreading information throughout the ciphertext, thus increasing unbreakability of the encryption algorithm. More precisely, diffusion encryption maps each bit of the plaintext to multiple bits in the ciphertext. Any change in a plaintext bit should correspond to variations in multiple bits of the ciphertext, successfully achieving a correlation between the plaintext and ciphertext. Permutation encryption encrypts the plaintext by replacing each bit with a different bit, which reorders the information of the plaintext to hide content, instead of altering values of each bit in the plaintext. The key process of the permutation encryption is to establish a one-to-one mapping between each bit of the plaintext and ciphertext, thus achieving a confusion effect of information.

In fact, the quantity of algorithms have been proposed for data encrypting based on DNA computing. For example, Khan et al. [8] propose DNAact Ran, a DNA computing-based sequence analysis engine, which could accurately detect a ransomware attack with their designed constraints and kmer frequency vector. Combining DNA computing with chaotic systems, Zou et al. [9] involve the DNA-based strand displacement strategy with the chaotic system, where the generated chaotic sequence greatly improves the unbreakability of their encryption system, owing to its capability to stand with statistical attacks. Most recently, Namasudra et al. [10] propose a DNA computing-based access control algorithm, where 1024 random keys are fed into the DNA computing system for the user's secret data encryption, significantly improving the security capability of the control model.

Inspired by the idea of utilizing DNA computing to encrypt, we aim to encrypt high-dimensional image data in this paper. Compared with text structure data, the image is characterized with high-dimensional and unstructured proptoses, which have two major difficulties in designing proper encryption algorithms. **First**, how to achieve an unbreakable capability for image data with less encoding complexity, since a more complex encoding strategy generally promotes encryption performance. **Second**, since the image is equipped with a natural property of high-dimensional complexity, how algorithms could involve such a property for for encoding. In other words, images are a natural source of randomness, where adjacent pixels marked with a considerable amount of redundant information demonstrate strong correlations. Therefore, the characteristic of smooth variation between pixels can be leveraged to speed up the image encryption process.

To address the above difficulties, we develop a novel hash encoding-based DNA computing algorithm to effectively encrypt high-dimensional image data, which consists of the DNA hash encoding module and content-aware encrypting module. The hash algorithm is unique in encrypting, which outputs a fixed-length output with data as complicated as possible. Moreover, it shares several impressive characteristics, such as a small computing burden with a simple function calculation; unidirectionality without possible ways to reverse the input data through hash results, tampering resistance where small modification would greatly vary the output results; and anti-collision capability, which guarantees that the unique output could be gained with different inputs. Inspired by these properties, we involve hash encoding into DNA computing inside the proposed DNA hash encoding module, thus boosting the encryption capability of high-dimensional data with a small computation burden. Considering high-dimensional image data as a natural source to encode with less complexity, a content-aware encrypting module is proposed to map between image content and encryption results with several simple but effective functions, thus offering new ways to encrypt images based on DNA computing.

To sum up, the contributions of this paper are three-fold:

- The proposed method combines the impressive characteristics of DNA computing and the hash function to realize an unbreakable image encryption with less computation burden.
- A novel DNA-based hash encoding module is proposed, which involves the hash function to construct mappings from high-dimensional image data to DNA bases.
- Considering correlations between adjacent pixels as natural sources of complexity, the proposed content-aware encrypting module sunccesfully generates random DNA sequences with chaos properties, which adopts non-linear functions originating from correlations of pixles as source of complexity.

The rest of this paper is organized as follows. Section 2 reviews the related work. Section 3 first presents an overview on algorithm structure, and then presents detailed steps of encryption and decryption. Section 4 conducts several experiments, including an analysis on the key space, histogram, pixel correlation, information entropy, sensitivity and computation cost. Finally, Section 5 concludes the paper and demonstrates the prospect.

2. Related Work

In this section, we provide a brief literature review of this paper, including an introduction to DNA computing and DNA computing-based image encryption.

2.1. Introduction to DNA Computing

With the development of DNA computing, the DNA strand displacement reaction has gradually become an important means to build complex digital circuits operating at room temperature. To solve the limitations of DNA logic circuits, i.e., slow computing speed and complex circuit design, Song et al. [11] propose a DNA logic circuit structure based on single strand logic gates, which improves the computing speed of DNA circuits and the number of DNA strands based on the biological performance of the strand replacement DNA polymerase. Combined with the commonly used cascading strategy, their proposed structure could aid to build large-scale logic circuits.

To build the DNA-based switch circuit, Wang et al. [12] developed a modular DNA molecular switch for the digital operation of any function, which solves the limitation of operation speed and signal-to-noise ratio comparing with the traditional logic gate circuit. In their experiments, their proposed DNA-based switch circuit not only controls the current signal transmission through the switch signal, but also controls the current signal transmission direction through the difference of molecular free energy in the reaction path. Essentially, the development of the DNA switch circuit has laid a solid foundation for molecular computers with decision-making abilities. Therefore, Thubagere et al. [13] successfully developed DNA-based robots, which could autonomously transport molecular-level goods to a designated location without any energy supply.

To improve the security of the physical layer, DNA computing is designed for data encryption during transmission. For example, Liu et al. [14] propose an image encryption scheme, where the DNA-based image compression are used to ensure the security of transmission. Later, Xiao et al. [15] propose a chaotic OFDM (Orthogonal Frequency Division Multiplexing) hybrid security method to ensure the security of the physical layer, which not only dynamically adjusts parameters of DNA-based encoding rules and scrambling methods, but also controls the positional and traversal direction of the helix through chaotic sequences.

2.2. DNA Computing-Based Image Encryption

DNA computing builds the encryption system based on modular arithmetic cryptography. Moreover, a chaotic strategy has been proved to be well matched with DNA ciphers, due to its characteristics of pseudo randomness, unpredictability and so on [16]. For example, Babaei et al. [17] realize a theoretically unbreakable data encryption algorithm, where they build a chaos strategy based on the difference between the original and optimized DNA-based encryption messages. Later, Wang et al. [18] first use piecewise linear chaotic mapping (PWLCM) and logistic mapping to generate encryption parameters, and then use DNA computing for information encoding. Specifically, their PWLCM could make each small segment of the original data be paired with an element of the pseudo-random sequence, thereby increasing the complexity of the encryption algorithm. Afterwards, Samiullah et al. [19] use three chaotic systems including PWLCM, Lorenz and 4D Lorenz to build a multi-level security system, which could largely increase the non-repeatability of the key.

To reduce the computational cost of image encryption, Malik et al. [20] use a tent map to select key streams generated by their proposed confusing and diffusing channels. Later, Khan et al. [21] propose a DNA-based image encryption method to effectively reduce memory usage, which selects the most informativepart from the visual appearance analyzed by their proposed dependent chaotic system. Afterwards, Ravichandran et al. [22] propose a medical image encryption algorithm based on the combination of the integer wavelet transform (IWT), DNA and chaos system. Their strategy of utilizing the advantages of different encryption methods ensures the security of electronic health records during network transmission. Recently, Zhang et al. [23] proposed a multi-image encryption algorithm, which not only resists conventional attacks to protect image content, but also improves the transmission speed via Internet by reducing image size by encoding. Most recently, Wu et al. [24] designed a random encoding, sequencing and diffusion strategy based on content-aware DNA computing, which greatly improves the difficulty of cracking even if attackers obtain partial or all information of the transmitted image. Most relevant to our method, Chen et al. [25] propose an image encryption scheme that incorporates adaptive diffusion permutation and DNA random coding. We further improve their idea by involving the hash function and image content-based encoding, thus greatly improving the capability to resist different types of attacks.

3. Methods

We provide a detailed description of the proposed method with three parts, i.e., the overall framework, DNA hash encoding module and content-aware encryption module.

3.1. Overall Framework

We begin with an introduction to a basis of DNA computing in encryption. DNA bases are classified into four types, namely, adenine (A), thymine (T), cytosine (C) and guanine (G). It is worth noting that the former two types are complementary pairs, and so are the latter two types. We then describe rules for DNA-based sequence encoding, which uses a binary encoding idea to represent the input sequence under rules shown in Table 1 with four bases, i.e., A, C, G and T. With the aid of encoding rules of Table 1, we can convert binary sequence into a DNA-based encoding form for further computation. For example, under rule 2, A, C, G and T are represented as 00, 01, 10, and 11, respectively. If a pixel in the input image refers to 179 in gray levels and its binary representation is 10110011, its corresponding DNA-based encoding sequence should be G-T-A-T under rule 2. During decoding, if a DNA-based encoding sequence is G-A-T-C under rule 2, its corresponding binary value should be 10001101.

Table 1. DNA-based binary encoding rules, which uses different rules to represent binary values with DNA bases.

	Rule 1	Rule 2	Rule 3	Rule 4	Rule 5	Rule 6	Rule 7	Rule 8
A	00	00	11	11	10	01	10	01
C	10	01	10	01	00	00	11	11
G	01	10	01	10	11	11	00	00
T	11	11	00	00	01	10	01	10

We then demonstrate the overall framework in Figure 1, which consists of sender and receiver structures. Specifically, the whole process of encryption inside the sender can be described as follows:

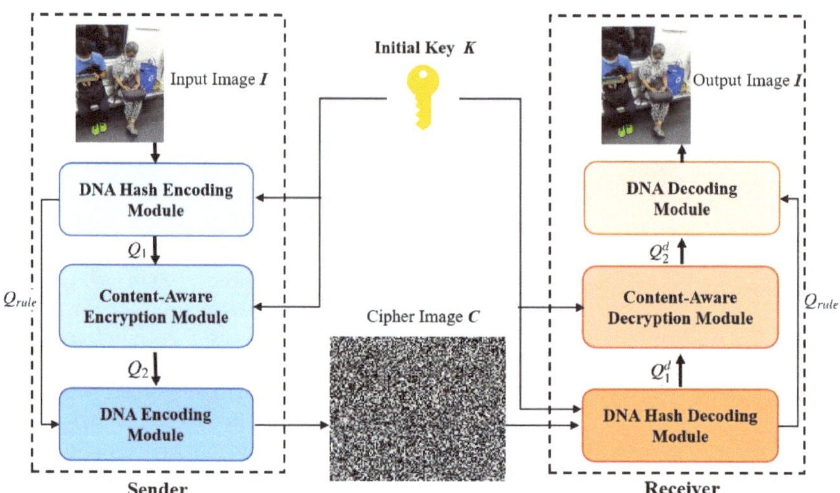

Figure 1. The overall structure of the proposed method, including sender and receiver. They both contain a DNA hash encoding module and a DNA decoding module. In fact, content-aware decryption module can be regarded as the inverse version of content-aware encryption module.

Step 1. Feed the initial key K and the input image I into the DNA hash encoding module, which computes the rule selection sequence Q_{rule} and DNA-encoded Sequence Q_1.

Step 2. K and Q_1 are regarded as input of Content-aware Encryption module, which computes new sequence Q_2 with permutation and diffusion, thus encoding complexity for high unbreakability.

Step 3. Q_{rule} and Q_2 are fed into the DNA encoding module, which encodes Q_2 by utilizing Q_{rule} to generate a binary bitstream under DNA encoding rules. Afterwards, we transfer the generated bitstream as a ciphered image C with $H' \times W'$.

Step 4. Transmit C from the sender to the receiver via Internet.

The whole process of decryption inside the receiver can be described as follows:

Step 5. Regarding C as input, the receiver sends it to DNA hash decoding module with the initial key K, thus generating the rule selection sequence Q_{rule} and a DNA-based decoding sequence Q_1^d. It is noted that we keep the rule selection sequence exactly the same to ensure the consistency of decryption content.

Step 6. Feed K and Q_1^d into the content-aware decryption module for the reversed permutation and diffusion, thus generating a reversed DNA-based decoding sequence Q_2^d.

Step 7. Feed Q_2^d and Q_{rule} into the DNA decoding module, which uses Q_{rule} as the rule selector to decode Q_2^d into the binary bitstream I_{bit}. Finally, we transfer I_{bit} into the output image I for computation. Note that the input and output image should remain the same due to the general principles of encryption.

3.2. Design of DNA Hash Encoding Module

The DNA hash encoding module is designed to transform image data into a DNA-encoded form with different DNA bases, which could involve significant properties of the hash function, i.e., fixed-length output, small computing burden, tampering resistance and anti-collision ability. Such property help improve the unbreakability of the proposed method with low computing cost.

As shown in Table 1, there only exist eight encoding patterns that satisfy the "complementary" rule, where each two bits satisfys the condition that there XOR results should be true. For example, if 00 and 01 are encoded as A and C, respectively, 11 and 10 should be encoded as T and G, respectively. There are only eight encoding rules obtained through this combination, and they are all listed in Table 1. Based on the analysis of the "complementary" rule, we convert image data into a bit stream with the size of 3 bits to describe 8 different mapping rules, where the 2 bit is mapped as one DNA base for the encoding basis. Meanwhile, two more 8-bit streams are designed at the beginning of the stream to record the length and width of the original image, which could help restore the input image after encryption in a proper way.

Essentially, the proposed DNA hash encoding module plays a crucial role in encryption, which randomly employs various sets of complementary rules to encode each pixel. We demonstrate the overall encoding process of it during encryption in Figure 2. The detailed steps of the encoding and decoding process inside the DNA hash encoding module are also listed with pseudo codes in Algorithms 1 and 2. First, we adopt the SHA256 algorithm to generate two important parameters, i.e., initial key value and initializing factor. To effectively improve the security of the cryptosystem, we then feed the computing factor p into the PWLCM, generating the rule selection sequence with the following equation:

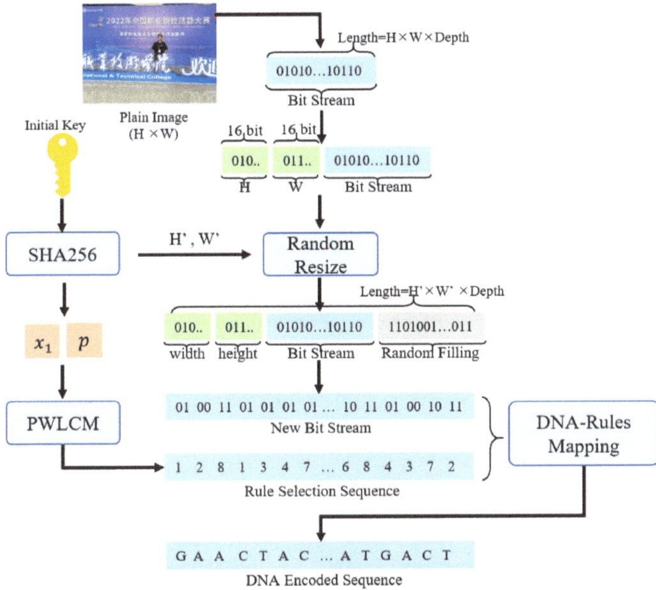

Figure 2. Encoding process inside DNA hash encoding module.

$$x_{n+1} = \begin{cases} \dfrac{x_n}{p}, & 0 \leq x < p \\ \dfrac{x_n - p}{0.5 - p}, & p < x \leq 0.5 \\ F(1 - x_n, p), & 0.5 < x \leq 1 \end{cases} \tag{1}$$

It's noted that $p \in (0, 0.5]$, which enables variety in outcome with little modification in key values, whi ch is highly resistant to attack. Finally, we use the rule selection sequence to help encode the bitstream, which serves as DNA rules to encode pixels with DNA-based rules as shown in Table 1.

Algorithm 1: DNA hash encoding process

Input: Initial Key K, Plain Image I
Output: Rule Selection Sequence Q_{rule}, DNA encoded Sequence Q_1

1   H, W = original height and original width of I;
2   Randomly generate H' and W', where $H < H' < 2 \times H$ and $W < W' < 2 \times W$;
3   Convert the integer numbers H' and W' into a 16 bit binary bitstream HW_{bit} ;
4   Convert the plain image I into a binary bitstream I_{bit};
5   Randomly generate a binary bitarray R_{bit} whose length is
    $H' \times W' \times 8 - H \times W \times 24 - 16$;
6   $I_{bit} = HW_{bit} \oplus I_{bit} \oplus R_{bit}$ (where $\oplus$ indicates concatenate operation);
7   $length = \text{Length}(I_{bit})$;
8   $HASH_K = \text{SHA256}(K)$;
9   $H_1 = HASH_K(0:63)$;
10   $H_2 = HASH_K(64:127)$;
11   $x_1 = modf\left(H_1/10^{15}\right)$;
12   $p = modf\left(H_2/10^{15}\right)/2$;
13   **For** (t = 0 **to** $length/2$);
14      $x_{t+1} = PWLCM(x_t, p)$;
15   $X = [x_1, x_2, ..., x_n]$;
16   $Q_{rule} = mod\left(floor\left(X \times 10^{15}\right), 8\right)$;
17   Introduce the DNA encoding table T ;
18   **For**(t = 0 **to** $length/2$);
19      $Q_1(t) = T(Q_{rule}(t), I_{bit}(2t, 2t+1))$;
20   Output Rule Selection Sequence Q_{rule}, DNA encoded sequence Q_1;

Algorithm 2: DNA Decoding Process

Input: Rule Selection Sequence Q_{rule}, DNA Encoded Sequence Q_2
Output: Cipher Image C

1   $length = \text{Length}(Q_2)$;
2   $T = \text{DNA Encoding Table}$;
3   **For** (t = 0 **to** $length$);
4      $C_{bit}(2t, 2t+1) = T(S_{rule}(t), S_2(t))$;
5   Change cipher bit array C_{bit} into cipher image C;
6   Output Cipher Image C;

For readers' convenience, we further offer a step-by-step explanation of hash encoding in Figure 3, which performs an encryption process on a sample image, as follows:

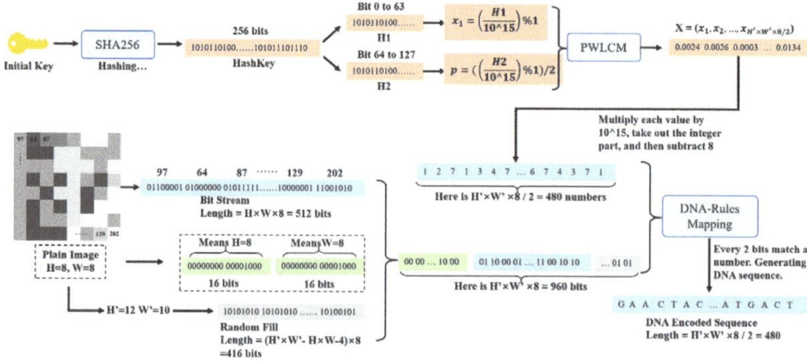

Figure 3. An example of DNA hash encoding by inputting an image with size 8×8.

Step 1. Obtain the height and width of the original image, where H and W equals 8 for the sample image. Convert the original image to a bitstream containing 512 bits of data.

Step 2. Save the original height and width as 16-bit data. For example, if H is 8, its corresponding bitstream is 00000000 00001000. Therefore, we use 32 bits of data to save the original height and width for further restoration.

Step 3. Randomly generate a new height and width, where both values should be larger but twice as small as the the original ones. For display, we set the new height H' and width W' to 12 and 10, respectively.

Step 4. It is noted that there exist $H' \times W' \times 8 = 960$ bits of data for encoding, where the first 32 bits represent the original height and width, and the other 512 bits refer to image content. It's noted that 416 bits remain unused, where they are randomly filled to make up a total 960-bit bitstream for further computation.

Step 5. We generate a 256-bit bit sequence HashKey through the SHA256 hash algorithm based on the input initial key. Note that $H1$ and $H2$ can be viewed as 64-bit numbers for computing.

Step 6. With $H1$ and $H2$, we calculate $x_1 = (H1/10^{15})\%1$ and $p = ((H1/10^{15})\%1)/2$.

Step 7. Then, x_1 and p are fed into the PWLCM function to generate a sequence, where $(H' \times W' \times 8)/2 = (12 \times 10 \times 8)/2 = 480$ numbers are generated. Each of them is multiplied by 10^{15}, modulared with 8, achieving results with a rule sequence containing 480 numbers.

Step 8. We further match 960 bits of original data with rule sequence, where we match every 2 bits of data with 1 rule number based on DNA mapping rules in Table 1. After converting, we could obtain results as a DNA hash-encoded sequence, which would be sent to the content-aware encryption module for permutation and diffusion.

3.3. Content-Aware Encryption Module

The content-aware encryption enhancing module is designed to reorganize the data structure by highly non-linear functions, originating from correlations between adjacent pixels and patches. It is noted that DNA operations could vary to form more diversified expressions, thus greatly improving the variability of encrypted sequences. Moreover, the inherent complexity of DNA operations can be enhanced by utilizing the relevance between neighboring pixels, borrowing non-linear equations from neighboring pixels and patches. Therefore, the proposed module is designed to naturally involve the complexity of the image, greatly boosting the difficulty of the cracking.

We demonstrate the structure design of the proposed module in Figure 4, where the DNA-encoded sequence and rule selection sequence are regarded as inputs for the module. Essentially, we design a reversible permutation and diffusion algorithm to combine both forms of complexity for better unbreakability. Both algorithms could be directly applied on the input image without additional transmission cost via Internet or LAN. Furthermore, we have designed several computation operations that satisfy the commutative law, namely ADD, SUB and XOR, which not only ensures the variance of DNA sequences, but also shares the same parameters to reduce unnecessary transmission costs in LAN.

Specifically, we describe the calculation process of the DNA Sequence Permutation algorithm in Algorithm 3, which is built on the following four parameters:

$$\begin{cases} \dfrac{dx_1}{dt} = x_{n+1} = a(y_n - x_n) + w_n \\ \dfrac{dy_1}{dt} = cx_n - y_n - x_n z_n \\ \dfrac{dz_1}{dt} = x_n y_n - b_n z_n \\ \dfrac{dw_1}{dt} = -y_n z_n + \gamma w_n \end{cases} \tag{2}$$

where a, b, c and γ are preset parameters equaling 10, 3/8, 28 and $[-1.52, -0.06]$, respectively. It is noted that such a calculation process ensures the chaos property of the proposed module. Afterwards, four parameters are sent into the Hyper Chaotic Lorenz System (HCLS) to generate the permutation control sequence Q_p, where HCLS is a highly nonlinear dynamical system, being sensitive to the initial values for unpredictability.

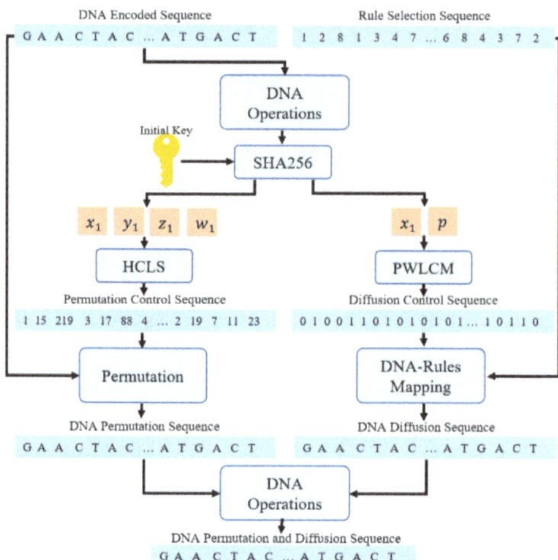

Figure 4. Structure design of the proposed content-aware encryption module.

Algorithm 3: DNA Sequence Permutation Process

Input: Initial Key K, DNA Encoded Sequence Q_1
Output: DNA Permutation Sequence Q_1^p
1   $TA = $ DNA ADD Table;
2   $TX = $ DNA XOR Table;
3   $length = $ Length(Q_1);
4   $AR = Q_1(0); XR = Q_1(0)$;
5   **For** $(i = 1$ **to** $length)$;
6       $AR = TA(AR, Q_1(i))$;
7       $XR = TX(XR, Q_1(i))$;
8   $HASH_D = $ SHA256$([AR, XR])$;
9   $HASH_K = $ SHA256(K);
10   $HASH_{DK} = HASH_D \oplus HASH_K$;
11   $A_1 = HASH_{DK}(0:63); A_2 = HASH_{DK}(64:127); A_3 = HASH_{DK}(128:191);$
      $A_4 = HASH_{DK}(192:255)$;
12   $x_1 = \left(\text{mod}\left(\text{fix}\left(A_1/10^8 \right), 80 \right) - 40 \right) + \left(A_1/10^{14} - fix\left(A_1/10^{14} \right) \right)$;
13   $y_1 = \left(\text{mod}\left(\text{fix}\left(A_2/10^8 \right), 80 \right) - 40 \right) + \left(A_2/10^{14} - fix\left(A_2/10^{14} \right) \right)$;
14   $z_1 = \left(\text{mod}\left(\text{fix}\left(A_3/10^8 \right), 80 \right) + 1 \right) + \left(A_3/10^{14} - fix\left(A_3/10^{14} \right) \right)$;
15   $w_1 = \left(\text{mod}\left(\text{fix}\left(A_4/10^8 \right), 500 \right) - 250 \right) + \left(A_4/10^{14} - fix\left(A_4/10^{14} \right) \right)$;
16   Put x_1, y_1, z_1, w_1 into HCLS to generate a sequence $X = [x_1, x_2, ..., x_n]$ by iterating;
17   $Q_p = mod\left(floor\left(X \times 10^{15} \right), length \right)$;
18   **If** (This is the encryption process);
19       $i = 0, j = length/2$;
20   **Else**;
21       $i = length/2, j = 0$;
22   **For** $(k = i$ **to** $j)$;
23       $Q_1(Q_p(k)) \leftrightarrow Q_1(Q_p(length - k)$;
24   $Q_1^p = Q_1$;
25   Output DNA Permutation Sequence Q_1^p;

Similarly, we describe the calculation process of the DNA Sequence diffusion algorithm in Algorithm 4. Specifically, we first calculate key parameters X_1 and p based on values of H_1 and H_2 transmitted from the last module. Then, we input these parameters into the PWLCM algorithm to generate a diffusion control sequence. With the input rule selection sequence, we adopt DNA rules to map from numbers to DNA bases, thus generating DNA diffusion sequence. Finally, we apply DNA-based calculations on the generated permutation sequence and diffusion sequence, thus obtaining the merged sequence as the final output of the DNA permutation and diffusion sequence.

Algorithm 4: DNA Sequence Diffusion Process

Input: Initial Key K, Rule Selection Sequence Q_{rule}, Permutated Sequence Q_1^p
Output: Permutated and Diffused Sequence S_2

1   Convert the plain image I into a binary bitstream I_{bit};
2   length = Length(S_p);
3   $HASH_K = $ SHA256(K);
4   $H_1 = HASH_K(128:191)$; $H_2 = HASH_K(192:255)$;
5   $x_1 = modf\left(H_1/10^{15}\right)$;
6   $p = modf\left(H_2/10^{15}\right)$;
7   **For** $(t = 0$ **to** $length/2)$;
8      $x_{t+1} = PWLCM(x_t, p)$;
9   $X = [x_0, x_1, ..., x_{length/2}]$;
10   $Y = mod\left(floor\left(\mathbf{X} \times 10^{15}\right), 256\right)$;
11   $T =$ DNA Encoding Table;
12   **For** $(i = 0$ **to** $length/2)$;
13      $S_{key}(i) = T(S_{rule}(i), Y(2i, 2i+1))$;
14   $TA = $ DNA ADD Operation Table;
15   $TS = $ DNA SUB Operation Table;
16   $TX = $ DNA XOR Operation Table;
17   **If** (in encryption process);
18      $D(0) = TA(Q_p(0), S_{key}(0))$;
19      $D(0) = TX(D(0), S_{key}(0))$;
20   **For** $(i = 0$ **to** $length)$;
21   **If** $mod(i, 2) == 1$;
22          $D(i) = TX(Q_p(i), S_{key}(i))$; $D(i) = TX(D(i), D(i-1))$;
23   **Else**;
24          $D(i) = TA(Q_p(i), S_{key}(i))$; $D(i) = TX(D(i), D(i-1))$;
25   **Elif** (in decryption process);
26   **For** $(i = length$ **to** $0)$;
27   **If** $mod(i, 2) == 1$;
28          $D(i) = TX(Q_p(i), S_{key}(i))$; $D(i) = TX(D(i), Q_p(i-1))$;
29   **Else**;
30          $D(i) = TA(Q_p(i), Q_p(i))$; $D(i) = TX(D(i), S_{key}(i-1))$;
31      $D(0) = TA(Q_p(0), S_{key}(0))$;
32      $D(0) = TX(D(0), S_{key}(0))$;
33   $Q_2 = D(0 : length - 1)$ Output Permutated and Diffused Sequence Q_2;

---

4. Experiment and Analysis

We firstly introduce the experiment settings and implementation details. Then, we perform variants of the analysis to verify the resistance against different attacks. Afterwards, we perform ablation and computation analysis to evaluate performance. Finally, the comparison experiments are conducted.

4.1. Experiment Settings and Implementation Details

We deploy the proposed encryption system in TCP/IP environment. We connected two computers through the Tenda-AC71200M router, thus forming a simple wireless LAN environment. Specifically, the original image is provided by a laptop (sender) and encrypted into a cipher image by the proposed method. Then, the cipher image is transmitted to another laptop (receiver) via IP Messenger protocol and decrypted into a plain image.

It is noted that we use two computers with the same configurations to simulate the sender and receiver. Both computers are equipped with an Intel Core i5-12400F CPU, NVIDIA RTX 3060Ti GPU, and 16GB memory. Additionally, the encryption method is implemented based on Python 3.8, and the initial key is set to 'GOOD-LUCK'.

4.2. Key Space Analysis

Key space is defined as the size of a key range, i.e., the number of keys, where different encryption algorithms have their own specific key space. Essentially, we should ensure that the key space is large enough to withstand brute force attacks. Specifically, we can calculate that the size of the key space of the proposed method is $S = (0.5 \times 10^{15})^2 \times (1 \times 10^{15})^2 \times (80 \times 10^{14})^3 \times (500 \times 10^{14}) = 6.410^{127} \approx 2^{418}$. Obviously, such a large key space ensures the capability to resist brute force attacks.

4.3. Histogram Analysis

The statistical analysis attack refers to the fact that attackers can obtain a quantity of information by analyzing the distribution of image pixel values. To verify the effectiveness of the proposed method against the statistical analysis attack, we compare the pixel value distribution of the original and cipher image. As shown in Figure 5a–e, the histogram of the original image fluctuates greatly, where attackers can easily construct effective attack strategies. Figure 5f–j shows the corresponding cipher images and pixel histograms, which proves that the encrypted image can effectively hide image information to resist statistical analysis attacks.

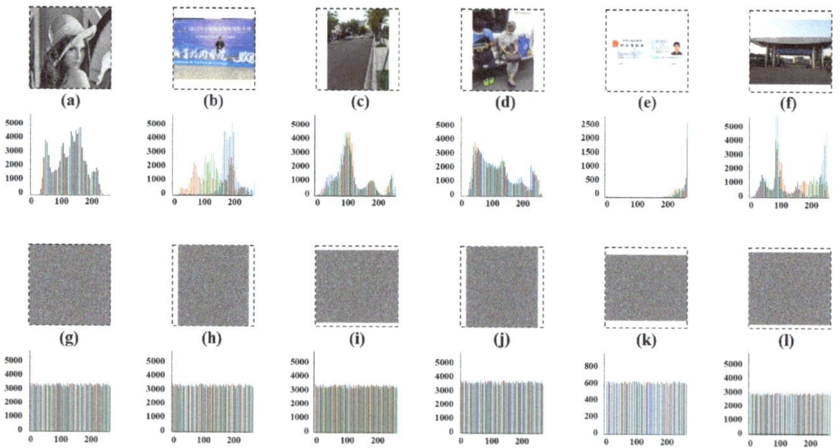

Figure 5. The corresponding histograms for different plain and cipher images, where (**a**–**f**) are input plain images of 'Lena', 'Poster', 'Street', 'Subway', 'ID Card' , 'Hohai University', and (**g**–**l**) are their corresponding cipher images.

4.4. Pixel Correlation Analysis

Permutation and diffusion operations are used to scramble adjacent pixels between images, thus reducing the correlation between adjacent pixels and improving the security after image encryption. We can calculate the correlation coefficient between a and b to evaluate such correlation effects

$$\begin{cases} r_{ab} = \dfrac{\mathrm{cov}(a,b)}{\sqrt{D(a)}\sqrt{D(b)}} \\ \mathrm{cov}(a,b) = \frac{1}{N}\sum_{i=1}^{N}(a_i - E(a))(b_i - E(b)) \\ D(a) = \frac{1}{N}\sum_{i=1}^{N}(a_i - E(a))^2 \\ E(a) = \frac{1}{N}\sum_{i=1}^{N} a_i \end{cases} \tag{3}$$

where $\mathrm{cov}(a,b)$, $D(a)$ and $E(a)$ are defined as the covariance, mean square and expected error, respectively.

Table 2 shows the correlation coefficients of the plain and cipher image, where we can observe that the pixel correlation has been significantly reduced by comparing between both images. Correspondingly, Figure 6 shows the comparison of image correlation before and after encryption, which proves that the proposed method can effectively reduce the correlation in three directions.

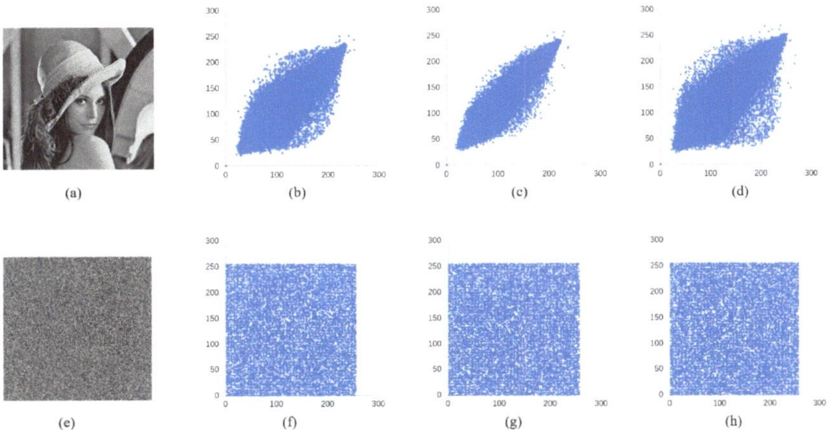

Figure 6. Correlation coefficients distributions of 'Lena' (**a**) and its corresponding cipher image (**e**), where (**b**,**f**) are distributions in horizontal direction, (**c**,**g**) are distributions in vertical direction, and (**d**,**h**) are distributions in diagonal direction.

Table 2. Results of correlation coefficients analysis on several testing images.

Images	Plain Image			Cipher Image		
	V	**H**	**D**	**V**	**H**	**D**
Baboon	0.8813	0.7524	0.7638	0.0039	0.0037	0.0048
Bridge	0.9344	0.9531	0.9812	−0.0089	0.0049	−0.0017
Chest	0.9674	0.9892	0.9816	−0.0037	0.0045	−0.0081
Lena	0.9832	0.9729	0.9631	0.0017	0.0026	0.0033
Hohai University	0.8681	0.8724	0.9076	−0.0018	−0.0089	0.0094

4.5. Information Entropy Analysis

Information entropy is used to measure the uncertainty of data, where higher information entropy refers to larger complexity for cracking. Supposing the number of keys of an encryption algorithm is K-bit, the ideal information entropy should be K. To evaluate the uncertainty of the data, i.e., the diffusion performance, information entropy can be calculated as follows:

$$H = -\sum_{i=1}^{256} p(e_i) \log_2 p(e_i) \tag{4}$$

where $p(e_i)$ represents the probability of the pixel being i.

The information entropy of five test images is 7.987421 (Poster), 7.996523 (Street), 7.989923 (Subway), 7.998615 (ID Card) and 7.988991 (Hohai University), where all the information entropy is close to 8, implying that the proposed method performs a pixel diffusion with an optimized performance.

4.6. Sensitivity Analysis

By modifying values of a few pixels and comparing results, differential attacks could help attackers to easily obtain the rules of encryption. Therefore, the cryptosystem is required to have a certain sensitivity to resist pixel modifications. We adopt a unified average changing intensity (UACI) and number of pixels change rate (NPCR) to measure such sensitivity:

$$\begin{cases} NPCR = \frac{1}{H \times W} \sum_{i=1}^{H} \sum_{j=1}^{W} D_{ij} \times 100\% \\ UACI = \frac{1}{H \times W} \sum_{i=1}^{H} \sum_{j=1}^{W} \left(\frac{I(i,j) - I'(i,j)}{255} \right) \times 100\% \end{cases} \tag{5}$$

where $D_{ij} = \begin{cases} 1, I(i,j) \neq I'(i,j) \\ 0, I(i,j) = I'(i,j) \end{cases}$. I and I' are the original and the modified image, respectively.

As shown in Figure 7, we choose an 8-bit image as the plain image and use the initial key 'GOOD-LUCK' to encrypt the image, obtaining the 'pseudo initialize factor' ($x_1 = 0.401894441844344, p = 0.33533929244635197$). Afterwards, we change the pseudo-initialized factor to generate a new cipher image E_0. Moreover, we define $x_1 = x_1 + 10^{-14}$ and $p = p + 10^{-14}$ to generate another two cipher images E_2 and E_3.

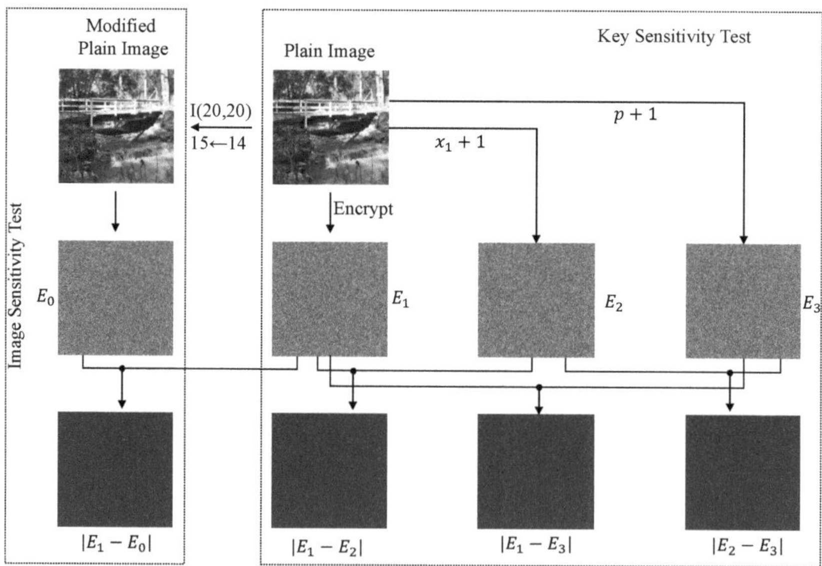

Figure 7. We demonstrate differential cipher images achieved by the proposed method, offering samples on sensitivity and key sensitivity analysis.

The average NPCR and UACI values for different images are shown in Table 3. In fact, Wu et al. [26] state that the expected UACI value should be 33.4635% for an image with ranging values from 0 to 255. It is noted that all results are close to the expected UACI values, which proves that the proposed is capable of resisting differential attacks.

Table 3. Results of NPCR and UACI achieved by the proposed method to evaluate sensitivity against differential attacks.

Images	NPCR	UACI
Poster	99.6732	33.4836
Street	99.5890	33.4821
Subway	99.6016	33.5104
ID Card	99.6232	33.4302
Hohai University	99.6278	33.4886
Key: $E_1 \leftrightarrow E_2$	99.5982	33.3821
Key: $E_1 \leftrightarrow E_3$	99.6107	33.4442
Key: $E_2 \leftrightarrow E_3$	99.5198	33.4250

4.7. Ablation Analysis

To verify the effectiveness of the proposed content-aware encryption module, we make an ablation analysis with "Lena" image as input, which is considered as the most commonly used test image [27]. As shown in Table 4, the performance without the proposed module is relatively poor, since simply adopting DNA encoding without content modeling fails in encoding as muchinformation of the input image. In fact, the proposed content-aware encryption module uses a permutation and a diffusion process to describe informative information remaining in the encoded data, greatly improving the complexity for breaking.

Table 4. Results of ablation analysis on "Lena" image, where CAEM indicates content-aware encryption module.

Ablation Methods	Correlation			Entropy	Sensitivity	
	V	H	D		NPCR	UACI
Ours (without CAEM)	0.1425	0.0986	0.0483	7.0145	93.2143	32.4218
Ours (Full)	0.0032	0.0019	0.0004	7.9982	99.6442	33.4732

4.8. Comparison Experiments

For comparative experiments, we follow the rules in Wu et al. [27], which are currently used by the vast majority of image encryption articles. Table 5 shows the experimental results with the comparing methods. It's noted that results achieved by other methods are directly obtained from their papers.

Table 5. Results achieved by the proposed method and several comparison methods.

Cryptosystems	Correlation			Entropy	Sensitivity	
	V	H	D		NPCR	UACI
Ours	0.0032	0.0019	0.0004	7.9982	99.6442	33.4732
Chen et al. [25]	−0.0064	0.0003	0.0110	7.9993	99.6218	33.5084
Zhang et al. [23]	0.0000016	−0.000003	−0.0000001	-	99.5009	33.4408
Aouissaoui et al. [28]	0.0240	0.0014	−0.0014	7.9978	99.6552	33.5871
Yan et al. [29]	−0.0056	−0.0012	−0.0020	7.9994	99.62	33.55

Essentially, correlation refers to the capability of resistance to statistical attack. Information entropy represents the capability of resistance to entropy attack; meanwhile, sensitivity represents the capability of resistance to the differential attack. It can be observed from Table 5 that the proposed method cannot achieve the best performance on all measurements. For example, Zhang et al. [23] has the best performance in correlation, but the worst in NPCR. Such a phenomenon means their method cannot bear a differential attack. Aouissaoui et al. [28] has the best performance in NPCR, but the worst in entropy and correlation. Such an observation means it would fail under an entropy attack and statistical attack. The proposed method has achieved a balanced performance in three measurements, which represents it does not have as much weaknesses as the other two comparison methods by obtaining enough capability of resistance to any type of attacks. Therefore, it can be concluded that our method is competitive among these methods.

4.9. Computation Cost Analysis

We compare the computing cost of the proposed method and other DNA computing-based methods in Table 6, thus proving its relatively low computing burden. It's noted that we use color image with 1024×1024 as input. All programs are deployed with C language and tested on the same devices (two laptops with i7-9750H CPU and 8GB DDR3 RAM), thus ensuring the fairness of comparisons. It is noted that the encryption and decryption process would cost 0.491 s and 0.507 s with the proposed method, which has certain advantages when comparing to other DNA computing-based methods.

Table 6. Computation cost comparisons among the proposed method and several DNA computing-based methods.

Cryptosystems	Encryption Time	Decryption Time
Ours	0.491 s	0.507 s
Chen et al. [25]	0.525 s	0.534 s
Zhang et al. [23]	0.572 s	0.568 s
Aouissaoui et al. [28]	0.554 s	0.549 s
Yan et al. [29]	0.598 s	0.602 s

5. Conclusions

The proposed method consists of two modules, i.e., DNA hash encoding and content-aware encrypting module. Specifically, the former one builds the quantity of hash mappings from image pixels to DNA computing bases, thus integrating the advantages of the hash function and DNA computing. The latter one considers strong correlations between adjacent pixels as sources of complexity, thus reorganizing the data structure by highly non-linear functions extracted from neighboring pixels and patches. Experimental results demonstrate that the proposed method could achieve a superior performance in various security measurements, compared with several of the latest comparison methods. Our future work includes DNA computing design on medical images a.

Author Contributions: Conceptualization, H.L. and L.Z.; methodology, H.L. and L.Z.; software, H.C.; validation, H.L., L.Z. and L.Z.; formal analysis, Y.W.; investigation, Y.W.; resources, H.L.; data curation, L.Z.; writing—original draft preparation, H.L. and L.Z.; writing—review and editing, H.L. and Y.W.; visualization, H.L. and L.Z.; supervision, Y.W.; project administration, Y.W.; funding acquisition, H.L. and Y.W. All authors have read and agreed to the published version of the manuscript.

Funding: This research was funded by Provincial Ideological and Political Education Research Projects in the Education Department of Zhejiang Province, the Fundamental Research Funds for the Central Universities under Grant B220202074, the Fundamental Research Funds for the Central Universities, JLU.

Informed Consent Statement: Not applicable.

Data Availability Statement: The data presented in this study are available on request from the corresponding author. The data are not publicly available due to privacy.

Conflicts of Interest: The authors declare no conflict of interest.

References

1. Wu, Y.; Cao, H.; Yang, G.; Lu, T.; Wan, S. Digital Twin of Intelligent Small Surface Defect Detection with Cyber-Manufacturing Systems. *ACM Trans. Internet Technol.* **2022**. [CrossRef]
2. Shu, J.J.; Wang, Q.W.; Yong, K.Y. DNA-based computing of strategic assignment problems. *Phys. Rev. Lett.* **2011**, *106*, 188702. [CrossRef]
3. Adleman, L.M. Molecular computation of solutions to combinatorial problems. *Science* **1994**, *266*, 1021–1024. [CrossRef] [PubMed]

4. Zhang, C.; Wang, Z.; Liu, Y.; Yang, J.; Zhang, X.; Li, Y.; Pan, L.; Ke, Y.; Yan, H. Nicking-assisted reactant recycle to implement entropy-driven DNA circuit. *J. Am. Chem. Soc.* **2019**, *141*, 17189–17197. [CrossRef] [PubMed]
5. Yang, J.; Wu, R.; Li, Y.; Wang, Z.; Pan, L.; Zhang, Q.; Lu, Z.; Zhang, C. Entropy-driven DNA logic circuits regulated by DNAzyme. *Nucleic Acids Res.* **2018**, *46*, 8532–8541. [CrossRef] [PubMed]
6. Zhang, C.; Zhao, Y.; Xu, X.; Xu, R.; Li, H.; Teng, X.; Du, Y.; Miao, Y.; Lin, H.C.; Han, D. Cancer diagnosis with DNA molecular computation. *Nat. Nanotechnol.* **2020**, *15*, 709–715. [CrossRef]
7. Ma, Q.; Zhang, C.; Zhang, M.; Han, D.; Tan, W. DNA Computing: Principle, Construction, and Applications in Intelligent Diagnostics. *Small Struct.* **2021**, *2*, 2100051. [CrossRef]
8. Khan, F.; Ncube, C.; Ramasamy, L.K.; Kadry, S.N.; Nam, Y. A Digital DNA Sequencing Engine for Ransomware Detection Using Machine Learning. *IEEE Access* **2020**, *8*, 119710–119719. [CrossRef]
9. Zou, C.; Wei, X.; Zhang, Q.; Zhou, C.; Zhou, S. Encryption algorithm based on DNA strand displacement and DNA sequence operation. *IEEE Trans. Nanobiosci.* **2021**, *20*, 223–234. [CrossRef]
10. Namasudra, S. Fast and Secure Data Accessing by Using DNA Computing for the Cloud Environment. *IEEE Trans. Serv. Comput.* **2022**, *15*, 2289–2300. [CrossRef]
11. Song, T.; Eshra, A.; Shah, S.; Bui, H.; Fu, D.; Yang, M.; Mokhtar, R.; Reif, J. Fast and compact DNA logic circuits based on single-stranded gates using strand-displacing polymerase. *Nat. Nanotechnol.* **2019**, *14*, 1075–1081. [CrossRef]
12. Wang, F.; Lv, H.; Li, Q.; Li, J.; Zhang, X.; Shi, J.; Wang, L.; Fan, C. Implementing digital computing with DNA-based switching circuits. *Nat. Commun.* **2020**, *11*, 121. [CrossRef]
13. Thubagere, A.J.; Li, W.; Johnson, R.F.; Chen, Z.; Doroudi, S.; Lee, Y.L.; Izatt, G.; Wittman, S.; Srinivas, N.; Woods, D.; et al. A cargo-sorting DNA robot. *Science* **2017**, *357*, eaan6558. [CrossRef] [PubMed]
14. Liu, L.; Jiang, D.; Wang, X.; Zhang, L.; Rong, X. A Dynamic Triple-Image Encryption Scheme Based on Chaos, S-Box and Image Compressing. *IEEE Access* **2020**, *8*, 210382–210399.
15. Xiao, Y.; Chen, Y.; Long, C.; Shi, J.; Ma, J.; He, J. A novel hybrid secure method based on DNA encoding encryption and spiral scrambling in chaotic OFDM-PON. *IEEE Photonics J.* **2020**, *12*, 1–15. [CrossRef]
16. Wu, Y.; Guo, H.; Chakraborty, C.; Khosravi, M.; Berretti, S.; Wan, S. Edge Computing Driven Low-Light Image Dynamic Enhancement for Object Detection. *IEEE Trans. Netw. Sci. Eng.* **2022**, 1. [CrossRef]
17. Babaei, M. A novel text and image encryption method based on chaos theory and DNA computing. *Nat. Comput.* **2013**, *12*, 101–107. [CrossRef]
18. Wang, X.; Liu, C. A novel and effective image encryption algorithm based on chaos and DNA encoding. *Multim. Tools Appl.* **2017**, *76*, 6229–6245. [CrossRef]
19. Samiullah, M.; Aslam, W.; Nazir, H.; Lali, M.I.U.; Shahzad, B.; Mufti, M.R.; Afzal, H. An Image Encryption Scheme Based on DNA Computing and Multiple Chaotic Systems. *IEEE Access* **2020**, *8*, 25650–25663. [CrossRef]
20. Malik, M.G.A.; Bashir, Z.; Iqbal, N.; Imtiaz, M.A. Color Image Encryption Algorithm Based on Hyper-Chaos and DNA Computing. *IEEE Access* **2020**, *8*, 88093–88107. [CrossRef]
21. Khan, J.S.; Boulila, W.; Ahmad, J.; Rubaiee, S.; Rehman, A.U.; Alroobaea, R.; Buchanan, W.J. DNA and Plaintext Dependent Chaotic Visual Selective Image Encryption. *IEEE Access* **2020**, *8*, 159732–159744. [CrossRef]
22. Ravichandran, D.; Banu, S.A.; Murthy, B.K.; Balasubramanian, V.; Fathima, S.; Amirtharajan, R. An efficient medical image encryption using hybrid DNA computing and chaos in transform domain. *Med. Biol. Eng. Comput.* **2021**, *59*, 589–605. [CrossRef] [PubMed]
23. Zhang, Q.; Han, J.; Ye, Y. Multi-image encryption algorithm based on image hash, bit-plane decomposition and dynamic DNA coding. *IET Image Process.* **2021**, *15*, 885–896. [CrossRef]
24. Wu, Y.; Zhang, L.; Berretti, S.; Wan, S. Medical Image Encryption by Content-Aware DNA Computing for Secure Healthcare. *IEEE Trans. Ind. Inform.* **2023**, *19*, 2089–2098. [CrossRef]
25. Chen, J.; Zhu, Z.; Zhang, L.; Zhang, Y.; Yang, B. Exploiting self-adaptive permutation-diffusion and DNA random encoding for secure and efficient image encryption. *Signal Process.* **2018**, *142*, 340–353. [CrossRef]
26. Wu, Y.; Noonan, J.P.; Agaian, S. NPCR and UACI randomness tests for image encryption. *Cyber J. Multidiscip. J. Sci. Technol. J. Sel. Areas Telecommun. (JSAT)* **2011**, *1*, 31–38.
27. Wu, Y.; Zhou, Y.; Noonan, J.P.; Agaian, S. Design of image cipher using latin squares. *Inf. Sci.* **2014**, *264*, 317–339. [CrossRef]
28. Aouissaoui, I.; Bakir, T.; Sakly, A. Robustly correlated key-medical image for DNA-chaos based encryption. *IET Image Process.* **2021**, *15*, 2770–2786. [CrossRef]
29. Yan, X.; Wang, X.; Xian, Y. Chaotic image encryption algorithm based on arithmetic sequence scrambling model and DNA encoding operation. *Multim. Tools Appl.* **2021**, *80*, 10949–10983. [CrossRef]

Review

UAV Detection and Tracking in Urban Environments Using Passive Sensors: A Survey

Xiaochen Yan [1], Tingting Fu [2], Huaming Lin [3,*], Feng Xuan [3], Yi Huang [3], Yuchen Cao [4], Haoji Hu [4] and Peng Liu [2]

[1] HDU-ITMO Joint School, Hangzhou Dianzi University, Hangzhou 310018, China; yanxiaochen0117@163.com
[2] School of Computer Science and Technology, Hangzhou Dianzi University, Hangzhou 310018, China; ftt@hdu.edu.cn (T.F.)
[3] Hangzhou Security and Technology Evaluation Center, Hangzhou 310020, China; hyi_1985@163.com (Y.H.)
[4] College of Information Science and Electronic Engineering, Zhejiang University, Hangzhou 310027, China; yuchen_cao@zju.edu.cn (Y.C.); haoji_hu@zju.edu.cn (H.H.)
* Correspondence: lhm18@live.cn

Abstract: Unmanned aerial vehicles (UAVs) have gained significant popularity across various domains, but their proliferation also raises concerns about security, public safety, and privacy. Consequently, the detection and tracking of UAVs have become crucial. Among the UAV-monitoring technologies, those suitable for urban Internet-of-Things (IoT) environments primarily include radio frequency (RF), acoustic, and visual technologies. In this article, we provide a comprehensive review of passive UAV surveillance technologies, encompassing RF-based, acoustic-based, and vision-based methods for UAV detection, localization, and tracking. Our research reveals that certain lightweight UAV depth detection models have been effectively downsized for deployment on edge devices, facilitating the integration of edge computing and deep learning. In the city-wide anti-UAV, the integration of numerous urban infrastructure monitoring facilities presents a challenge in achieving a centralized computing center due to the large volume of data. To address this, calculations can be performed on edge devices, enabling faster UAV detection. Currently, there is a wide range of anti-UAV systems that have been deployed in both commercial and military sectors to address the challenges posed by UAVs. In this article, we provide an overview of the existing military and commercial anti-UAV systems. Furthermore, we propose several suggestions for developing general-purpose UAV-monitoring systems tailored for urban environments. These suggestions encompass considering the specific requirements of the application scenario, integrating detection and tracking mechanisms with appropriate countermeasures, designing for scalability and modularity, and leveraging advanced data analytics and machine learning techniques. To promote further research in the field of UAV-monitoring systems, we have compiled publicly available datasets comprising visual, acoustic, and radio frequency data. These datasets can be employed to evaluate the effectiveness of various UAV-monitoring techniques and algorithms. All of the datasets mentioned are linked in the text or in the references. Most of these datasets have been validated in multiple studies, and researchers can find more specific information in the corresponding papers or documents. By presenting this comprehensive overview and providing valuable insights, we aim to advance the development of UAV surveillance technologies, address the challenges posed by UAV proliferation, and foster innovation in the field of UAV monitoring and security.

Keywords: anti-UAV; UAV detection; UAV tracking; UAV dataset; anti-UAV systems

Citation: Yan, X.; Fu, T.; Lin, H.; Xuan, F.; Huang, Y.; Cao, Y.; Hu, H.; Liu, P. UAV Detection and Tracking in Urban Environments Using Passive Sensors: A Survey. *Appl. Sci.* **2023**, *13*, 11320. https://doi.org/10.3390/app132011320

Academic Editors: Yirui Wu and Shaohua Wan

Received: 15 September 2023
Revised: 6 October 2023
Accepted: 10 October 2023
Published: 15 October 2023

1. Introduction

With the rapid advancement of technology, unmanned aerial vehicles (UAVs) have found numerous urban IoT (Internet of Things) applications in fields such as rescue operations [1], surveillance [2,3], edge computing [4,5], disaster area reconstruction [6,7], aerial base stations [8,9], intelligent transportation [10,11], wireless power transfer [12], environmental monitoring [13], and more [14,15]. According to market forecasts, the UAV market will experience a compound annual growth rate (CAGR) of 7.9%, expanding from USD 26.2 billion in 2022 to USD 38.3 billion in 2027 [16]. However, the increasing use of UAVs also raises concerns about security, public safety, and privacy. To address these issues, researchers have been exploring various UAV monitoring technologies. UAV surveillance can be achieved through four primary technologies: radar, radio frequency (RF), visual, and acoustic surveillance.

Radar surveillance is a well-established and widely adopted method for airspace monitoring and reconnaissance. This technology can be used to detect UAVs that lack radio communication or with full autonomy. Doppler radar is a powerful technology that uses the Doppler effect to detect the velocity of moving objects. Micro-Doppler radar is an advanced version of Doppler radar that can detect speed and motion differences inside objects. Micro-Doppler radar is an ideal method for detecting UAVs because their propellers create large linear velocity differences [17]. Other radar technologies that may be applicable include non-line-of-sight radar, ultra-wideband radar, millimeter-wave radar [18], chaotic mono-static, bi-static [19], and multi-static radars [20]. Because of their low radar cross section [21] and slow speeds, detecting UAVs with radar is difficult and complex [22]. Experimental results indicate that the detection range of radar rarely surpasses 10,000 m [23].

UAVs emitting RF wave signals can be intercepted and analyzed to track and locate them [24]. In many cases, manually operated UAVs communicate with a ground station and a GNSS (Global Navigation Satellite System) for operation, making it possible to intercept signals and obtain information such as coordinates and video feeds. However, autonomous UAVs that rely solely on onboard sensors for operation may not emit RF signals, making it more challenging to locate and track them [17]. RF surveillance exhibit the capability to detect and locate UAVs within a range of 5000 m. However, their performance can be influenced by factors such as multipath and non-line-of-sight propagation.

In recent years, computer vision-based methods have become increasingly popular for detecting and monitoring UAVs. By leveraging deep learning techniques, models can be trained to automatically extract appearance and motion features from datasets of UAV images and videos, enabling the identification and tracking of these vehicles using video surveillance from cameras [25]. In addition, infrared cameras can be used to detect and identify UAVs in low ambient light conditions [26]. By combining computer vision techniques with infrared cameras, UAVs can be detected and monitored effectively in various lighting and environmental conditions. Vision-based methods encounter challenges in distinguishing UAVs from birds, particularly when the UAVs are situated at a considerable distance. In fact, identifying UAVs beyond a range of 1000 m becomes exceedingly arduous, if not nearly impossible.

Sound waves emitted by the power unit and propeller blades during UAV navigation can be detected using an acoustic microphone, which converts the pressure of these sound waves into electrical signals. These sound waves create an individualized "audio fingerprint" for each UAV, enabling individual identification. However, detecting these sound waves can be challenging in practice due to factors such as ambient noise and sound wave attenuation. Deep learning models have been applied to improve the effectiveness of this method for UAV detection. By training these models with datasets of acoustic signatures, they can learn to distinguish UAV signals from noise and other interference, enhancing the accuracy and reliability of this approach [27]. Nevertheless, acoustic monitoring is highly susceptible to ambient noise and possesses a constrained detection range. As per the conducted tests, the maximum detection range for UAVs does not exceed 300 m.

Table 1 provides a comprehensive overview of the aforementioned monitoring techniques. It is crucial to acknowledge that the detection distances presented are derived from existing literature and systems, and may exhibit variations based on factors such as UAV type, hardware parameters, and associated algorithms. To improve the effectiveness of UAV recognition, UAV detection systems typically utilize a combination of two to three of the aforementioned technologies [22].

Table 1. Comparison of UAV surveillance technologies.

Surveillance Technology	Localization/Tracking Method	Detection Range	Challenges
Radar	Doppler-based tracking Delay-based localization	10,000 m	Low radar cross section Low speed and altitude
Acoustic	TDOA\AOA-based localization	0–300 m	High ambient noise
Vision	Motion-based tracking	100–1000 m	Confuse with birds Indistingushable small objects
RF	RSS\AOA-based localization	5000 m	Ambient RF noise Multipath Non-line of sight

In the past, UAV-monitoring systems were primarily deployed in critical military and civilian facilities such as airports and military bases. However, with the increasing popularity of UAVs, the need for UAV-monitoring systems has expanded to a wider range of settings, including construction sites, communities, shopping malls, schools, and other locations. This has created a demand for UAV-monitoring systems that are more cost-effective, scalable, and responsive. To meet this challenge, researchers have been exploring ways to detect UAVs using lower-cost and passive sensors [28]. The development of software-defined radio has greatly reduced the cost of RF detection, making it more accessible to a broader range of users. In recent years, neural network-enhanced RF-based detection, visual-based detection, and acoustic-based detection have emerged as promising options for general UAV-monitoring systems in urban environments. With the development of lightweight models, there are already some models that can obtain acceptable results with very little computing resources, which makes it possible to use edge computing for UAV detection. While radar surveillance is highly effective in detecting aircraft, its use is limited to specific locations due to the high cost and radiation associated with the technology. As a result, it may not be suitable for detecting illegal UAVs in urban areas.

Our findings indicate that within the research community, there is a notable divergence in focus regarding anti-UAV research. While some researchers prioritize the development of low-cost and lightweight anti-UAV solutions, the majority of researchers show a greater interest in enhancing the effectiveness of UAV detection and tracking. Our survey aims to address this disparity and raise awareness among researchers regarding the broader anti-UAV requirements in urban environments. We believe there is a pressing need for further exploration of low-cost and passive sensor-based anti-UAV systems. By directing attention towards these areas, we hope to foster increased research and innovation in developing comprehensive anti-UAV solutions that are both cost-effective and capable of meeting the specific challenges posed by urban environments.

This paper provides a comprehensive review of passive UAV surveillance technologies, encompassing RF-based, acoustic-based, and vision-based methods for UAV detection, localization, and tracking. Our research reveals that certain lightweight UAV depth detection models have been effectively downsized for deployment on edge devices, facilitating the integration of edge computing and deep learning. In the city-wide anti-UAV, the integration of numerous urban infrastructure monitoring facilities presents a challenge in achieving a centralized computing center due to the large volume of data. To address this, calculations can be performed on edge devices, enabling faster UAV detection. Currently, there is a wide range of anti-UAV systems that have been deployed in both commercial and military sectors to address the challenges posed by UAVs. In this article, we provide an overview of the existing military and commercial anti-UAV systems. Furthermore, we propose several suggestions for developing general-purpose UAV-monitoring systems tailored for urban environments. These suggestions encompass considering the specific requirements of the application scenario, integrating detection and tracking mechanisms with appropriate countermeasures, designing for scalability and modularity, and leveraging advanced data analytics and machine learning techniques. To promote further research in the field of

UAV-monitoring systems, we have compiled publicly available datasets comprising visual, acoustic, and radio frequency data. These datasets can be employed to evaluate the effectiveness of various UAV-monitoring techniques and algorithms. All of the datasets mentioned are linked in the text or in the references. Most of these datasets have been validated in multiple studies, and researchers can find more specific information in the corresponding papers or documents. By presenting this comprehensive overview and providing valuable insights, we aim to advance the development of UAV surveillance technologies, address the challenges posed by UAV proliferation, and foster innovation in the field of UAV monitoring and security.

The rest of the paper is organized as follows. Section 2 gives the comprehensive review of UAV detection and identification methods. Section 3 describes ways of UAV localization and tracking. Anti-UAV system is introduced in Section 4. Finally, the conclusion is presented in Section 5.

2. UAV Detection and Identification

In this section, we will conduct a comprehensive analysis and comparison of RF-based, acoustic-based, and vision-based methods for the detection and identification of UAVs in urban IoT environments. We will explore these methods from various perspectives, including traditional methods, deep learning methods, and the available public and semi-public datasets. Additionally, we will discuss the specific challenges associated with identifying UAV intrusions in IoT environments and review recent research focused on leveraging edge devices for UAV detection. Table 2 provides a performance comparison of various UAV detection methods, with the data obtained from literature research. The vision method based on deep learning demonstrates good accuracy even when experiments are conducted on public datasets that are not specifically optimized for UAVs. However, recent studies in the literature have shown that after fine-tuning, optimization, and training on specialized UAV datasets, the accuracy rate can exceed 90%, indicating a high level of competitiveness.

Table 2. Comparison of different UAV detection methods.

Category	Method	Accuracy
Traditional RF-based Detection	Fractal dimension (FD) [29]	100%
	axially integrated bispectra (AIB) [30]	98%
	square integrated bispectra (SIB) [31]	96%
	Signal Spectrum (SFS) [32]	97.85%
	Wavelet energy entropy (WEE) [32]	93.75%
	Power spectral entropy (PSE) [32]	83.85%
RF-based Detection Using Deep Learning	Y. Mo et al. [33]	99%
	C. J. Swinney et al. [34]	100%
	S. Lu et al. [35]	98%
	Z. He et al. [36]	90.2%
	T. Li et al. [37]	98.57%
Traditional Acoustic-based Detection	Mel Frequency Cepstrum Coefficient (MFCC)	97%
Acoustic-based Detection Using Deep Learning	S. Al-Emadi et al. [27]	96.38%
	Q. Dong et al. [38]	99%
	İ Aydın et al. [39]	98.78%

Table 2. *Cont.*

Category	Method	Accuracy
Vision-based Detection Using Deep Learning	RCNN	58.50%
	SPPNet	59.20%
	Fast RCNN	70.00%
	Faster RCNN	73.20%
	YOLOv3	63.40%
	SSD	76.80%
	RetinaNet	59.10%

Traditional RF methods involving RF feature extraction have made progress in recent years. The RF approach based on deep learning uses technology to convert RF signals into images, facilitating feature extraction using deep networks. Traditional acoustic methods mainly rely on Mel-Frequency cepstrum coefficient (MFCC) feature extraction, often supplemented by linear predictive cepstrum coefficient (LPCC). Acoustic methods based on deep learning convert sound signals into spectrograms and extract relevant features. In recent years, vision-based detection methods, especially those based on deep learning, have gained prominence. The models listed in the study use public datasets that are not specifically optimized for UAV detection. However, when these models are optimized and trained on specialized UAV datasets, detection accuracy can often exceed 90%. The effectiveness of UAV detection methods depends on factors such as dataset quality, training techniques, and optimization for specific use cases.

2.1. RF-Based Detection

The process of RF-based UAV detection typically involves feature extraction from the received RF signal, followed by comparison of the extracted features with the UAV RF feature database to achieve UAV identification. In recent years, learning-based approaches have gained significant traction in the realm of UAV detection, leading to significant changes in RF-based UAV detection methods. To enhance UAV detection using learned methods, researchers have employed various techniques. Generally, these approaches can be divided into two categories. The first category involves extracting signal features using signal processing methods such as Fourier transform, followed by classification using SVM, decision trees, and other methods [29,32]. The second category involves applying simple processing to the signal, followed by feature extraction using deep neural networks for UAV signal identification [33–37]. This section will provide an overview of both traditional RF-based detection methods and deep learning-based approaches for UAV detection. Additionally, commonly used datasets for UAV detection will also be introduced.

2.1.1. Traditional RF-Based Detection

RF detection involves identifying both the flight control signal and the image transmission signal that are exchanged between the control station and the UAV. Its principle is to analyze the frequency, symbol rate, modulation type, signal jump, channel bandwidth and other information of these signals based on collecting the original RF signal, and extract the 'fingerprint' feature that identifies its uniqueness. To improve the accuracy of identification, researchers often choose to extract multiple features to serve as the 'fingerprint' information for UAVs. In this section, we shall examine the methodologies employed for the extraction of UAV spectral features in recent years.

The Fractal dimension (FD) [29] is a statistical metric that characterizes the complexity of a signal, and can also be used to quantify its roughness or irregularity. There are two methods for calculating the FD, namely the time domain method and the phase space domain method. Some commonly used FD include the correlation dimension, Hausdorff dimension, box dimension [40], and so on. The Higuchi algorithm is a highly efficient method for calculating fractal dimensions [41]. The FD has been shown to be able to achieve identification accuracies of up to 100%.

Bispectrum is a powerful tool for analyzing nonlinear, non-Gaussian, non-minimal phase stationary random signals. The bispectrum is a powerful tool for reducing the impact of white Gaussian noise (WGN). For better online signal classification and object recognition, several bispectral-based integrated feature extraction methods have been developed [42], including the axially integrated bispectra (AIB) [30] and the square integrated bispectra (SIB) [31]. The AIB feature and SIB feature exhibit accuracies of 98% and 96%, respectively.

Signal Spectrum (SFS) [32] is a graphical representation of the frequency distribution of a signal, which can be used to analyze its frequency content. Throughout the entire flight process of a UAV, the continuous rotation of its propellers induces vibrations that propagate throughout the fuselage. By employing STFT to analyze the received wireless signal, it is possible to distinguish between UAV and non-UAV signals based on differences in their respective SFS. The accuracy of this approach can reach 97.85%.

Wavelet energy entropy (WEE) [32] can analyze the change of signal entropy level after wavelet transform. The wavelet transform can capture the sudden short-term signal changes that will occur when the UAV is restored due to the effects of wind and the UAV's movement in three-dimensional space during the flight. The accuracy of the WEE method is 93.75%.

Power spectral entropy (PSE) [32] is a signal feature that describes the relationship between power spectrum and entropy rate, and can effectively differentiate between different signals. Existing literature indicates that under the same acquisition environment, UAV signals exhibit significantly higher PSE values compared to non-UAV signals. With respect to PSE, an accuracy of 83.85% has been achieved.

As shown in Table 3, the accuracy of FD, AIB, and SIB is higher than SFS, WEE, and PSE. However, the former ones (FD, AIB, and SIB) can only accurately identify known UAV types, and the accuracy of identification is low for UAVs that are unknown. The latter ones (SFS, WEE, and PSE) identify UAVs based on the physical characteristics of their flight, and maintains a high accuracy rate for UAVs that are not visible yet. Simultaneously utilizing multiple fingerprint features can enhance the accuracy of UAV identification.

Table 3. Comparison of traditional RF-based detection. the processed RF features are input into the machine learning algorithm (KNN, SVM, RandF) to identify the UAV, and the accuracy rate is optimized.

Fingerprint	Accuracy
Fractal dimension (FD) [29]	100%
Axially integrated bispectra (AIB) [30]	98%
Square integrated bispectra (SIB) [31]	96%
Signal Spectrum (SFS) [32]	97.85%
Wavelet energy entropy (WEE) [32]	93.75%
Power spectral entropy (PSE) [32]	83.85%

2.1.2. RF-Based Detection Using Deep Learning

As learning methods continue to evolve, researchers are exploring new ways to improve the effectiveness of signal feature extraction. A wide range of possibilities for signal feature extraction using learning methods have been explored in the literature. Recent research results [43,44] have demonstrated that convolutional neural networks (CNN) are highly suitable for RF-based detection and classification, highlighting the potential of deep learning methods in this field. By leveraging the ability of CNN to automatically extract relevant features from raw signal data, researchers have achieved significant improvements in UAV detection and classification performance.

As shown in Ref. [33], CNN can directly extract features from compressed sensing signals without requiring the use of reconstruction algorithms to restore the original signal. Extracting features in this way is a little easier than using traditional methods. The accuracy of the final classification can reach about 99%. In Ref. [34], the authors fed the spectrogram directly into the CNN to extract features. With this method, the prediction accuracy can reach up to 100%. S. Lu et al. [35] tried to use a time-frequency waterfall map directly to extract features by utilizing a network architecture comprising EESP cells (according to the description in this literature, EESP may refer to the ESP, efficient spatial pyramid [21]) and employing the VGG feature extraction method, and achieved good results. This method can achieve an accuracy of over 98%. In Ref. [36], after preprocessing the compressive signal using wavelet transform, the signal was mapped to a two-channel image, which was fed into a VGG-16 model to detect and classify UAVs. Using bispectral feature extraction based on two-dimensional Fourier transform, they were able to extract more frequency domain features. This method is 100% accurate in predicting whether UAVs are present or absent. When the prediction is classified as different types of UAVs, the accuracy rate is 98.9%. When the prediction classification includes different operating modes of various types of UAVs, the accuracy rate is 90.2%. Li et al. [37] directly converted the two frequency domain dimensions (f_1, f_2) and frequency domain amplitude $|B(f_1, f_2)|$ to grayscale images and undertook embedding into a Siamese network, supported by an image augmentation strategy. The accuracy rates for UAV type detection and operation mode detection are 98.57% and 92.31%, respectively.

These deep learning-based methods typically convert radio frequency signals into images such as spectrograms, and then use deep learning networks to extract features and identify UAVs. Table 4 presents a comparison of the aforementioned methods.

Table 4. Comparison of RF-based detection using deep learning.

Authors	Accuracy
Y. Mo et al. [33]	99%
C. J. Swinney et al. [34]	100%
S. Lu et al. [35]	98%
Z. He et al. [36]	90.2%
T. Li et al. [37]	98.57%

Recent literature on RF-based UAV detection has identified two major development directions. The first direction involves combining UAV physical characteristics with signal processing techniques to extract more effective spectral features, such as waveform energy entropy (WEE) and spectral flatness measure (SFM). This approach aims to boost the discriminative capability of the extracted features by leveraging unique physical properties of UAV signals. The second direction involves developing network models that can extract RF features faster and more accurately. This approach focuses on designing deep learning architectures that can effectively capture the relevant features from the RF signal data. By improving the efficiency and accuracy of feature extraction, these approaches aim to enhance the overall performance of RF-based UAV detection systems.

2.1.3. Datasets and Metrics for RF-Based Detection

In this section, public and semi-public datasets in recent years will be reviewed in detail.

There are several publicly available RF-based UAV datasets that researchers can use to develop and evaluate RF-based UAV detection methods. These include the DroneRF dataset [45], Drone remote controller RF signal dataset [46], Radio-Frequency Control and Video Signal Recordings of Drones [47], Dronesignals Dataset [48], DroneDetect Dataset [49], VTIDroneSETFFT [50], and the CARDINAL RF [51]. These datasets contain recorded segments from 3, 17, 10, 9, 7, 3, and 6 different UAVs, respectively, with VTIDroneSETFFT being the only dataset to contain records of the simultaneous operation of multiple UAVs. Unlike vision-based UAV datasets, there are relatively few publicly available RF-based UAV datasets, and RF datasets are frequently quite sizable due to the considerable amount of data acquired. Despite these challenges, RF-based UAV detection methods have shown promising results in detecting and classifying UAVs based on their RF emissions. Researchers interested in accessing these datasets can find the corresponding links in the references section.

2.2. Acoustic-Based Detection

Audio-based UAV detection is another commonly used technical solution for UAV detection. While sound waves and electromagnetic waves are different, audio-based detection and radio-frequency detection share some similarities due to their wave nature. Acoustic-based detection can be broadly categorized into two main types, i.e., traditional acoustic-based detection and acoustic-based detection using deep learning.

2.2.1. Traditional Acoustic-Based Detection

Cepstrum analysis is a widely used technique for analyzing audio signals, and it entails computing the inverse Fourier transform of the logarithmic power spectrum values. Many feature extraction methods for audio signals are developed based on cepstrum coefficients. The commonly used cepstrum coefficients include Linear Predictive Cepstrum Coefficient (LPCC) and Mel Frequency Cepstrum Coefficient (MFCC). By applying these techniques to audio-based UAV detection, researchers have achieved promising results in identifying UAVs based on their unique acoustic features.

The Mel frequency is a non-linear frequency scale that was designed to model the way in which the human ear perceives sound. It does not have a linear correspondence with Hz frequency. The MFCC is a spectrum feature that is computed based on the Mel frequency scale, and it has found extensive use in the domain of speech recognition. However, the accuracy of MFCC calculation decreases with increasing frequency. As a result, in practical applications, only low-frequency MFCCs are typically used, while medium to high frequency MFCCs are discarded. Despite this limitation, MFCC remains the most widely employed audio feature in current UAV recognition tasks, as it has demonstrated effectiveness in identifying UAVs based on their unique acoustic features [38,52–55]. The accuracy when using MFCC can be over 97%.The audio signal is initially segmented into frames and windowed. This involves dividing the signal into shorter time intervals and applying a window function to each interval to minimize spectral leakage. Each frame undergoes the application of the Fast Fourier Transform (FFT) to derive its linear spectrum. The linear spectrum is subsequently filtered using a bank of Mel filters that are spaced according to the Mel frequency scale. The logarithm of the filtered spectrum is then computed to obtain the LogMel spectrogram. The LogMel spectrogram is subsequently subjected to the Discrete Cosine Transform (DCT), an instrumental method utilized for the transformation of spectral data from the time domain to the frequency domain. A desired number of DCT coefficients are then retained to obtain the MFCC, which can be used as a feature for audio-based UAV detection.

Linear Predictive Cepstral Coefficients (LPCC) is a commonly used method for feature extraction in audio signal analysis [56] and is a variant of MFCC. Similar to MFCC, LPCC extracts features by preprocessing, framing, filter bank processing, Mel cepstrum, DCT, and other steps on speech signals. However, LPCC uses linear prediction (LP) filters instead of triangular filter banks in the filter bank processing stage to extract resonance peak information of speech signals. The LP filters used by LPCC in filter bank processing can more accurately extract the formant information of speech signals, making LPCC more effective than MFCC in the presence of Gaussian white noise. However, the high complexity of the LP filters and the corresponding increase in computational complexity must be balanced in practical applications. In the literature, LPCC is usually used as a supplement to MFCC, and the amalgamation of these two features is commonly employed to bolster the performance of UAV recognition [53].

2.2.2. Acoustic-Based Detection Using Deep Learning

Acoustic-based UAV detection has two primary directions of research. The first direction involves accurately and quickly separating sound sources and identifying UAV acoustic signals in noisy environments, which can be achieved using techniques such as ICA and fastICA [57]. The second direction focuses on exploring UAV detection in time-varying scenarios, which is important due to the rapid movement of UAVs and the need to track their acoustic signatures over time [57]. While acoustic-based UAV detection is less commonly used in commercial and military anti-UAV systems due to its short detection range, this approach still has significant potential for application in urban IoT environments such as residential areas, schools, and commercial areas. By leveraging the unique acoustic signatures of UAVs, this approach can help to identify and track UAVs in real-time, providing valuable information for UAV detection and countermeasure systems. Future research in this area may focus on developing more accurate and robust acoustic-based UAV detection methods that can function across a range of acoustic environments and under various operating conditions of UAVs.

Independent Component Analysis (ICA) is a computational method that can be employed to decompose multivariate signals into additive subcomponents. This is done under the assumption that the subcomponents are non-Gaussian and statistically independent of one another. ICA is a special case of blind source separation, and it can be used to separate acoustic signals from mixed recordings. ICA's ability to separate signals that are mixed with strong interference sources makes it a promising technique for detecting single or multiple UAVs when the signals are contaminated with noise [57]. In practice, ICA can be used as a part of signal preprocessing and combined with other analysis methods. For example, in Refs. [53,57], a combination of ICA and MFCC is used for UAV identification, with ICA used to preprocess the audio signals and extract independent acoustic sources before applying MFCC for feature extraction.

In addition to traditional signal processing techniques, learning-based methods are also widely used for feature extraction in audio-based UAV detection. Recent research [38,39] has shown that deep learning models can also be used to extract features from UAV acoustic signals for UAV detection. In Ref. [27], the authors converted audio clips into spectrograms and used them as input for deep learning algorithms. To train deep learning models, the algorithm extracted multiple features from the spectrograms generated by the audio signals. In Ref. [39], the received audio signal was converted into the MEL spectrum by short-term Fourier transform, recorded as an image, and then input into a specially designed Light-Weight Convolutional Neural Network to extract signal features. In Ref. [38], Log Mel spectrograms and MFCC were used as inputs to CNN to classify the acoustic signals as either indicative of UAV activity or not. Table 5 presents the accuracy of these methods. It showcases the effectiveness of the approaches described earlier in detecting UAVs using acoustic signals.

Table 5. Comparison of acoustic-based detection using deep learning.

Authors	Accuracy
S. Al-Emadi et al. [27]	96.38%
Q. Dong et al. [38]	99%
İ Aydın et al. [39]	98.78%

2.2.3. Datasets and Metrics for Acoustic-Based Detection

To evaluate the performance of acoustic-based detection, the following two datasets and metrics are usually used, i.e., DroneAudioDataset [58] (can be found from GitHub repository: https://github.com/saraalemadi/DroneAudioDataset, accessed on 1 June 2023) and the Casabianca's Dataset [59] (can be found from GitHub repository: https://github.com/pcasabianca/Acoustic-UAV-Identification, accessed on 1 June 2023). The DroneAudioDataset contains recordings of UAV propeller noise in an indoor environment, with the audio clips categorized as either 'Drone' or 'Unknown'. The 'Unknown' category consists of white noise and silence audio clips that do not contain any UAV signals. In contrast, the Casabianca's Dataset contains recordings of a variety of UAVs in a range of background sound situations, including airplanes, helicopters, traffic, thunderstorms, wind, and quiet environments. These datasets provide a valuable resource for researchers to develop and evaluate audio-based UAV detection methods under different acoustic conditions.

Since the number of acoustic UAV datasets is limited, researchers often have to create their own datasets by recording UAV acoustic signals from publicly available videos or by conducting their own experiments. This involves using appropriate recording equipment, such as microphones, and selecting suitable recording locations and environmental conditions to capture a diverse range of UAV acoustic signatures. Despite the challenges associated with creating and using self-collected datasets, this approach allows researchers to tailor their experiments to specific application scenarios and to evaluate the performance of their proposed methods under realistic conditions. However, it is important to ensure that the datasets are properly annotated and validated to ensure the accuracy and reliability of the results.

2.3. Vision-Based Detection

Object detection is a crucial task in the domain of computer vision that involves the detection of instances of certain types of visual objects (e.g., animals, people, or plants) in an image. By creating computational models and techniques to detect objects, computer vision applications can provide valuable information: what object is where? Object detection is widely recognized as a fundamental challenge, and it provides the basis for numerous other associated tasks. For instance, methods for object detection can be extended to instance segmentation, image captioning, and target tracking tasks. From an application standpoint, research on object detection can be broadly categorized into two primary topics: "general object detection", which aim to detect various types of objects within a unified framework, mimicking the visual perception and recognition capabilities of human beings, and "detection applications", which focuses on designing specialized approaches tailored to specific application scenarios, such as pedestrian detection, face detection, and other relevant domains. Over the past few years, rapidly advancing deep learning techniques have injected a new dynamic into object detection, bringing significant breakthroughs. Object detection has become increasingly popular across a diverse spectrum of real-world applications, including path planning, robot vision, video surveillance, and many others. The following section will introduce more typical target detection methods, including traditional object detection methods and deep learning-based approaches, as well as the commonly used UAVs object detection datasets at present. Table 6 shows the results of some state-of-the-art methods on different datasets.

Table 6. Accuracy of object detection on VOC07, VOC12 and MS-COCO datasets. Detectors in the table: RCNN, SPPNet, Fast RCNN, Faster RCNN, YOLOv3, SSD, RetinaNet.

mAp	VOC07 mAP	VOC12 mAP	COCO mAP@.5	COCO mAP@[.5, .95]
RCNN	58.50%	53.70%	-	-
SPPNet	59.20%	-	-	-
Fast RCNN	70.00%	68.40%	35.90%	19.70%
Faster RCNN	73.20%	70.40%	42.70%	21.90%
YOLOv3	63.40%	57.90%	-	-
SSD	76.80%	74.90%	46.50%	26.80%
RetinaNet	-	-	59.10%	39.10%

2.3.1. Traditional Object Detection

Viola Jones Detector

The VJ detector [60] relies on a simple yet effective object detection technique known as sliding windows. As part of this method, the window's position and scale are scanned in order to determine if there is a human face in any of the windows. Although the sliding windows technique used by the VJ detector appears to be straightforward, the computational demands of this approach were beyond the capabilities of computers at the time of its development. This method combines three essential techniques: "integral image", "feature selection", and "detection cascade", which significantly enhance the speed of detection.

HOG Detector

HOG [61] can be regarded as a significant improvement over previous techniques such as scale-invariant feature transformations and shape contexts. It is designed to be computed on a dense grid of uniformly spaced cells in order to balance feature invariance, which includes translation, scaling, illumination, and other factors, as well as nonlinearity and the need to distinguish between different object classes. The classification accuracy is further improved by applying overlapping local contrast normalization to the "blocks". In spite of the fact that HOG is capable of detecting objects of different classes, it is primarily motivated by pedestrian detection problems. HOG detectors resize input images several times while maintaining the detection window size to detect objects of different sizes. In various computer vision applications for many years, the HOG detector has been a key component of many object detectors.

2.3.2. Deep Learning Based Object Detection
RCNN

It is easy to understand how RCNN (Regions with CNN features) [62] works. Initially, selective search is used to extract an object candidate frame set. An ImageNet-trained CNN model is then applied to each candidate frame and rescaled to a fixed-size image, such as AlexNet, to extract features. The final step is to determine whether objects are present or absent within each region and to identify object classes using a linear SVM classifier. For instance, RCNN has demonstrated a substantial increase in mean Average Precision (mAP) 8.5%.

SPPNet

SPPNet (Spatial Pyramid Pooling Network) was proposed by K. He et al. [63] in 2014, which addressed the limitations of fixed-size inputs in CNN. A major contribution of SPP-Net is its Spatial Pyramid Pooling (SPP) layer, which allows CNN to generate fixed length representations without rescaling depending on the image size. This approach avoids the need for repeated computation of convolutional features by computing the feature mapping only once from the entire image. When used for object detection, SPPNet enables the generation of a fixed-length representation for any region, making it unnecessary to rescale the image or region of interest. Using this approach, SPPNet can detect objects at 20 times faster speeds than RCNN without compromising detection accuracy (VOC07 mAP = 59.2%).

Fast RCNN

In 2015, Fast RCNN [64] detectors improve upon RCNN and SPPNet according to R. Girshick. By using the Fast RCNN detector, both the bounding box regressor and the object detector are trained using the same network configuration, which is an improvement over the previous two-stage approach used by RCNN and SPPNet. With its streamlined architecture, Fast RCNN provides a significant speedup over previous methods, detecting objects more than 200 times faster than RCNN, while still achieving high accuracy. Fast RCNN improved the mAP from 58.5% (RCNN) to 70.0% on the VOC07 dataset.

Faster RCNN

S. Ren et al. introduced a faster RCNN [65] detector shortly after the Fast RCNN detector made its debut in 2015. The Faster RCNN detector is considered to be the first to detect deep learning end-to-end and in near real-time. It achieved a COCO mAP@.5 of 42.7%, COCO mAP@[.5, .95] of 21.9%, VOC07 mAP of 73.2%, and VOC12 mAP of 70.4%. By integrating proposal detection, feature extraction, and bounding box regression into a single framework, Faster RCNN detector provides a complete solution for object detection. Today, faster RCNN is widely used for object detection due to its enhanced speed and accuracy.

YOLO

YOLO (You Only Look Once) was proposed by R. Joseph et al. in 2015 to detect objects in one stage [66]. YOLO is known for its remarkable speed, with the fast version of YOLO running at 155 fps (VOC07 mAP = 52.7%), while the enhanced version runs at 45 fps (VOCO7 mAP = 63.4%, VOC12 mAP = 57.9%). Unlike previous object detection methods that used a "proposal detection + verification" paradigm, one neural network is applied to the whole input image in YOLO. An image is divided into a grid of cells, with each cell containing its bounding box and probability. However, when it comes to localizing small objects, its accuracy falls short compared to that of the two-stage detector. YOLO is very popular in the field of UAV detection because of its extremely fast speed and relatively high accuracy, and there have been many studies on UAV detection based on the YOLO algorithm in recent years [67,68].

SSD

W. Liu et al. presented SSD [69] in 2015 as the second level 1 detector in deep learning. This invention introduces multi-reference and multi-resolution detection techniques that greatly enhance detection accuracy, especially for small objects. SSD outperforms previous detectors in terms of speed and accuracy, with VOC07 mAP of 76.8%, VOC12 mAP of 74.9%, COCO mAP@.5 of 46.5%, mAP@[.5, .95] of 26.8%, and a fast version running at 59 fps. SSD detects objects at 5 different scales at different layers of the network, as opposed to previous detectors that detected at only the top layer.

RetinaNet

RetinaNet [70], proposed by Lin et al. in 2018, is a one-level target detector that overcomes the challenge of training a single-level detector with an extreme foreground-background class imbalance. While two-level target detectors can achieve higher accuracy by proposing dense candidate locations, single-level detectors have been limited by difficulties in training an unbalanced set of positive and negative examples. A novel loss function called "focal loss" was proposed to address this problem, which alters the standard cross-entropy loss function so that the detector can focus more on difficult, misclassified examples during training, mitigating the loss resulting from well-classified or simple examples. The use of focal loss greatly improves the accuracy of RetinaNet, allowing it to achieve COCO mAP@.5 of 59.1% and mAP@[.5, .95] of 39.1%, while maintaining the speed of previous single-stage detectors.

2.4. Datasets and Metrics for Vision-Based Detection

In this section, public and semi-public datasets in recent years will be reviewed.

2.4.1. Well-Known Datasets

The construction of larger datasets with smaller biases is crucial for the development of advanced computer vision algorithms. During the last decade, numerous benchmarks and datasets have been released for object detection, including datasets such as the PASCAL VOC challenge [71,72], the MS-COCOCO detection challenge [73], and others. Table 6 shows the detection accuracies of the above deep learning methods on the VOC07, VOC12, and MS-COCO datasets.

2.4.2. Object Detection Dataset for UAVs

The main image-based UAV datasets are the Real World dataset [74], the Det-Fly dataset [75], the MIDGARD dataset [76], the USC-Drone dataset [77], and the DUT-Anti-UAV [78]. These 5 datasets contain 56,821, 13,271, 8775, 18,778, and 10,109 images, respectively, and their respective resolutions are 640 × 480, 3840 × 2160, 752 × 480, 1920 × 1080 pixels, and various resolutions. Their samples are shown in Figure 1. Compared to other UAV datasets, the Real World dataset stands out for its diversity in terms of UAV types and environments, despite the low-resolution quality of the images. This is because all the data in the Real World dataset are obtained from YouTube videos, whereas other datasets are typically collected by the researchers themselves. On the other hand, the Det-Fly dataset addresses the shortcomings of single-view UAV data by providing multi-view data that covers a wider range of angles and perspectives. The MIDGARD dataset and USC-Drone dataset also contain data from a single type of UAV and a richer environment, but they are limited by the disadvantage of having a single viewpoint. The DUT-Anti-UAV contains more than 35 types of UAVs with various background. The background in the dataset includes a wide variety of environments, such as the sky, black clouds, jungle, high-rise building, residential building, farmland, and playground. In addition, various lighting conditions (including dawn, day, dusk and night) as well as different weather conditions are taken into account. The datasets referenced, namely the Real World dataset, the Det-Fly dataset, the MIDGARD dataset, the USC-Drone dataset, and the DUT-Anti-UAV, can be accessed through the following provided links: https://github.com/Maciullo/DroneDetectionDataset, https://github.com/Jake-WU/Det-Fly, https://mrs.felk.cvut.cz/midgard, https://github.com/chelicynly/A-Deep-Learning-Approach-to-Drone-Monitoring and https://github.com/wangdongdut/DUT-Anti-UAV, accessed on 1 June 2023.

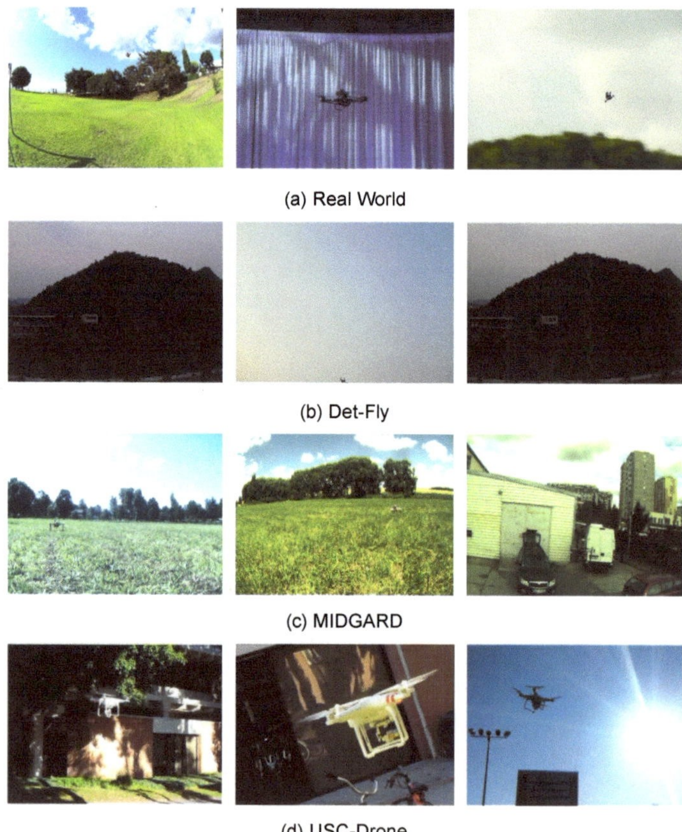

Figure 1. Some example images in (**a**) Real World, (**b**) Det-Fly, (**c**) MIDGARD, and (**d**) USC-Drone.

There are several publicly available video-based UAV datasets datasets, the Anti-UAV [79] and the USC-GRAD-STDdb [80]. The DUT-Anti-UAV also contain 24,804 videos, and their resolutions are mostly 1920 × 1080 pixels. There are 318 RGB-T pairs in the Anti-UAV dataset, each of which consists of an RGB video and a thermal video. A 25 FPS frame rate was used to record these videos, which feature various UAVs flying in the sky. It is publicly available at https://github.com/ucas-vg/Anti-UAV, accessed on 1 June 2023. Meanwhile, in USC-GRAD-STDdb dataset, there are 115 video segments, with a total of over 25,000 annotated frames in HD 720p resolution. The USC-GRAD-STDdb dataset is available at https://citius.usc.es/t/usc-grad-stddb, accessed on 1 June 2023.

2.5. Anti-UAV in IoT Environments

With the development of UAV technology, UAVs have become an important part of the modern IoT. Anonymous UAVs may hijack, steal, or disrupt data transmission within the Internet of Things. Therefore, Anti-UAV in IoT Environments face some new application scenarios, such as detecting only UAVs that attempt to invade the IoT, rather than all UAVs that invade areas. In the IoT, there are a great number of edge devices, and the detection and tracking of UAVs with the help of edge computing is also a research direction. This part reviews some research on counter-UAVs in the IoT environment.

2.5.1. Detect and Identify UAVs Trying to Hack IoT

The anti-UAVs mentioned above are all researched on UAVs invading areas of interest. In the IoT environment, people sometimes only want to detect and identify anonymous UAVs that attempt to invade the IoT. A blockchain-based UAV authentication network is an effective solution [81,82].

BANDA [81] (Blockchain-Assisted Network for Drone Authentication) is a system that leverages blockchain technology to enhance the authentication and security of UAVs. It operates by accepting real-time inputs about UAV information and add-on packages through cloud-based infrastructure. These inputs are then verified using proof-of-authority algorithms and smart contracts. The decentralized nature of BANDA is achieved through the deployment of systems on UAVs, ground control stations, and UAV defense systems. This distributed network allows for secure authentication and prevents malicious attacks and intrusions from anonymous UAVs. By utilizing a blockchain-based approach, BANDA ensures the integrity and authenticity of UAV information, enabling reliable identification and verification processes. This helps in mitigating security risks associated with unauthorized or malicious UAV activities. The optimized application of BANDA enables robust protection against potential threats and enhances the overall security of UAV operations.

RBFNNs [82], which stands for Radial Basis Function Neural Networks, is a blockchain-based model that enhances data integrity and storage capabilities, enabling intelligent decision-making across different Internet of Drone (IoD) environments. By leveraging blockchain technology, this model facilitates decentralized predictive analytics and the application of deep learning methods in a distributed manner. It proves to be a feasible and effective approach for the IoD environment. The utilization of blockchain in developing decentralized predictive analytics and models ensures data integrity and enables secure sharing of deep learning methods. This approach aligns well with the requirements and constraints imposed by network intrusion detection, making RBFNNs a suitable choice for developing classifiers in such scenarios. By combining the strengths of blockchain and deep learning, RBFNNs provide a robust and secure framework for intelligent decision making in the IoD environment. It offers the potential for reliable and efficient network intrusion detection while maintaining data integrity and compliance with the constraints of the system.

2.5.2. Anti-UAV with Edge Computing

Edge computing is used to store and process data on edge devices. It has the characteristics of fast data processing and analysis speed and strong real-time performance. However, edge devices have limited computing power, so they can only perform lightweight operations. Currently, edge computing's application in UAV detection and tracking primarily revolves around sensor data fusion. This approach effectively reduces data storage and bandwidth requirements while enhancing latency and response time. An example of a commercially available multi-sensor fusion networking device is Droneshield's SmartHub Mk2. Furthermore, the literature has explored lightweight deep network models that enable quick and accurate UAV detection and tracking within the constraints of limited computational resources. These studies shed light on the potential of leveraging edge computing for UAV detection and tracking. Carolyn J. Swinney et al. [34] introduces a cost-effective early warning system for UAV detection and classification. The system is composed of a BladeRF software-defined radio (SDR), a wideband antenna, and a Raspberry Pi 4, which together form an edge node. Remarkably, this setup is designed to be affordable, with a total cost of under USD 540. This produced overall accuracy for a two-class detection system at 100% and 90.9% for UAV type classification on the UAVs tested. The inference times for two-class detection in this system range from 15 to 28 s, while for the six-class UAV type classification system, the inference times range from 18 to 28 s. RF-UAVNet [83] is a lightweight convolutional neural network based on RF. Its grouped convolution layer can significantly reduce network size and computing cost; multi-level skip connections and multi-gap mechanisms can effectively improve accuracy. Notably, it achieves remarkable performance with an accuracy of approximately 99.9% for UAV detection, 98.6% for UAV classification, and 95.3% for operation recognition. What sets RF-UAVNet apart is its low complexity, boasting a mere 11,000 parameters. TIB-Net [84] introduces a cyclic pathway in the iterative backbone to keep the model size lightweight while utilizing low-level feature information, and the integrated spatial attention module further improves the performance. TIB-Net stands out not only for its compact size, but also for its efficiency. With a model size of less than 700 Kb and a remarkably low number of parameters at 0.1 million, TIB-Net demonstrates its ability to achieve notable results (approximately 89.2% for UAV detection) while maintaining a lightweight structure. In addition, there are other lightweight models available, such as the visual-based MOB-YOLO [85], which can be deployed on edge nodes. However, it is worth noting that the accuracy of MOB-YOLO is relatively lower at 49.62%. Nonetheless, it is reasonable to assume that this model can achieve higher accuracy if trained on a dedicated UAV dataset. It is essential to consider further research and training to potentially enhance the accuracy of MOF-YOLO for improved performance in UAV detection and classification tasks. According to our observation, the development of these lightweight models mainly considers taking certain measures to reduce the size and calculation amount of the model while retaining low-level information as much as possible, and using some mechanisms to improve accuracy.

In the future, city-wide anti-UAV systems are poised to become integral to the development of smart cities. However, the vast amount of surveillance data generated in the city's airspace poses a significant challenge for a centralized computing center. To address this challenge, deploying detection models directly on edge devices and utilizing edge computing for UAV detection on these devices, can greatly reduce data transmission requirements and expedite UAV detection processes.

3. UAV Localization and Tracking

UAV localization and tracking play a crucial role in anti-UAV research. This section provides a review of UAV localization and UAV tracking methods. UAV localization research primarily concentrates on utilizing RF and acoustic-based methods. These methods are preferred due to their ability to accurately determine the location of UAVs. Regarding UAV tracking, we categorize the studies into filter-based approaches and deep Siamese networks approaches. Filter-based methods utilize various filters for tracking UAVs, while deep Siamese networks approaches leverage deep learning techniques for tracking. The advancements made in these research areas are thoroughly discussed, highlighting the progress and innovations achieved in UAV tracking.

3.1. UAV Localization

RF-based positioning technology has become increasingly mature, leading to numerous studies on RF-based UAV positioning. RF sensors are the only technology that can locate both the UAV and the pilot. On the other hand, acoustic-based UAV positioning technology is feasible for short-distance detection [22]. Although acoustic-based UAV localization technology started later than other methods, it has achieved remarkable results in recent years. In contrast, vision-based UAV tracking technology has its advantages, but due to the rapid movement of UAVs, the effect of visual positioning technology is not ideal. This section focuses on reviewing the RF and acoustic-based UAV localization technologies.

3.1.1. RF-Based Localization

There are a variety of localization methods based on radio frequency, such as Angle of Arrival measurement (AOA), Time of Arrival measurement (TOA), Time Difference of Arrival measurement (TDOA), and RSS measurement. To perform TOA localization, both transmitter and receiver clocks must be synchronized, which is typically difficult to accomplish in anti-UAV scenarios. Similarly, the TDOA positioning method relies on calculating the time difference when the signal arrives at each receiver, which can be challenging to achieve with widely spaced receivers in an anti-UAV system. Additionally, the accuracy of the positioning based on RSS measurement is generally low, and this method has limited applications in UAV positioning. In UAV localization tasks, most studies have focused on localization methods using AOA [29,86]. Despite the challenges associated with RF-based UAV positioning, these methods have shown promising results and have the advantage of being able to locate both the UAV and the pilot. Future research may focus on developing more accurate and robust RF-based UAV localization methods that can operate in a range of environments and under various UAV operating conditions.

In recent years, various parametric models have been proposed to compute AOA, such as beamforming, subspace-based methods, cross-correlation-based methods, maximum likelihood methods, and sparse array processing methods. These methods are based on pre-established models for position estimation, which can be cumbersome and computationally intensive. Figure 2 illustrates the specific positioning model. To address these issues, many studies have begun to use machine learning approaches to solve the AOA estimation problem [87,88]. From experimental results, machine learning-based approaches have shown superiority over traditional methods in terms of processing speed and accuracy. However, these methods may only achieve satisfactory results when the distribution of the training set and the test set are the same, which can be difficult to achieve in practical applications where it is challenging to cover all possible scenarios in the training set. To address these challenges, researchers are now using deep neural networks for emission source localization [86,89]. Deep Convolutional Neural Networks have been used for enhancing angle classification accuracy, as they allow high-level features to be learned from high-dimensional unstructured data at multiple scales [90]. Compared to traditional MD-based and ML-based methods, CNN-based methods have demonstrated superior accuracy levels, according to simulation results. Table 7 [90] displays the advantages and disadvantages of the methods.

Table 7. Comparison of AOA estimation techniques.

Approach	Model	Advantage	Disadvantage
Model-driven	Beamforming [91]	Easy installation	High computational complexity Vulnerable at high SNR
	MUSIC [92]	High accuracy	Performance degradation of far field High installation cost
	ESPRIT [93]	Array calibration is not required More complex than MUSIC	Lower accuracy in non-ideal sensor design High expensive system
Machine learning	CWSVR [94]	Independent on array geometrics	Degrade performance at non identical distributions
	GCA [95]	Less expensive than MD approach	Unable to process high density data
Deep learning	Spectrum CNN [89]	Construct complicated propagation model	Low accuracy at low SNR
	Multi-speaker CNN [96]	Easy to perform at non-identical	High computational expense

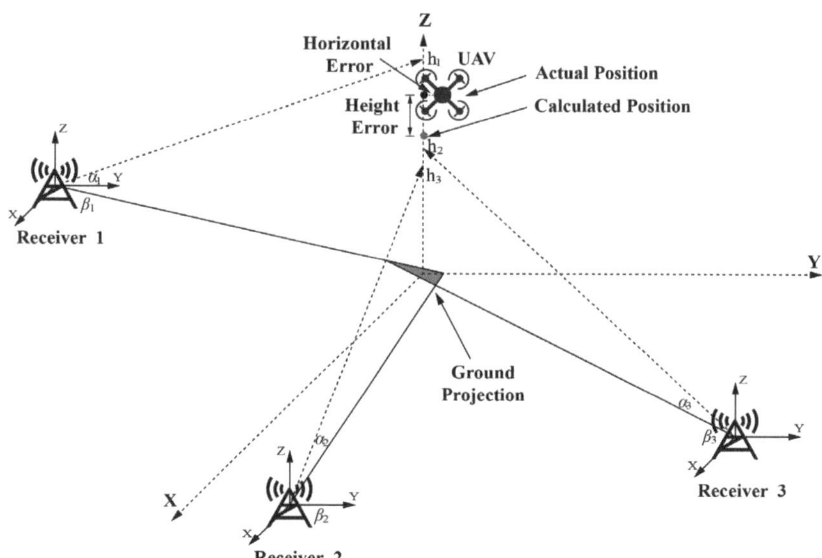

Figure 2. The positioning model of AOA involves projecting the position of the UAV onto the ground using the AOA of the LOS (line of sight) path β_i selected by the i-th receiver. The height h_i of the UAV can then be calculated using the pitch angle α_i of each receiver and the projected position pos of the UAV on the ground. By averaging the height calculated from multiple receivers, the overall height of the UAV h can be obtained. Finally, the UAV's space position $posd$ can be determined by combining the height h and the projected position pos on the ground.

3.1.2. Acoustic-Based Localization

There are two primary strategies for acoustic-based UAV positioning: AOA and TDOA [97]. The AOA method has been introduced in the previous section, and in this section, we focus on the TDOA method. The TDOA-based sound source localization method is implemented using a microphone array to find the three-dimensional position of the sound source. The TDOA source estimation involves a two-step procedure. First, the time delay between pairs of microphones in the array is estimated using the generalized cross-correlation function. This function measures the time difference between the arrival of a sound wave at different microphones in the array. Once the time delays have been estimated, they can be used as input to an equation that relates the time delays to the source location. Inverting this equation yields an estimate of the source location. The accuracy of the estimation depends on the quantity and distribution of microphones in the array, as well as the characteristics of the sound wave being detected [98,99]. The time complexity of modern TDOA-SSL methods is generally low for three-dimensional estimation of the source position, but multipath propagation and environmental noise can affect its positioning process. To address these challenges, some researchers have proposed using machine learning methods to limit these effects [100], while others have used bionic methods to improve accuracy [101].

Acoustic-based UAV localization has a maximum detection range of about 100 m under the condition that the SNR is lower than −5 dB [100]. However, for far-field conditions, the acoustic localization effect may not meet expectations. In such cases, RF-based localization can complement this issue. Therefore, the use of multiple technologies to compose an anti-UAV system is necessary to achieve effective UAV detection and tracking capabilities.

3.2. UAV Tracking

UAV tracking is a type of object tracking task that involves estimating the state and trajectory of the UAV. Object tracking (OT) is a fundamental open problem that has been extensively studied in the past decade. Two prominent paradigms for OT are Filters-based methods and deep Siamese Networks (SNs) [102]. In this section, we review recent literature on UAV tracking and discuss the advancements made in this field.

3.2.1. Filters-Based Tracker

The core of the prediction problem is to solve the distribution, which is particularly challenging when only part of the data is observed. In such cases, it is necessary to solve a conditional distribution, also known as the posterior distribution. In time-shifting systems, the current posterior distribution is often referred to as a filter. For instance, in object tracking, an inference is made based on the current state of the system, which involves a filtering task. Bayesian filtering is a widely used framework for solving filtering problems, based on a recursive method. Kalman filter is a popular Bayesian filter that is based on Gaussian process modeling. It uses a recursive method to solve linear filtering problems. The small memory footprint and high speed of the Kalman filter make it well-suited for real-time applications and embedded systems, as it only needs to retain the previous state and can efficiently update the estimated state in real-time. Different types of Kalman filters are among the most important technologies for moving target tracking [103–105]. In these studies, the Kalman filter is typically used for tracking after UAV detection and recognition. For instance, in Ref. [105], the authors focus on tracking a large number of targets simultaneously using Kalman filter.

In anti-UAV systems and UAV detection and tracking tasks, it is almost a consensus to utilize different types of sensors as they complement each other and greatly improve the detection effect. The Kalman filter is also known for its extraordinary advantages in multi-sensor data fusion, which is one of the reasons why it is widely used in UAV tracking tasks. By integrating information from multiple sensors, the Kalman filter can provide more accurate and reliable estimates of UAV states and trajectories. Therefore, the Kalman filter plays a crucial role in improving the effectiveness of multi-sensor-based anti-UAV systems.

Discriminative Correlation Filters (DCF) is a widely used method in visual tracking tasks. The DCF-based tracking approach involves training correlation filters online over regions of interest by minimizing a least-squares loss. The trained filters are then convolved by Fast Fourier Transform (FFT) to detect objects in consecutive frames [102]. DCF has also been shown to be a highly competitive method in vision-based UAV tracking tasks [106–108]. Through the exploitation of discriminative information between the target object and the background, DCF-based methods can achieve high tracking accuracy and robustness against challenging scenarios, such as occlusion and deformation. Therefore, DCF-based methods have gained significant attention in the field of UAV tracking and have been shown to outperform many other state-of-the-art methods.

3.2.2. Deep Siamese Networks

Deep Siamese Networks (SNs) have become a popular approach for learning similarities between target images and search image regions in visual tracking tasks. With SNs, real-time applications can fully leverage end-to-end learning and overcome the limitations of pretrained CNN. Siamese trackers are instructed to deal with various tracking challenges, such as rotations, perspective changes, and lighting changes, by using offline training videos. This enables the tracker to accurately localize target objects in consecutive frames [102]. Over the years, several advanced Siamese-based trackers have been proposed, such as SiamFC (Fully-Convolutional Siamese Networks) [109], SiamRPN++ (Siamese region proposal network plus plus) [110], and LTMU (High-Performance Long-Term Tracking with Meta-Updater) [111]. The robustness of these trackers against challenging scenarios, combined with their state-of-the-art performance on several benchmark datasets, has established them as highly competitive methods in the area of UAV tracking.

SiamFC [109]

A fully convolutional Siamese network is used in this classic tracking algorithm to perform cross-correlations between the template slice and the search area in order to locate the target. Additionally, a multi-scale strategy is employed to determine the appropriate scale of the object being tracked. By leveraging the discriminative information learned from the offline training videos, the Siamese-based tracker can accurately estimate the position and scale of the target object in real-time. On several benchmark datasets, this method has shown to achieve state-of-the-art performance and has gained considerable attention in the UAV tracking field.

SiamRPN++ [110]

SiamRPN++ is an advanced tracking algorithm based on the Siamese architecture that incorporates a region proposal network (RPN) into the Siamese network, enabling a very deep backbone network. The overall framework consists of two branches: a classification branch, which selects the optimal anchor, and a regression branch, which predicts the offsets of the anchor. By leveraging the RPN mechanism, SiamRPN++ can generate high-quality proposals for the target object, which enhances the robustness and speed of the tracking algorithm. SiamRPN++ is more robust and faster than SiamFC, thanks to the introduction of the RPN mechanism and the removal of the multi-scale strategy. SiamRPN++ has demonstrated state-of-the-art performance on several benchmark datasets and has become one of the most competitive tracking methods in the field of UAV tracking.

LTMU [111]

This method is not only a short-term tracker but also a long-term tracker. The main contribution of this method is its offline-trained meta-updater, which is utilized to determine whether the tracker needs to be updated in the current frame or not, resulting in improved robustness. Furthermore, a long-term tracking framework is designed that utilizes a SiamRPN-based re-detector, an online verifier, and an online local tracker in conjunction with the proposed meta-updater. By integrating these components, the tracker can maintain accurate tracking of the target object over an extended period, even in scenarios where the object undergoes significant changes, such as occlusion or appearance variations. In both short- and long-term tracking benchmarks, this method has demonstrated excellent discriminative capability and robustness, making it an extremely competitive method.

While the Kalman filter is widely used in UAV tracking tasks due to its low computational cost, achieving high accuracy results in practical applications is often challenging [112]. Recently, DCF methods and Siamese tracking algorithms have shown excellent performance in object tracking, thanks to their use of end-to-end offline learning. However, this has only been possible with the availability of large-scale training datasets [102]. Despite these advancements, achieving low-cost, high-precision real-time UAV tracking remains a challenge in anti-UAV research. The development of more efficient algorithms and the availability of more diverse and comprehensive datasets for training and testing are essential for further advancing the field of UAV tracking.

4. Anti-UAV System

In recent years, there has been a significant amount of literature on anti-UAV systems, and several anti-UAV systems have been used commercially and militarily.

One such system is DedroneTracker [113], a multi-sensor platform (RF, PTZ camera, radar) with countermeasure capabilities (jamming) released by Dedrone. The system can be extended and customized to meet specific field requirements and can automatically capture a portfolio of forensic data, including UAV make, model, time and length of UAV activity, and video verification. Dedrone also offers RF sensors and jammers, with a detection range of up to 1.5 km (up to 5 km in special cases) and a maximum jamming range of about 2 km. Droneshield [114] is another system that provides anti-UAV defense solutions, offering a range of standalone portable products as well as rapidly deployable fixed-site solutions. It employs a variety of surveillance technologies, including radar, audio, video, and radio frequency, to detect UAVs and provide effective countermeasures. Droneshield's jammer can immediately cease video transmission back to the UAS operator and allow the UAV to respond to a live vertical controlled landing or return to the operator controller or starting point. Another RF-based UAV detection system is ARDRONIS [115], developed by Rohde and Schwarz. ARDRONIS performs detection for frequency-spread spectrum (FHSS) signals and WLAN signals, with a detection range of up to 7 km for commercial off-the-shelf remote signals and up to 5 km for UAVs such as the DJI Phantom 4 under ideal conditions. The AUDS [116] system, on the other hand, uses radar technology, video, and thermal imaging to detect and track UAVs and has directional radio frequency inhibition capabilities. With its Air Security Radar, the range of detection can be up to 10 km. The ORELIA Drone-Detector system [117] consists of an acoustic sensor and software for protected object monitoring, target tracing, and sensor adjustment. The detection range is about 100 m, and multiple acoustic detectors can be installed to detect UAVs in all directions. Falcon Shield [118] is a rapidly deployable, scalable, and modular system that combines electro-optics, electronic surveillance, and radar sensors. ELTA Systems [119] is another solution that combines radar, RF, and photoelectric sensors to detect and track UAVs more than 5 km away and to take soft and hard kill measures against them.

Each system uses different surveillance techniques and implements various functions, as summarized in Table 8. The examples of these systems are shown in Figure 3. These systems have demonstrated their effectiveness in UAV detection and tracking. To enhance their detection capabilities, the majority of anti-UAV systems opt to utilize multiple types of sensors. These systems primarily find application in scenarios such as airport operations and military protection. However, Dedrone stands out as a rare system provider that specifically addresses urban settings, including sports events and outdoor concerts. DeDrone and ARDRONIS are two anti-UAV systems that offer video evidence recording capabilities, although they employ different methods. DeDrone utilizes the visual monitor within the system to directly record video evidence when a UAV is detected. On the other hand, ARDRONIS obtains the UAV's field of view by intercepting the visual transmission signal of the UAV. It is important to note that while DeDrone can record and save evidence whenever a UAV is detected, ARDRONIS can only achieve video recording when it successfully cracks the visual transmission signal of the UAV. Dedrone's products and design concepts serve as a valuable source of inspiration for tackling the challenges of anti-UAV systems in urban environments. Their emphasis on scalable software platforms and leveraging existing infrastructure offers valuable insights for research and development in this field. By incorporating such approaches, it is possible to enhance the effectiveness and adaptability of anti-UAV systems in urban settings.

Table 8. Comparison of anti-UAV systems.

System	Technology				Function			Video Recording	Range
	Radar	Audio	Visual	RF	Detection	Tracking	Defense		
DedroneTracker	✔	✘	✔	✔	✔	✔	✔	✔	5000 m
Dronesshield	✔	✔	✔	✔	✔	✔	✔	✘	-
ARDRONIS	✘	✘	✘	✔	✔	✔	✔	✔	5000 m
AUDS	✔	✘	✔	✘	✔	✔	✔	✘	10,000 m
ORELIA	✘	✔	✘	✘	✔	✔	✘	✘	100 m
Falcon Shield	✔	✘	✔	✔	✔	✔	✔	✘	-
ELTA	✔	✘	✔	✔	✔	✔	✔	✘	-

Figure 3. Examples of anti-UAV systems: (**a**) DeDrone, (**b**) Droneshield, (**c**) ARDRONIS, (**d**) AUDS, (**e**) ORELIA, (**f**) Falcon Shield, (**g**) ELTA. The aforementioned images are sourced from multiple official websites of anti-UAV systems.

Based on our observation, the more mature anti-UAV systems are currently mainly focused on military applications, and these systems are often implemented using a combination of multiple sensors. RF monitoring, radar, and vision sensors are the three most commonly used types of multi-sensors. However, radar is not suitable for use in certain locations, such as urban environments, due to its strong radiation characteristics [22]. In urban environments, RF surveillance emerges as a crucial method for locating UAV pilots, while visual surveillance plays a pivotal role in capturing UAV flight videos as evidence of intrusions. This combination of RF and visual surveillance techniques provides a highly competitive solution for UAV detection in urban settings. Although the acoustic-based method has shortcomings such as small detection range and being easily affected by noise, it can also provide an effective supplement for UAV detection under certain conditions, such as poor visual conditions and complex electromagnetic environments.

Inspired by the above anti-UAV systems, we propose several suggestions for developing effective and scalable general anti-UAV systems. First, it is essential to consider the specific needs and requirements of the application scenario when selecting and combining sensors. Indeed, various environmental factors can impact the effectiveness of different surveillance methods in urban environments. For instance, visual surveillance may face challenges in locations with low visibility caused by heavy fog, sand, or dust. Similarly, complex electromagnetic environments can negatively affect the performance of RF surveillance. Additionally, acoustic surveillance may not be suitable in areas with strong winds or high levels of ambient noise. It is essential to consider these factors when selecting the most appropriate surveillance method for UAV detection, ensuring optimal performance in diverse urban scenarios.

Second, the detection and tracking components of the system should be integrated with appropriate countermeasure capabilities. One of the most prevalent countermeasure equipment against intruding UAVs is the use of RF jamming guns. These devices are designed to disrupt the communication and control signals of the UAVs, forcing them to either land or return to their point of origin. RF jamming guns are widely employed as a common UAV countermeasure. Another commonly utilized countermeasure involves deploying UAVs to capture intruding UAVs. This approach involves using specially designed UAVs equipped with nets, cables, or other mechanisms to physically intercept and capture the unauthorized UAVs. Both RF jamming guns and UAV capture methods serve as effective countermeasures in mitigating the risks and potential threats posed by unauthorized UAV activities.

Third, the system should be designed to be scalable and modular, allowing for easy deployment and adaptation to changing conditions. The design concept put forth by Dedrone Company offers valuable inspiration. It emphasizes the importance of creating scalable anti-UAV systems with a platform at the core. Such a design enables easier integration and maintenance of sensors in the future. By adopting a scalable approach, anti-UAV systems can adapt to evolving threats and technological advancements, ensuring flexibility and efficiency in the long run.

Fourth, data analytics and machine learning techniques can be employed to enhance the accuracy and efficiency of UAV detection and tracking. Indeed, machine learning techniques have already found extensive application in the field of UAV detection. The continuous advancements in machine learning algorithms and the availability of more comprehensive datasets are key factors in enhancing UAV detection capabilities. By leveraging more efficient learning methods and utilizing diverse and representative datasets, the accuracy and effectiveness of UAV detection systems can be significantly improved. This ongoing development in machine learning holds promise for further advancements in UAV detection technology.

Fifth, although accuracy is an important evaluation indicator, lightweight anti-UAV systems that sacrifice part of the accuracy seem to be more in demand in some non-important scenarios. In mobile scenarios or situations where budget constraints exist, lightweight anti-UAV systems that can operate on portable devices offer a more suitable solution. These systems, designed to be lightweight and portable, can be easily deployed and utilized in various environments. They provide flexibility and cost-effectiveness, making them a practical choice for scenarios where mobility and budget considerations are important factors. By leveraging lightweight anti-UAV systems, organizations can enhance their capabilities for UAV detection and mitigation while maintaining operational efficiency.

Furthermore, compatibility with existing sensors, such as ubiquitous video surveillance equipment, could significantly reduce the cost of anti-UAV systems. This suggestion of leveraging existing video surveillance equipment, inspired by Dedrone's products, is indeed valuable. By ensuring compatibility and utilization of the already deployed video surveillance infrastructure, the deployment costs of anti-UAV systems can be significantly reduced. This approach aligns with the concept of using edge computing to assist in UAV detection, as previously discussed. By calibrating the physical locations of these monitoring devices, it becomes possible to detect and track illegally intruding UAVs effectively. This application scenario showcases the potential for cost-effective and efficient UAV detection by leveraging existing resources and edge computing capabilities.

5. Conclusions

In this paper, we have summarized the technical classification and implementation methods of UAV detection and tracking in urban IoT environment. We have also reviewed the performance of edge computing and deep learning empowered anti-UAV systems, highlighting their strengths and limitations. Furthermore, we have proposed several suggestions for developing effective and scalable general anti-UAV systems, such as considering the specific needs of the application scenario, integrating detection and tracking components with appropriate countermeasures, designing for scalability and modularity, employing data analytics and machine learning techniques, and ensuring compliance with relevant regulations.

To facilitate further research in the field of anti-UAV systems, we have also presented publicly available visual, acoustic, and radiofrequency datasets that can be used to evaluate the performance of different anti-UAV techniques and algorithms. We hope that these datasets will be useful for researchers in developing and testing new anti-UAV systems and techniques.

Author Contributions: Conceptualization, T.F., P.L. and H.H.; validation, F.X., H.L. and H.H.; formal analysis, P.L., Y.C. and T.F.; investigation, F.X., Y.H. and H.H.; resources, X.Y., Y.H. and H.H.; data curation, X.Y., H.L. and F.X.; writing—original draft preparation, X.Y. and Y.C.; writing—review and editing, X.Y. and P.L.; supervision, P.L. and T.F. All authors have read and agreed to the published version of the manuscript.

Funding: This research was funded by Zhejiang Public Information Industry Co., Ltd.

Institutional Review Board Statement: Not applicable.

Informed Consent Statement: Not applicable.

Data Availability Statement: Not applicable.

Conflicts of Interest: The authors declare no conflict of interest.

References

1. Tian, S.; Li, Y.; Zhang, X.; Zheng, L.; Cheng, L.; She, W.; Xie, W. Fast UAV path planning in urban environments based on three-step experience buffer sampling DDPG. *Digit. Commun. Netw.* **2023**. [CrossRef]
2. Ding, G.; Wu, Q.; Zhang, L.; Lin, Y.; Tsiftsis, T.A.; Yao, Y.D. An amateur drone surveillance system based on the cognitive Internet of Things. *IEEE Commun. Mag.* **2018**, *56*, 29–35. [CrossRef]
3. Shen, Y.; Pan, Z.; Liu, N.; You, X. Performance analysis of legitimate UAV surveillance system with suspicious relay and anti-surveillance technology. *Digit. Commun. Netw.* **2022**, *8*, 853–863. [CrossRef]
4. Lin, N.; Tang, H.; Zhao, L.; Wan, S.; Hawbani, A.; Guizani, M. A PDDQNLP Algorithm for Energy Efficient Computation Offloading in UAV-assisted MEC. *IEEE Trans. Wirel. Commun.* **2023**. [CrossRef]
5. Deng, X.; Wang, L.; Gui, J.; Jiang, P.; Chen, X.; Zeng, F.; Wan, S. A review of 6G autonomous intelligent transportation systems: Mechanisms, applications and challenges. *J. Syst. Archit.* **2023**, *142*, 102929. [CrossRef]
6. Liu, J.; Peng, J.; Xu, W.; Liang, W.; Liu, T.; Peng, X.; Xu, Z.; Li, Z.; Jia, X. Maximizing Sensor Lifetime via Multi-node Partial-Charging on Sensors. *IEEE Trans. Mob. Comput.* **2022**, *22*, 6571–6584. [CrossRef]
7. Xu, W.; Xie, H.; Wang, C.; Liang, W.; Jia, X.; Xu, Z.; Zhou, P.; Wu, W.; Chen, X. An Approximation Algorithm for the h-Hop Independently Submodular Maximization Problem and Its Applications. *IEEE/ACM Trans. Netw.* **2023**, *31*, 1216–1229. [CrossRef]
8. Cheng, Z.; Liwang, M.; Chen, N.; Huang, L.; Guizani, N.; Du, X. Learning-based user association and dynamic resource allocation in multi-connectivity enabled unmanned aerial vehicle networks. *Digit. Commun. Netw.* **2022**, *in press*. [CrossRef]
9. Heidari, A.; Jafari Navimipour, N.; Unal, M.; Zhang, G. Machine learning applications in internet-of-drones: Systematic review, recent deployments, and open issues. *ACM Comput. Surv.* **2023**, *55*, 1–45. [CrossRef]
10. Wang, L.; Deng, X.; Gui, J.; Jiang, P.; Zeng, F.; Wan, S. A review of Urban Air Mobility-enabled Intelligent Transportation Systems: Mechanisms, applications and challenges. *J. Syst. Archit.* **2023**, *141*, 102902. [CrossRef]
11. Iftikhar, S.; Asim, M.; Zhang, Z.; Muthanna, A.; Chen, J.; El-Affendi, M.; Sedik, A.; Abd El-Latif, A.A. Target Detection and Recognition for Traffic Congestion in Smart Cities Using Deep Learning-Enabled UAVs: A Review and Analysis. *Appl. Sci.* **2023**, *13*, 3995. [CrossRef]
12. Shi, J.; Cong, P.; Zhao, L.; Wang, X.; Wan, S.; Guizani, M. A two-stage strategy for UAV-enabled wireless power transfer in unknown environments. *IEEE Trans. Mob. Comput.* **2023**, *in press*. [CrossRef]
13. Abbas, N.; Abbas, Z.; Liu, X.; Khan, S.S.; Foster, E.D.; Larkin, S. A Survey: Future Smart Cities Based on Advance Control of Unmanned Aerial Vehicles (UAVs). *Appl. Sci.* **2023**, *13*, 9881. [CrossRef]
14. Motlagh, N.H.; Taleb, T.; Arouk, O. Low-altitude unmanned aerial vehicles-based internet of things services: Comprehensive survey and future perspectives. *IEEE Internet Things J.* **2016**, *3*, 899–922. [CrossRef]
15. Weng, L.; Zhang, Y.; Yang, Y.; Fang, M.; Yu, Z. A mobility compensation method for drones in SG-eIoT. *Digit. Commun. Netw.* **2021**, *7*, 196–200. [CrossRef]
16. Markets. *UAV Market by Point of Sale, Systems, Platform (Civil & Commercial, and Defense & Government), Function, End Use, Application, Type (Fixed Wing, Rotary Wing, Hybrid), Mode of Operation, Mtow, Range & Region-Global Forecast to 2027*; Technical Report; Research and Markets: Dublin, Ireland, 2022.
17. Chamola, V.; Kotesh, P.; Agarwal, A.; Naren; Gupta, N.; Guizani, M. A Comprehensive Review of Unmanned Aerial Vehicle Attacks and Neutralization Techniques. *Ad Hoc Netw.* **2021**, *111*, 102324. [CrossRef]
18. Li, S.; Chai, Y.; Guo, M.; Liu, Y. Research on Detection Method of UAV Based on micro-Doppler Effect. In Proceedings of the 2020 39th Chinese Control Conference (CCC), Shenyang, China, 27–29 July 2020; pp. 3118–3122. [CrossRef]
19. Pappu, C.S.; Beal, A.N.; Flores, B.C. Chaos based frequency modulation for joint monostatic and bistatic radar-communication systems. *Remote Sens.* **2021**, *13*, 4113. [CrossRef]
20. Abd, M.H.; Al-Suhail, G.A.; Tahir, F.R.; Ali Ali, A.M.; Abbood, H.A.; Dashtipour, K.; Jamal, S.S.; Ahmad, J. Synchronization of monostatic radar using a time-delayed chaos-based FM waveform. *Remote Sens.* **2022**, *14*, 1984. [CrossRef]
21. Mehta, S.; Rastegari, M.; Caspi, A.; Shapiro, L.; Hajishirzi, H. Espnet: Efficient spatial pyramid of dilated convolutions for semantic segmentation. In Proceedings of the European Conference on Computer Vision (ECCV), Munich, Germany, 8–14 September 2018, pp. 552–568.
22. Shi, X.; Yang, C.; Xie, W.; Liang, C.; Shi, Z.; Chen, J. Anti-Drone System with Multiple Surveillance Technologies: Architecture, Implementation, and Challenges. *IEEE Commun. Mag.* **2018**, *56*, 68–74. [CrossRef]

23. Farlik, J.; Kratky, M.; Casar, J.; Stary, V. Radar cross section and detection of small unmanned aerial vehicles. In Proceedings of the 2016 17th International Conference on Mechatronics-Mechatronika (ME), Prague, Czech Republic, 7–9 December 2016; pp. 1–7.

24. Zhang, J.; Liu, M.; Zhao, N.; Chen, Y.; Yang, Q.; Ding, Z. Spectrum and energy efficient multi-antenna spectrum sensing for green UAV communication. *Digital Commun. Netw.* **2022**, *9*, 846–855. [CrossRef]

25. Unlu, E.; Zenou, E.; Riviere, N.; Dupouy, P.E. Deep learning-based strategies for the detection and tracking of drones using several cameras. *IPSJ Trans. Comput. Vis. Appl.* **2019**, *11*, 7. [CrossRef]

26. Fang, H.; Xia, M.; Zhou, G.; Chang, Y.; Yan, L. Infrared small UAV target detection based on residual image prediction via global and local dilated residual networks. *IEEE Geosci. Remote. Sens. Lett.* **2021**, *19*, 7002305 . [CrossRef]

27. Al-Emadi, S.; Al-Ali, A.; Mohammad, A.; Al-Ali, A. Audio based drone detection and identification using deep learning. In Proceedings of the 2019 15th International Wireless Communications & Mobile Computing Conference (IWCMC), Tangier, Morocco, 24–28 June 2019; pp. 459–464.

28. Dumitrescu, C.; Minea, M.; Costea, I.M.; Cosmin Chiva, I.; Semenescu, A. Development of an acoustic system for UAV detection. *Sensors* **2020**, *20*, 4870.

29. Nie, W.; Han, Z.C.; Zhou, M.; Xie, L.B.; Jiang, Q. UAV detection and identification based on WiFi signal and RF fingerprint. *IEEE Sensors J.* **2021**, *21*, 13540–13550.

30. Tugnait, J.K. Detection of non-Gaussian signals using integrated polyspectrum. *IEEE Trans. Signal Process.* **1994**, *42*, 3137–3149.

31. Yao, Y.; Yu, L.; Chen, Y. Specific Emitter Identification Based on Square Integral Bispectrum Features. In Proceedings of the 2020 IEEE 20th International Conference on Communication Technology (ICCT), Nanning, China, 28–31 October 2020; pp. 1311–1314. [CrossRef]

32. Nie, W.; Han, Z.C.; Li, Y.; He, W.; Xie, L.B.; Yang, X.L.; Zhou, M. UAV detection and localization based on multi-dimensional signal features. *IEEE Sensors J.* **2021**, *22*, 5150–5162.

33. Mo, Y.; Huang, J.; Qian, G. UAV Tracking by Identification Using Deep Convolutional Neural Network. In Proceedings of the 2022 IEEE 8th International Conference on Computer and Communications (ICCC), Chengdu, China, 9–12 December 2022; pp. 1887–1892.

34. Swinney, C.J.; Woods, J.C. Low-Cost Raspberry-Pi-Based UAS Detection and Classification System Using Machine Learning. *Aerospace* **2022**, *9*, 738. [CrossRef]

35. Lu, S.; Wang, W.; Zhang, M.; Li, B.; Han, Y.; Sun, D. Detect the Video Recording Act of UAV through Spectrum Recognition. In Proceedings of the 2022 IEEE International Conference on Artificial Intelligence and Computer Applications (ICAICA), Dalian, China, 24–26 June 2022; pp. 559–564.

36. He, Z.; Huang, J.; Qian, G. UAV Detection and Identification Based on Radio Frequency Using Transfer Learning. In Proceedings of the 2022 IEEE 8th International Conference on Computer and Communications (ICCC), Virtual, 9–12 December 2022; pp. 1812–1817.

37. Li, T.; Hong, Z.; Cai, Q.; Yu, L.; Wen, Z.; Yang, R. Bissiam: Bispectrum siamese network based contrastive learning for uav anomaly detection. *IEEE Trans. Knowl. Data Eng.* **2021**. [CrossRef]

38. Dong, Q.; Liu, Y.; Liu, X. Drone sound detection system based on feature result-level fusion using deep learning. *Multimed. Tools Appl.* **2023**, *82*, 149–171.

39. Aydın, İ.; Kızılay, E. Development of a new Light-Weight Convolutional Neural Network for acoustic-based amateur drone detection. *Appl. Acoust.* **2022**, *193*, 108773.

40. Schweinhart, B. Persistent homology and the upper box dimension. *Discret. Comput. Geom.* **2021**, *65*, 331–364.

41. Higuchi, T. Approach to an irregular time series on the basis of the fractal theory. *Phys. D Nonlinear Phenom.* **1988**, *31*, 277–283.

42. Zhang, X.D.; Shi, Y.; Bao, Z. A new feature vector using selected bispectra for signal classification with application in radar target recognition. *IEEE Trans. Signal Process.* **2001**, *49*, 1875–1885. [CrossRef]

43. Al-Sa'd, M.F.; Al-Ali, A.; Mohamed, A.; Khattab, T.; Erbad, A. RF-based drone detection and identification using deep learning approaches: An initiative towards a large open source drone database. *Future Gener. Comput. Syst.* **2019**, *100*, 86–97.

44. Mo, Y.; Huang, J.; Qian, G. Deep Learning Approach to UAV Detection and Classification by Using Compressively Sensed RF Signal. *Sensors* **2022**, *22*, 3072. [PubMed]

45. Allahham, M.S.; Al-Sa'd, M.F.; Al-Ali, A.; Mohamed, A.; Khattab, T.; Erbad, A. DroneRF dataset: A dataset of drones for RF-based detection, classification and identification. *Data Brief* **2019**, *26*, 104313. [CrossRef] [PubMed]

46. Ezuma, M.; Erden, F.; Anjinappa, C.K.; Ozdemir, O.; Guvenc, I. Drone Remote Controller RF Signal Dataset. 2020. Available online: https://ieee-dataport.org/open-access/drone-remote-controller-rf-signal-dataset (accessed on 1 June 2023). [CrossRef]

47. Vuorenmaa, M.; Marin, J.; Heino, M.; Turunen, M.; Riihonen, T. Radio-Frequency Control and Video Signal Recordings of Drones. 2020. Available online: https://zenodo.org/record/4264467 (accessed on 1 June 2023). [CrossRef]

48. Basak, S.; Rajendran, S.; Pollin, S.; Scheers, B. Drone classification from RF fingerprints using deep residual nets. In Proceedings of the 2021 International Conference on COMmunication Systems & NETworkS (COMSNETS), Bangalore, India, 5–9 January 2021; pp. 548–555. [CrossRef]

49. Swinney, C.J.; Woods, J.C. DroneDetect Dataset: A Radio Frequency dataset of Unmanned Aerial System (UAS) Signals for Machine Learning Detection & Classification. 2021. Available online: https://ieee-dataport.org/open-access/dronedetect-dataset-radio-frequency-dataset-unmanned-aerial-system-uas-signals-machine (accessed on 1 June 2023).

50. Sazdić-Jotić, B.; Pokrajac, I.; Bajčetić, J.; Bondžulić, B.; Obradović, D. Single and multiple drones detection and identification using RF based deep learning algorithm. *Expert Syst. Appl.* **2022**, *187*, 115928. [CrossRef]

51. Medaiyese, O.; Ezuma, M.; Lauf, A.; Adeniran, A. Cardinal RF (CardRF): An Outdoor UAV/UAS/Drone RF Signals with Bluetooth and WiFi Signals Dataset. 2022. Available online: https://ieee-dataport.org/documents/cardinal-rf-cardrf-outdoor-uavuasdrone-rf-signals-bluetooth-and-wifi-signals-dataset (accessed on 1 June 2023). [CrossRef]

52. Svanström, F.; Alonso-Fernandez, F.; Englund, C. Drone Detection and Tracking in Real-Time by Fusion of Different Sensing Modalities. *Drones* **2022**, *6*, 317.

53. Uddin, Z.; Qamar, A.; Alharbi, A.G.; Orakzai, F.A.; Ahmad, A. Detection of Multiple Drones in a Time-Varying Scenario Using Acoustic Signals. *Sustainability* **2022**, *14*, 4041.

54. Jamil, S.; Rahman, M.; Ullah, A.; Badnava, S.; Forsat, M.; Mirjavadi, S.S. Malicious UAV detection using integrated audio and visual features for public safety applications. *Sensors* **2020**, *20*, 3923.

55. Guo, J.; Ahmad, I.; Chang, K. Classification, positioning, and tracking of drones by HMM using acoustic circular microphone array beamforming. *EURASIP J. Wirel. Commun. Netw.* **2020**, *2020*, 1–19.

56. Gupta, H.; Gupta, D. LPC and LPCC method of feature extraction in Speech Recognition System. In Proceedings of the 2016 6th International Conference-Cloud System and Big Data Engineering (Confluence), Noida, India, 14–15 January 2016; pp. 498–502.

57. Uddin, Z.; Altaf, M.; Bilal, M.; Nkenyereye, L.; Bashir, A.K. Amateur Drones Detection: A machine learning approach utilizing the acoustic signals in the presence of strong interference. *Comput. Commun.* **2020**, *154*, 236–245.

58. Al-Emadi, S.A.; Al-Ali, A.K.; Al-Ali, A.; Mohamed, A. Audio Based Drone Detection and Identification using Deep Learning. In Proceedings of the IWCMC 2019 Vehicular Symposium (IWCMC-VehicularCom 2019), Tangier, Morocco, 24–28 June 2019.

59. Casabianca, P.; Zhang, Y. Acoustic-based UAV detection using late fusion of deep neural networks. *Drones* **2021**, *5*, 54.

60. Viola, P.; Jones, M.J. Robust real-time face detection. *Int. J. Comput. Vis.* **2004**, *57*, 137–154.

61. Dalal, N.; Triggs, B. Histograms of oriented gradients for human detection. In Proceedings of the CVPR'05, San Diego, CA, USA, 20–26 June 2005; Volume 1, pp. 886–893.

62. Girshick, R.; Donahue, J.; Darrell, T.; Malik, J. Rich feature hierarchies for accurate object detection and semantic segmentation. In Proceedings of the CVPR, Columbus, OH, USA, 23–28 June 2014; pp. 580–587.

63. He, K.; Zhang, X.; Ren, S.; Sun, J. Spatial pyramid pooling in deep convolutional networks for visual recognition. *IEEE Trans. Pattern Anal. Mach. Intell.* **2015**, *37*, 1904–1916.

64. Girshick, R. Fast r-cnn. In Proceedings of the ICCV, Santiago, Chile, 13–16 December 2015; pp. 1440–1448.

65. Ren, S.; He, K.; Girshick, R.; Sun, J. Faster r-CNN: Towards real-time object detection with region proposal networks. *IEEE Trans. Pattern Anal. Mach. Intell.* **2017**, *39*, 1137–1149.

66. Redmon, J.; Divvala, S.; Girshick, R.; Farhadi, A. You only look once: Unified, real-time object detection. In Proceedings of the CVPR, Las Vegas, NV, USA, 27–30 June 2016; pp. 779–788.

67. Ajakwe, S.O.; Ihekoronye, V.U.; Kim, D.S.; Lee, J.M. DRONET: Multi-Tasking Framework for Real-Time Industrial Facility Aerial Surveillance and Safety. *Drones* **2022**, *6*, 46.

68. Wang, J.; Hongjun, W.; Liu, J.; Zhou, R.; Chen, C.; Liu, C. Fast and Accurate Detection of UAV Objects Based on Mobile-Yolo Network. In Proceedings of the 2022 14th International Conference on Wireless Communications and Signal Processing (WCSP), Virtually, 1–3 November 2022; pp. 1–5.

69. Liu, W.; Anguelov, D.; Erhan, D.; Szegedy, C.; Reed, S.; Fu, C.Y.; Berg, A.C. Ssd: Single shot multibox detector. In Proceedings of the ECCV, Amsterdam, The Netherlands, 11–14 October 2016; pp. 21–37.

70. Lin, T.Y.; Goyal, P.; Girshick, R.; He, K.; Dollár, P. Focal loss for dense object detection. In Proceedings of the ICCV, Venice, Italy, 22–29 October 2017; pp. 2980–2988.

71. Everingham, M.; Van Gool, L.; Williams, C.K.; Winn, J.; Zisserman, A. The pascal visual object classes (voc) challenge. *Int. J. Comput. Vis.* **2010**, *88*, 303–338.

72. Everingham, M.; Eslami, S.; Van Gool, L.; Williams, C.K.; Winn, J.; Zisserman, A. The pascal visual object classes challenge: A retrospective. *Int. J. Comput. Vis.* **2015**, *111*, 98–136.

73. Lin, T.Y.; Maire, M.; Belongie, S.; Hays, J.; Perona, P.; Ramanan, D.; Dollár, P.; Zitnick, C.L. Microsoft coco: Common objects in context. In Proceedings of the ECCV, Zurich, Switzerland, 6–12 September 2014; pp. 740–755.

74. Pawełczyk, M.; Wojtyra, M. Real world object detection dataset for quadcopter unmanned aerial vehicle detection. *IEEE Access* **2020**, *8*, 174394–174409.

75. Zheng, Y.; Chen, Z.; Lv, D.; Li, Z.; Lan, Z.; Zhao, S. Air-to-air visual detection of micro-uavs: An experimental evaluation of deep learning. *IEEE Rob. Autom. Lett.* **2021**, *6*, 1020–1027.

76. Walter, V.; Vrba, M.; Saska, M. On training datasets for machine learning-based visual relative localization of micro-scale UAVs. In Proceedings of the ICRA, Paris, France, 31 May–31 August 2020; pp. 10674–10680.

77. Chen, Y.; Aggarwal, P.; Choi, J.; Kuo, C.C.J. A deep learning approach to drone monitoring. In Proceedings of the APSIPA ASC, Kuala Lumpur, Malaysia, 12–15 December 2017; pp. 686–691.

78. Zhao, J.; Zhang, J.; Li, D.; Wang, D. Vision-Based Anti-UAV Detection and Tracking. *IEEE Trans. Intell. Transp. Syst.* **2022**, *23*, 25323–25334.

79. Jiang, N.; Wang, K.; Peng, X.; Yu, X.; Wang, Q.; Xing, J.; Li, G.; Zhao, J.; Guo, G.; Han, Z. Anti-UAV: A large multi-modal benchmark for UAV tracking. *arXiv* **2021**, arXiv:2101.08466.

80. Bosquet, B.; Mucientes, M.; Brea, V. STDnet: A ConvNet for Small Target Detection. In Proceedings of the 29th British Machine Vision Conference, Newcastle, UK, 3–6 September 2018.

81. Ajakwe, S.O.; Saviour, I.I.; Kim, J.H.; Kim, D.S.; Lee, J.M. BANDA: A Novel Blockchain-Assisted Network for Drone Authentication. In Proceedings of the 2023 Fourteenth International Conference on Ubiquitous and Future Networks (ICUFN), Paris, France, 4–7 July 2023; pp. 120–125.

82. Heidari, A.; Navimipour, N.J.; Unal, M. A Secure Intrusion Detection Platform Using Blockchain and Radial Basis Function Neural Networks for Internet of Drones. *IEEE Internet Things J.* **2023**, *10*, 8445–8454.

83. Huynh-The, T.; Pham, Q.V.; Nguyen, T.V.; Da Costa, D.B.; Kim, D.S. RF-UAVNet: High-performance convolutional network for RF-based drone surveillance systems. *IEEE Access* **2022**, *10*, 49696–49707.

84. Sun, H.; Yang, J.; Shen, J.; Liang, D.; Ning-Zhong, L.; Zhou, H. TIB-Net: Drone detection network with tiny iterative backbone. *IEEE Access* **2020**, *8*, 130697–130707.

85. Liu, Y.; Liu, D.; Wang, B.; Chen, B. Mob-YOLO: A Lightweight UAV Object Detection Method. In Proceedings of the 2022 International Conference on Sensing, Measurement & Data Analytics in the Era of Artificial Intelligence (ICSMD), Harbin, China, 22–24 December 2022; pp. 1–6.

86. Golam, M.; Akter, R.; Naufal, R.; Doan, V.S.; Lee, J.M.; Kim, D.S. Blockchain Inspired Intruder UAV Localization Using Lightweight CNN for Internet of Battlefield Things. In Proceedings of the MILCOM 2022—2022 IEEE Military Communications Conference (MILCOM), Norfolk, VA, USA, 28 November–2 December 2022; pp. 342–349.

87. Barthelme, A.; Utschick, W. DoA estimation using neural network-based covariance matrix reconstruction. *IEEE Signal Process. Lett.* **2021**, *28*, 783–787.

88. Wang, M.; Yang, S.; Wu, S.; Luo, F. A RBFNN approach for DoA estimation of ultra wideband antenna array. *Neurocomputing* **2008**, *71*, 631–640.

89. Wu, L.; Liu, Z.M.; Huang, Z.T. Deep Convolution Network for Direction of Arrival Estimation With Sparse Prior. *IEEE Signal Process. Lett.* **2019**, *26*, 1688–1692. [CrossRef]

90. Akter, R.; Doan, V.S.; Huynh-The, T.; Kim, D.S. RFDOA-Net: An Efficient ConvNet for RF-Based DOA Estimation in UAV Surveillance Systems. *IEEE Trans. Veh. Technol.* **2021**, *70*, 12209–12214. [CrossRef]

91. Van Veen, B.D.; Buckley, K.M. Beamforming: A versatile approach to spatial filtering. *IEEE Assp. Mag.* **1988**, *5*, 4–24.

92. Schmidt, R. Multiple emitter location and signal parameter estimation. *IEEE Trans. Antennas Propag.* **1986**, *34*, 276–280.

93. Roy, R.; Kailath, T. ESPRIT-estimation of signal parameters via rotational invariance techniques. *IEEE Trans. Acoust. Speech, Signal Process.* **1989**, *37*, 984–995.

94. Wu, L.L.; Huang, Z.T. Coherent SVR learning for wideband direction-of-arrival estimation. *IEEE Signal Process. Lett.* **2019**, *26*, 642–646.

95. Sun, Y.; Chen, J.; Yuen, C.; Rahardja, S. Indoor sound source localization with probabilistic neural network. *IEEE Trans. Ind. Electron.* **2017**, *65*, 6403–6413.

96. Chakrabarty, S.; Habets, E.A. Multi-speaker DOA estimation using deep convolutional networks trained with noise signals. *IEEE J. Sel. Top. Signal Process.* **2019**, *13*, 8–21.

97. Blanchard, T.; Thomas, J.H.; Raoof, K. Acoustic localization and tracking of a multi-rotor unmanned aerial vehicle using an array with few microphones. *J. Acoust. Soc. Am.* **2020**, *148*, 1456–1467.

98. Alameda-Pineda, X.; Horaud, R. A geometric approach to sound source localization from time-delay estimates. *IEEE/ACM Trans. Audio Speech Lang. Process.* **2014**, *22*, 1082–1095.

99. Chen, J.; Zhao, Y.; Zhao, C.; Zhao, Y. Improved two-step weighted least squares algorithm for TDOA-based source localization. In Proceedings of the 2018 19th International Radar Symposium (IRS), Piscataway, NJ, USA, 20–22 June 2018; pp. 1–6.

100. Shi, Z.; Chang, X.; Yang, C.; Wu, Z.; Wu, J. An acoustic-based surveillance system for amateur drones detection and localization. *IEEE Trans. Veh. Technol.* **2020**, *69*, 2731–2739.

101. Heydari, Z.; Mahabadi, A. Real-time TDOA-based stationary sound source direction finding. *Multimedia Tools Appl.* **2023**, 1–32. [CrossRef]

102. Javed, S.; Danelljan, M.; Khan, F.S.; Khan, M.H.; Felsberg, M.; Matas, J. Visual object tracking with discriminative filters and siamese networks: A survey and outlook. *IEEE Trans. Pattern Anal. Mach. Intell.* **2023**, *45*, 6552–6574.

103. Jahangir, M.; Ahmad, B.I.; Baker, C.J. Robust drone classification using two-stage decision trees and results from SESAR SAFIR trials. In Proceedings of the 2020 IEEE International Radar Conference (RADAR), Washington, DC, USA, 28–30 April 2020; pp. 636–641.

104. Jouaber, S.; Bonnabel, S.; Velasco-Forero, S.; Pilte, M. NNAKF: A Neural Network Adapted Kalman Filter for Target Tracking. In Proceedings of the ICASSP 2021-2021 IEEE International Conference on Acoustics, Speech and Signal Processing (ICASSP), Toronto, ON, Canada, 6–11 June 2021; pp. 4075–4079.

105. Campbell, M.A.; Clark, D.E.; de Melo, F. An algorithm for large-scale multitarget tracking and parameter estimation. *IEEE Trans. Aerosp. Electron. Syst.* **2021**, *57*, 2053–2066.

106. Huang, Z.; Fu, C.; Li, Y.; Lin, F.; Lu, P. Learning aberrance repressed correlation filters for real-time UAV tracking. In Proceedings of the IEEE/CVF International Conference on Computer Vision, Seoul, Republic of Korea, 27 October–2 November 2019; pp. 2891–2900.

107. Li, Y.; Fu, C.; Ding, F.; Huang, Z.; Pan, J. Augmented memory for correlation filters in real-time UAV tracking. In Proceedings of the 2020 IEEE/RSJ International Conference on Intelligent Robots and Systems (IROS), Las Vegas, NV, USA, 25–29 October 2020; pp. 1559–1566.
108. Yuan, D.; Chang, X.; Li, Z.; He, Z. Learning adaptive spatial-temporal context-aware correlation filters for UAV tracking. *ACM Trans. Multimed. Comput. Commun. Appl. (TOMM)* **2022**, *18*, 1–18.
109. Bertinetto, L.; Valmadre, J.; Henriques, J.F.; Vedaldi, A.; Torr, P.H. Fully-convolutional siamese networks for object tracking. In Proceedings of the Computer Vision–ECCV 2016 Workshops: Amsterdam, The Netherlands, 8–10 October 2016; Part II 14, pp. 850–865.
110. Li, B.; Wu, W.; Wang, Q.; Zhang, F.; Xing, J.; Yan, J. Siamrpn++: Evolution of siamese visual tracking with very deep networks. In Proceedings of the IEEE/CVF Conference on Computer Vision and Pattern Recognition, Long Beach, CA, USA, 15–20 June 2019; pp. 4282–4291.
111. Dai, K.; Zhang, Y.; Wang, D.; Li, J.; Lu, H.; Yang, X. High-performance long-term tracking with meta-updater. In Proceedings of the IEEE/CVF Conference on Computer Vision and Pattern Recognition, Seattle, WA, USA, 13–19 June 2020; pp. 6298–6307.
112. Zitar, R.A.; Mohsen, A.; Seghrouchni, A.E.; Barbaresco, F.; Al-Dmour, N.A. Intensive Review of Drones Detection and Tracking: Linear Kalman Filter Versus Nonlinear Regression, an Analysis Case. *Arch. Comput. Methods Eng.* **2023**, *30*, 2811–2830.
113. DeDrone. Counter Drone Software. [EB/OL]. Available online: https://www.dedrone.com/products/counter-drone-software (accessed on 1 June 2023).
114. Dronesshield. DroneSentry-X Vehicle, Ship and Fixed Site C-UAS Detect-and-Defeat. [EB/OL]. Available online: https://www.droneshield.com/products/sentry-x (accessed on 1 June 2023).
115. Rohde&Schwarz. R&S®ARDRONIS for Effective Drone Defense. [EB/OL]. Available online: https://www.rohde-schwarz.com.cn/products/aerospace-defense-security/counter-drone-systems_250881.html (accessed on 1 June 2023).
116. blighterz. AUDS Anti-UAV Defence System. [EB/OL]. Available online: https://www.blighter.com/products/auds-anti-uav-defence-system/ (accessed on 1 June 2023).
117. dronebouncer. Orelia Drone-Detector. [EB/OL]. Available online: http://dronebouncer.com/en/orelia-drone-detector/ (accessed on 1 June 2023).
118. leonardo. Falcon Shield. [EB/OL]. Available online: https://uk.leonardo.com/en/innovation/falcon-shield/ (accessed on 1 June 2023).
119. IAI. Elta-Systems. [EB/OL]. Available online: https://www.iai.co.il/about/groups/elta-systems (accessed on 1 June 2023).

applied sciences

Article

Hybrid Deep Convolutional Generative Adversarial Network (DCGAN) and Xtreme Gradient Boost for X-ray Image Augmentation and Detection

Ahmad Hoirul Basori *, Sharaf J. Malebary and Sami Alesawi

Faculty of Computing and Information Technology in Rabigh, King Abdulaziz University, Jeddah 21589, Saudi Arabia; smalebary@kau.edu.sa (S.J.M.)
* Correspondence: abasori@kau.edu.sa

Abstract: The COVID-19 pandemic has exerted a widespread influence on a global scale, leading numerous nations to prepare for the endemicity of COVID-19. The polymerase chain reaction (PCR) swab test has emerged as the prevailing technique for identifying viral infections within the current pandemic. Following this, the application of chest X-ray imaging in individuals provides an alternate approach for evaluating the existence of viral infection. However, it is imperative to further boost the quality of collected chest pictures via additional data augmentation. The aim of this paper is to provide a technique for the automated analysis of X-ray pictures using server processing with a deep convolutional generative adversarial network (DCGAN). The proposed methodology aims to improve the overall image quality of X-ray scans. The integration of deep learning with Xtreme Gradient Boosting in the DCGAN technique aims to improve the quality of X-ray pictures processed on the server. The training model employed in this work is based on the Inception V3 learning model, which is combined with XGradient Boost. The results obtained from the training procedure were quite interesting: the training model had an accuracy rate of 98.86%, a sensitivity score of 99.1%, and a recall rate of 98.7%.

Keywords: COVID-19 pandemic; endemic; DCGAN; Inception V3; deep learning; chest X-ray

Citation: Basori, A.H.; Malebary, S.J.; Alesawi, S. Hybrid Deep Convolutional Generative Adversarial Network (DCGAN) and Xtreme Gradient Boost for X-ray Image Augmentation and Detection. *Appl. Sci.* **2023**, *13*, 12725. https://doi.org/10.3390/app132312725

Academic Editor: João M. F. Rodrigues

Received: 31 October 2023
Revised: 12 November 2023
Accepted: 23 November 2023
Published: 27 November 2023

1. Introduction

The respiratory infection known as COVID-19, colloquially referred to as the coronavirus, has had a substantial impact on a considerable global population. According to recent statistical data, the cumulative number of global infections has exceeded 243 million cases, with Saudi Arabia reporting a significant figure of over 801,000 cases. The fatality rate in Saudi Arabia stands at 1.15%, leading to a total of 9223 recorded deaths [1,2]. Despite the widespread administration of the vaccination to a majority of the population, the health authority emphasizes the continued importance of adhering to health precautions. COVID-19 manifests a range of symptoms, including elevated body temperature, emesis, and even gastrointestinal distress in the form of diarrhea. X-ray imaging can be utilized as a comprehensive diagnostic tool to assess the impact of viral infections on the human lungs and their potential to spread to other organs. The potential for utilizing X-ray images in the analysis of COVID-19 via visual inspection presents an opportunity to employ image processing and deep learning techniques for the comprehensive investigation of infection areas and the potential prediction of future spread. The utilization of convolutional neural networks (CNNs) for deep learning has been widely acknowledged as a robust method for the detection and recognition of medical images [3,4]. Deep learning has substantially assisted in the classification of COVID-19 via the use of X-rays. Other researchers have proposed an automatic detection process using a deep learning method. Their methods concern applying adaptive median filtering and histogram equalization [5]. Neha Gianchandani et al. proposed a method for rapid COVID-19 diagnosis using ensemble deep transfer learning models from chest radiographic images. They designed two deep learning models utilizing chest X-ray images to diagnose COVID-19 [6]. Abdullahi Umar Ibrahim et al. used a deep learning approach based on a pre-trained AlexNet model to classify COVID-19, pneumonia, and normal CXR scans obtained from different public databases [7]. Ebenezer Jangam et al. used a stacked ensemble of VGG 19 and DenseNet 169 models to detect COVID-19, achieving high accuracy and recall in five different datasets consisting of chest X-ray images and chest CT scans [7]. The advancement of technologies for diagnosing COVID-19 from X-ray images has several advantages, as shown in Table 1.

Table 1. Chest X-ray identification advantages.

Approach	Advantages
Chest X-ray diagnosis	The expeditious identification and assessment of a substantial patient population within a constrained timeframe.
Chest X-ray images	The aforementioned data may be obtainable from various hospitals and clinics responsible for their management. Individuals can be diagnosed with COVID-19.
Light chest X-ray detection system	Implementing the light chest X-ray detection system can potentially mitigate widespread infection via timely diagnosis. The physician may request that the patient engage in self-isolation.
Disease complication	Chest X-rays are a valuable diagnostic tool for investigating and monitoring numerous diseases, including those associated with COVID-19.

1.1. Motivation of Study

During the pandemic, chest X-rays and computer tomography (CT) had the potential for the clinical diagnosis and evaluation of COVID-19 patients. These approaches are useful for doctors performing diagnosis and therapy. Keymal and Şen used automatic detection with a bidirectional LSTM network and deep features [8]. They relied on the Bi-LSTM network inside the deep feature and then compared it with the deep neural network of the fivefold cross-validation approach. Their model was trained using 100 epochs and 32 batch sizes. The fourth section of their model was designed to handle the activation function, with two neurons representing the classes COVID-19 and no-finding [8]. A radiological study found a diagnostic tool for COVID-19. However, it can also be used for measuring the chronicity of diseases or complications in patients [9]. Most researchers rely on the visual interpretation of chest X-ray images, which requires an expert such as a radiologist or medical doctor. The densities of chest X-rays might vary depending on the severity of the disease. Figure 1 shows the denseness of chest X-ray images.

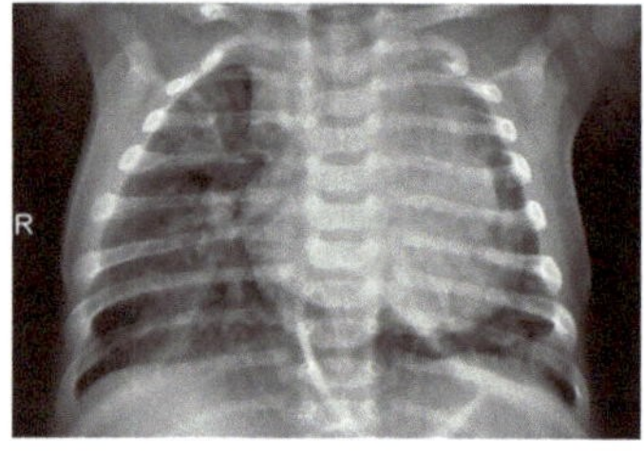
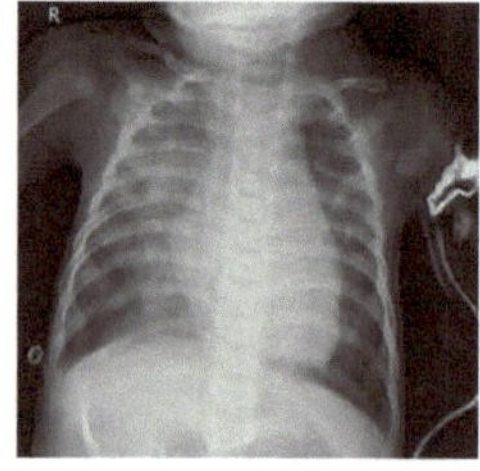
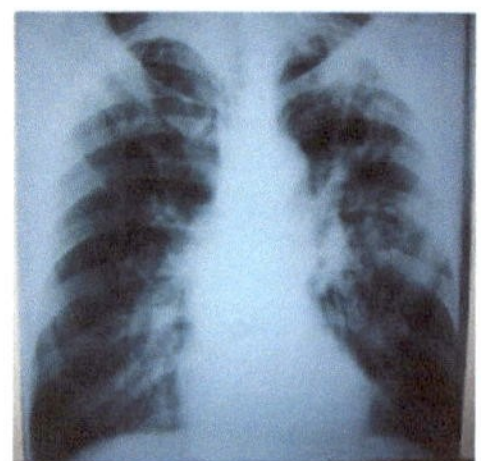

Figure 1. Chest X-ray images of patients with various diseases (COVID-19, pneumonia, and tuberculosis).

1.2. State of the Art

The utilization of artificial intelligence (AI) equipped with deep learning capabilities has become prevalent throughout the pandemic for assessing chest X-ray pictures. The present study aims to develop a research design for the identification of anomalies in the lung pictures of patients during the initial screening process. Another researcher also utilized artificial intelligence to identify COVID-19 based on cough sounds. The researchers directed their attention towards the analysis of cough sounds obtained from several datasets, with the objective of detecting and diagnosing respiratory diseases. This was achieved via the utilization of a support vector machine (SVM) classifier and linear regression techniques. The utilization of artificial neural networks (ANNs) and the random forest (RF) classifier has also been observed in the analysis of cough patterns and the subsequent determination of the severity of respiratory diseases in patients [10]. Respiratory conditions, such as asthma and cough, can be detected via the application of Wigner distribution methodologies within a low-noise setting [10].

The other study has focused on the problem of classifying COVID-19 patients from healthy people using a limited public dataset. They used two-step transfer learning models to train on small datasets [11]. They conducted a pre-trained deep residual network on a large pneumonia dataset in their first step. Satisfying results for COVID-19 detection in healthy and pneumonia-infected X-ray images were obtained. Their proposed model used two types of layers for pre-trained steps using ResNet. Afterwards, they used feature smoothing (FSL) based on 1×1 convolution to maintain the shape of tensors. They continued by smoothing pre-trained tensors to acquire the characteristics of the new image input. FSL will then be used with residual blocks for further processing with a feature extraction layer (FEL).

FEL has double channels and three operations to be combined with a feature map into a single dimension [11]. The input of the X-ray images is 224×224 pixels, using three channels in ResNet34. Then, it is fed into three residual blocks inside ResNet34. FSL is inserted before and after the second residual block. While FEL is implanted after the third residual block, the final training uses a $512 \times k$ layer (k refers to the number of classes) with limited X-ray images of COVID-19. The images are cropped using a ratio of 0.8–1.0. Then, it is rotated with a random rotation angle [−20, 20]. The result is an initial cross-entropy loss that is relatively high at around 0.67, with an average loss of 0.73. The significant decrease in entropy loss was sighted in the 100th batch at 0.47. Then, after the 1635th batch, cross-extropy loss can be pressed until 0.072; this number is very substantial since the initial loss is around 0.67. State-of-the-art COVID-19 automatic detection using chest X-rays varies depending on the data's characteristics and the convolutional network's structure. Table 2 summarizes related works on COVID-19 detection using chest X-ray images.

1.3. Contribution of the Proposed Work

The primary contributions of this study can be succinctly stated as follows: (1) the implementation of data augmentation using the deep convolutional generative adversarial network (DCGAN), resulting in enhanced dataset quality; (2) the utilization of a modified Inception V3 model, specifically designed to accommodate both the mixed DCGAN dataset and the original dataset, thereby achieving a greater accuracy rate compared to previous research efforts. The XGradient Boost algorithm enhances feature mapping. The framework of the paper is outlined as follows: Section 1 is dedicated to examining the current level of knowledge, and the underlying factors that drive research endeavors. Section 2 provides an overview of existing studies pertaining to the application of deep learning techniques in the detection of chest diseases. Section 3 of the document encompasses the content and proposed approach. Section 4 of the paper provides a comprehensive analysis of the experimental results and subsequent comments. Additionally, it evaluates the performance of the proposed model in comparison to earlier studies. Finally, Section 5 serves as the concluding section of the study.

Table 2. Summary of the previous related work.

Author	Proposed Model	Image Size	Pre-Trained	Achievement	Remarks
Rouchi, Z, et al. [11]	Two-Step transfer learning using XRayNet(2) and XRayNet(3)	224 × 224	Yes	AUC score of 99.8% on the training dataset and 98.6 on the testing dataset. Overall accuracy is 91.92%.	Small number of datasets
N, Kumar et al. [12]	Integration of previous pre-trained models (EfcientNet, GoogLeNet, and XceptionNe)	224 × 224	Yes	The AUC score is 99.2% for COVID-19, 99.3% for normal, 99.01% for pneumonia, and 99.2% for tuberculosis.	Medium number of datasets
S. Lafraxo M. el Ansari [5]	Integrated architecture using an adaptive median filter, convolutional neural network, and histogram equalization	256 × 256	No	The proposed system known as CoviNet is able to achieve an accuracy rate of 98.6% for binary and 95.8% for multiclass classification.	Medium number of datasets
A. Narin [13]	Based on ResNet50, with support vector machines (SVMs); quadratic and cubic	1024 × 1024	Yes	The input images are fed to a convolutional neural network (ResNet) and then three SVM models are used (linear, quadratic, and cubic). The result shows that SVM-Quadratic outperformed the others by 99% in terms of overall accuracy.	Medium number of datasets
Saif, A.F.M. et al. [14]	CapsCovNet capsule convolutional neural network with three blocks	128 × 128	Yes	There are two input images: US images extracted from the US video dataset and chest X-ray images. Their method has outperformed state-of-the art US images with increments around 3.12–20.2%.	Medium number of datasets

2. Related Work

The researchers observed that individuals with COVID-19 infection exhibited obvious ground-glass opacities in their lungs, which appeared darker in comparison to the surrounding area when compared to individuals without COVID-19 [15,16]. Consequently, scientists place significant reliance on utilizing chest X-ray pictures to detect and analyze COVID-19, as well as determining the subsequent course of action for patients. In practical contexts, the ability to analyze numerous cases simultaneously is necessary in order to address the limitations imposed by staff and testing kit availability. Hence, the utilization of image-based detection techniques using X-ray pictures holds significant promise in assisting hospitals and medical personnel by facilitating the identification of infected patients. The radiography system is widely accessible in most hospitals, and medical staff is generally better acquainted with interpreting X-ray images as opposed to utilizing the latest testing kits [15,16]. Furthermore, the generative adversarial network (GAN) framework has gained significant popularity in the field of medical image processing. In their study, Zhao et al. introduced a framework that utilizes the VGG16 model and a DCGAN-based model for generating synthetic lung pictures. The generated images are then used for classification purposes, employing the forward and backward GAN techniques [17]. The efficiency of generative adversarial networks (GANs) has also been demonstrated in the detection of anomalies in retinal images, specifically in the comparison between healthy and unhealthy retinal images. Hence, the absence of a sufficient dataset has compelled researchers to further investigate the available dataset. Consequently, the utilization of a generative adversarial network (GAN) has become imperative in order to address the limitations imposed by the dataset. The new study published in the *Journal of Radiology* has demonstrated the higher performance of chest X-ray imaging compared to laboratory testing methods, such as PCR or fast tests. As a result, numerous research studies have concluded that the utilization of chest radiography as the primary screening method is recommended for the detection of COVID-19 infection. The integration of artificial intelligence (AI) with radiographic imaging has the potential to facilitate extensive detection capabilities and streamline the responsibilities of medical professionals, enabling them to allocate their attention towards providing effective treatment to patients who have tested positive for a medical condition.

Computers have a big role in diagnosing diseases. Globally, doctors have detected viruses in the chest by looking at X-ray images of the patient's chest [18]. X-rays are commonly used to diagnose pneumonia, but it is still difficult to distinguish between lung infections caused by COVID-19 or pneumococcal pneumonia using X-ray images [19]. It is difficult for radiologists to distinguish COVID-19 from viral pneumonia because the images of various pneumonia viruses are similar and overlap with other lung diseases. Thus, to help diagnose COVID-19, artificial intelligence with deep learning that is able to provide fast, precise, and inexpensive results is needed [20]. With Mask_RCNN, diagnosing diseases using X-ray images will be more accurate and efficient for understanding the type and severity of patterns [21]. Deep learning has been widely used in the field of biomedical image processing, where the results have proven to be effective in classifying various diseases such as pneumonia, respiratory disorders, etc. [12]. Abdul Qayyum et al. detected COVID-19 on lung X-ray images using the depth-wise multilevel feature concatenated deep neural network method [22]. N. Kumar et al. [12] detected COVID-19 using a deep transfer learning-based ensemble model designed by integrating EffientNet, GoogLeNet, and XceptionNet for the early diagnosis of COVID-19 infection. Faisal Muhammaad Shah et al. [23] reviewed some newly emerging AI-based models that can detect COVID-19 from X-ray or CT images of the lungs.

Sourabh Shastri et al. designed a nested ensemble model using deep learning methods based on a long short-term memory (LSTM) model, which evaluated confirmed intensive care COVID-19 and death cases in India [24]. Kemal Akyol and Baha Sen propose deep learning using the Bi-LSTM network to detect COVID-19 and no-finding cases via chest X-ray images [16]. K. Shankar et al. developed a metaheuristic-based fusion model for COVID-19 diagnosis using chest X-ray images [25]. Medical image collection is a very expensive and tedious process that requires the participation of radiologists and researchers [26]. The ability of CNNs to perform such tasks is due to their large number of parameters and fine-tuning approach. Thus, even with limited datasets, it can carry out the best detection and recognition process [27–29]. The dataset of images for medical imaging is quite hard to collect and might be very expensive because it involves radiologists and researchers [30]. During the pandemic, collecting X-ray datasets for the chest was still difficult; however, another researcher has proposed a data augmentation method to augment the synthetic dataset. This method is used to encompass a modified dataset for training purposes.

Previous researchers have used several approaches for data augmentation, such as image transformation, color adjustment, distorting, or polishing. Currently, the advanced form of data augmentation is more capable than conventional data amplification. One of the methods that previous researchers used was known as the generative adversarial network, or the GAN. This approach is very innovative for producing synthetic images without supervision via the *min–max* game algorithm. The GAN uses two disparate networks, $G(z)$ and $D(x)$, that have different functions. $G(z)$ is used to generate a faithful copy of the image in order to fake another network with a certain discriminator, which is computed by $D(x)$ [30]. The infected patient with COVID-19 might show several symptoms, such as fever and a cough similar to the flu. However, serious cases show organ failure, breathing difficulty, and even death [31,32]. As a result of the exponential growth of COVID-19 infections, many countries are facing serious issues with their health systems, and most of them are at the point of downfall because their facilities are not capable of handling a large number of patients at the same time. The testing kit and ventilator stock decreased rapidly; therefore, most countries employ a lockdown policy, strictly ban any gathering, and ask everybody to stay at home. One of the critical steps in detecting the infection in patients is testing the method. Primarily, COVID-19 testing used polymerase chain reaction (rRT-PCR) tests [33,34]. This is also known as a swab test. By taking some liquid from the nose or throat, the result might be observed in a few hours or days. Furthermore, the other approach that can be used is to obtain X-ray radiography images of the patient's chest [34].

3. Research Method and Materials

This section describes the dataset, DCGAN, Xtreme Gradient Boost (XGBoost), and evaluation metrics and measurement.

3.1. Dataset

Historically, researchers have directed their attention towards employing deep learning techniques for the purpose of binary classification in the context of binary images. The primary objective of our research was to apply the generative adversarial network (GAN) methodology in conjunction with convolutional neural networks (CNNs) and Xtreme Gradient Boosting (XGBoost) techniques. Multiple datasets consisting of both normal and pneumonia patient data were obtained from the Paul Cohen dataset, encompassing a range of resolutions [35]. An alternative version of the dataset that Joseph Paul Cohen provided, which includes samples of COVID-19, further supports the dataset that we used [36]. We have chosen these two samples as they are among the initial datasets collected for chest X-ray analysis. Subsequently, the aforementioned dataset will be utilized as input for the DCGAN Network in order to generate synthetic images. Figure 2 illustrates an example of normal, pneumonia, and COVID-19 chest X-ray images.

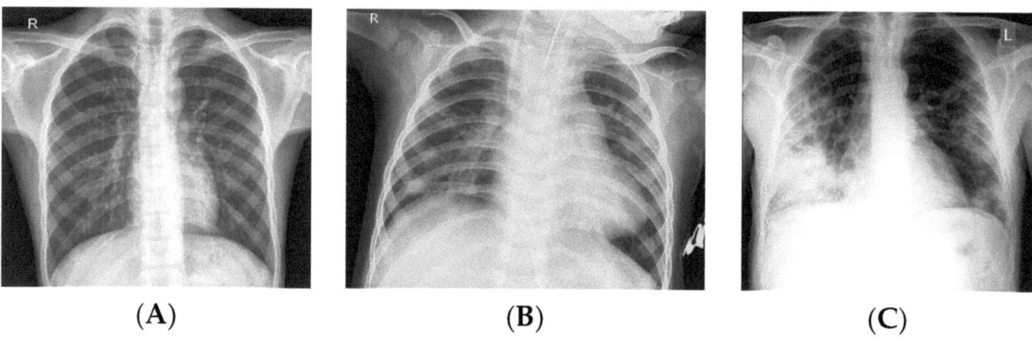

Figure 2. Sample of dataset: (**A**) normal, (**B**) pneumonia, and (**C**) COVID-19.

3.2. Proposed Methodology

The generative adversarial network (GAN) can produce images or samples of data that imitate the original dataset's feature distribution. There are two core components of the GAN, which are the generator and discriminator, that are trained simultaneously via a combative process. The discriminator learns how to differentiate the fake and authentic image datasets, while the generator discovers methods for producing images that look like authentic images [37]. The architecture of the GAN is depicted in Figure 3.

The generator utilizes latent space coordinates as an input to generate a novel image. Typically, a vector with 100 numerical values represents the latent space. During the training phase, the generator learns the process of mapping each individual point in order to generate an image. Once the model has been retrained, it will acquire new mapping capabilities.

The process of learning GANs is to perform real-time training on both the discriminator and generator networks, which is challenging and requires *min–max* adjustments for the discriminator and generator. This process is shown in Equation (1) [38].

$$\frac{min}{G}\frac{max}{D}\ V_{GAN}(D,G) = \mathbb{E}_{x\sim p_{data}(x)}[logD(x)] + \mathbb{E}_{z\sim p_z(z)}[log(1-D(G(z)))] \qquad (1)$$

$\mathbb{E}_{x\sim p_{data}(x)}$ is the expected value of real instances, while $\mathbb{E}_{z\sim p_z(z)}$ is the expected value of fake instances. $p_z(z)$ represents the random noise variable for a standard normal distribution. $G(z)$ denotes the generator function that is responsible for mapping the data to space, *x* denotes the original data, and *D(x)* denotes the probability of original data(*x*) and not the generated ones.

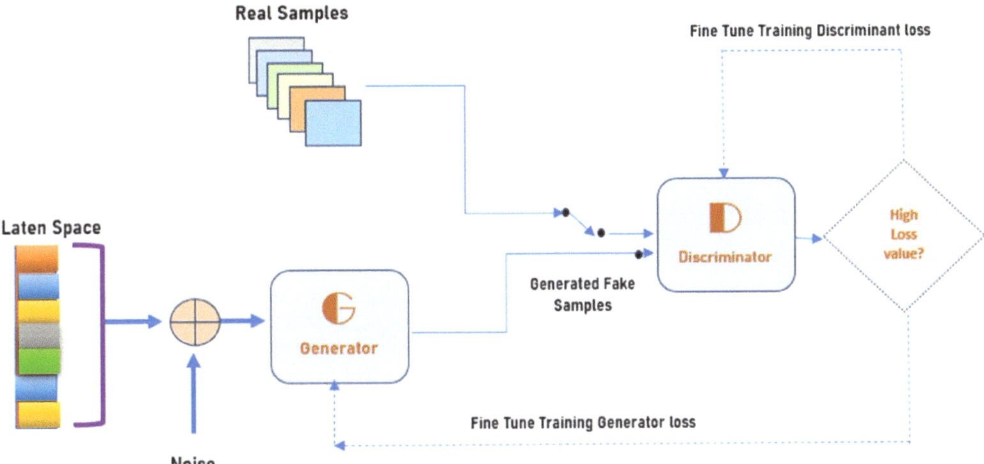

Figure 3. The architecture of the generative adversarial network (GAN) [37].

It is widely acknowledged that Inception V3 is a convolutional neural network design that belongs to the Inception family. This approach incorporates various enhancements, including label smoothing, factorized 7×7 convolution, and an auxiliary classifier. The Inception module initially consists of a 5×5-sized convolutional layer, which is then decreased by employing two 3×3 convolutions. The auxiliary classifier is employed in order to enhance the convergence of deep neural networks by mitigating the issue of vanishing gradients. Traditionally, the techniques of maximum and average polling have been employed in order to decrease the dimensions of the grid and feature map. The Inception V3 model reduces the size of the grid by performing a division operation on the current grid size, dividing it by 2. For instance, if a grid consists of a $d \times d$ grid with k filters, the process of reduction will yield a $d/2 \times d/2$ grid with $2k$ filters.

The core concept of this paper is to combine DCGAN with transfer learning (Inception_V3) to acquire better chest X-ray image results. The architecture of the Inception-V3 model is depicted in Figure 4, while the proposed methodology is portrayed in Figure 5.

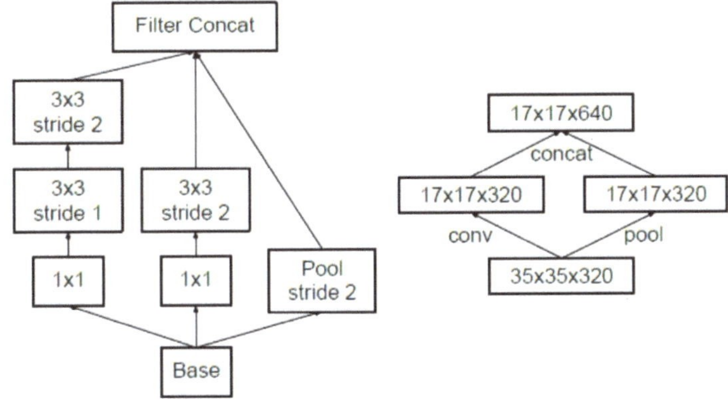

Figure 4. The architecture of Inception V3 Szegedy, C. et al. [39].

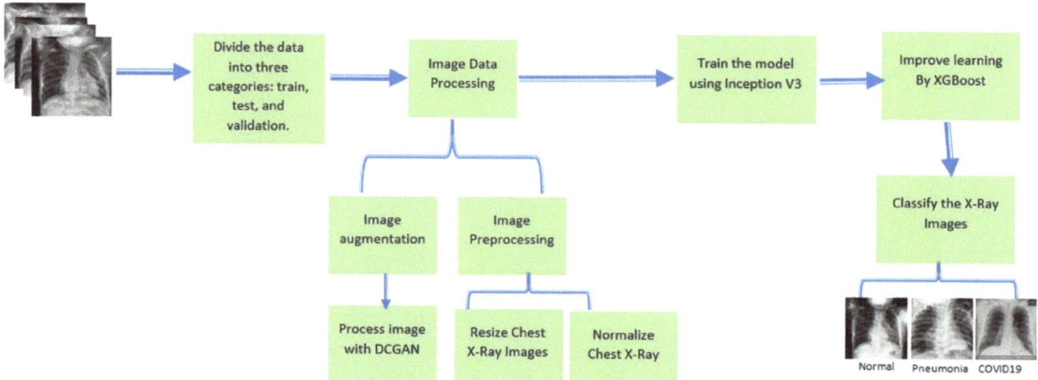

Figure 5. Proposed methodology for DCGAN-InceptionV3.

Image data processing is essential to the convolutional neural network (CNN) because it will determine the classification result. The initial process is an image-resizing process that resizes the image into 288 × 288 dimensions for training purposes, followed by image normalization. The next step will involve image augmentation using DCGAN. Then, it will continue the training process using Inception V3.

3.3. Xtreme Gradient Boost (XGBoost)

We have adopted the XgBoost algorithm from Chen, T. and Guestrin, C. [40]. XGBoost is designed for given datasets that have n examples and m features, and it is represented by $\mathcal{D}\{(x_i, y_i)\}$ $(|\mathcal{D}| = n, x_i \in \mathbb{R}^m, y_i \in \mathbb{R})$. By using the K additive function in Equation (2), it can be used for predicting the output:

$$\hat{y}_i = (x_i) = \sum_{k=1}^{K} f_k(x_i), f_k \in \mathcal{F} \tag{2}$$

where $\mathcal{F} = \left\{f(x) = \omega_{q(x)}\right\} (q : \mathbb{R}^m \to T, \omega \in \mathbb{R}^T)$ is the spatial regression of trees, q represents the tree structures, and T represents the number of leaves on the tree. The total score will be given by ω by adding up the corresponding leaves. The regularized objective for learning a set of functions is given by Equations (3) and (4).

$$\mathcal{L}() = \sum_i l(\hat{y}_i - y_i) + \sum_k \Omega(f_k) \tag{3}$$

$$where\ \Omega(f) = \gamma T + \frac{1}{2}\lambda \|\omega\|^2 \tag{4}$$

The given tree representing Equation (3) cannot be trained using Euclidian distances; therefore, Chen, T. and Guestrin, C. [40] used an additive approach, as shown in Equation (5). $\hat{y}_i^{(t)}$ represents the prediction of the *i-th* instance at the *t-th* iteration.

$$\mathcal{L}^{(t)} = \sum_{i=1}^{n} l\left(y_i, \hat{y}_i^{(t-1)} + f_t(x_i)\right) + \Omega(f_t) \tag{5}$$

Equation (5) can be improved further with a second-order approximation, as depicted in Equation (6):

$$\mathcal{L}^{(t)} \simeq \sum_{i=1}^{n} l\left(y_i, \hat{y}_i^{(t-1)}\right) + g_i f_t(x_i) + \frac{1}{2}h_i f_t^2(x_i) + \Omega(f_t) \tag{6}$$

where $g_i = \partial_{\hat{y}_i^{(t-1)}} l\left(y_i, \hat{y}_i^{(t-1)}\right)$ and $h_i = \partial^2_{\hat{y}_i^{(t-1)}} l\left(y_i, \hat{y}_i^{(t-1)}\right)$, and both g_i and h_i are loss functions for the first and second gradient statistics at step t. Equation (6) can be simplified more by removing the constant term; thus, the equation will transform into Equation (7) [40].

$$\widetilde{\mathcal{L}}^{(t)} = \sum_{i=1}^{n} \left[g_i f_t(x_i) + \frac{1}{2} h_i f_t^2(x_i) \right] + \Omega(f_t) \tag{7}$$

If we define $I_j = \{i|q(x_i) = j\}$ as an instance that symbolizes leaf j, then Equation (7) can be revised as described in Equations (8) and (9) by expanding Ω.

$$\widetilde{\mathcal{L}}^{(t)} = \sum_{i=1}^{n} \left[g_i f_t(x_i) + \frac{1}{2} h_i f_t^2(x_i) \right] + \gamma T + \frac{1}{2} \lambda \sum_{j=1}^{T} w_j^2 \tag{8}$$

$$= \sum_{j=1}^{T} \left[\left(\sum_{i \in I_j} g_i \right) w_j + \frac{1}{2} \left(\sum_{i \in I_j} h_i + \lambda \right) w_j^2 \right] + \gamma T \tag{9}$$

For the static value of $q(x)$, the optimum weight w_j^* of leaf j is given by Equations (10) and (11).

$$w_j^*(q) = -\frac{\sum_{i \in I_j} g_i}{\sum_{i \in I_j} h_i + \lambda} \tag{10}$$

$$\widetilde{\mathcal{L}}^{(t)}(q) = -\frac{1}{2} \sum_{j=1}^{T} \frac{\left(\sum_{i \in I_j} g_i \right)^2}{\sum_{i \in I_j} h_i + \lambda} \tag{11}$$

Algorithm 1 describes the proposed approach's process flow:

Algorithm 1: Data Augmentation and XGradientBoost

Input: Chest X-ray Image $I\,(x,y)$
Step 1: Preprocessing, convert inage to grayscale, then apply normalization.
$I : \{\mathbb{X} \subseteq \mathbb{R}^n\} \rightarrow \{Min, \ldots, Max\}$ become. $I_N : \{\mathbb{X} \subseteq \mathbb{R}^n\} \rightarrow \{newMin, \ldots, newMax\}$. $I_N = (newMax - newMin)\frac{1}{1+e^{-\frac{I-\beta}{\alpha}}} + newMin$
Step 2: Basic Augmentation image (scale, flip, rotate)
Scale, $I(x,y)x_{ratio} = \frac{old\_image.x}{new\_image.x}$
$\qquad y_{ratio} = \frac{old\_image.y}{new\_image.y}$
$\qquad I_{new}\left(floor(x * x_{ratio}), floor(y * y_{ratio})\right)$
$\qquad$ Horizontal Flip, $I_{new}(width - x - 1, y)$
$\qquad$ Rotation $I_{new} = (x_r, y_r)$
$x_r = (x - center\_x) * \cos(angle) - (y - center\_y) * \sin(angle) + center\_x$
$y_r = (x - center\_x) * \sin(angle) + (y - center\_y) * \cos(angle) + center\_y$
Step 3: Advanced augmentation image(DCGAN)
Compute the expected value of real $\mathbb{E}_{x \sim p_{data}(x)}$ and fake instance
$\mathbb{E}_{z \sim p_z(z)}$ using Equation (1)
Step 4: Features Extraction
Push the image $I_{new} = (x_r, y_r)$, into convolution network, then calculate output of Pixel
Value $V = \frac{\sum_{i=1}^{q}\left(\sum_{j=1}^{q} f_{ij}d_{ij}\right)}{F}$
Step 5: Features Mapping with XGradient Boost
Add the output to Xgradient boost tree by predicting the optimum weight of the leaf j, described in Equation (6).
Output: $I_{new} = (x_r, y_r)$

3.4. Evaluation Metrics

We have followed the standard and most common metrics for evaluation: precision, recall, and F1 score. The standard definitions are defined in Equations (12)–(14):

$$Precision = \frac{TP}{TP + FP'} \tag{12}$$

$$Recall\ or\ True\ Positive\ Rate = \frac{TP}{TP + FN'} \tag{13}$$

where *FP*, *FN*, *TP*, and *TN* are false-positive, false-negative, true-positive, and true-negative values, correspondingly. The F1 score is used to measure the model's accuracy, and it can be computed using Equation (4):

$$F1 = \frac{2 \times precision \times recall}{precision + recall} = \frac{2TP}{2TP + FP + FN'} \tag{14}$$

4. Results and Discussion

This section describes the implementation details of the proposed DCGAN-Inception V3 and Xtreme Gradient Boost.

4.1. Generating the DCGAN Image

The DCGAN method is used to generate a new dataset based on the original data by considering factors from the original image. There are two training parts used here:

- Part 1: the discriminator is trained to maximize the probability of correctly classifying the given input as either real or fake.
- Part 2: the generator is trained by minimizing *log (1-D(G-(Z)))* to generate a better fake image.

The process starts by generating fake images based on the DCGAN approach. Initially, the picture looks like a blank canvas. In the 252nd iteration, the shadow on chest X-ray images gradually appeared. Following the 504th iteration, the chest portion became more precise. Once the 1000th iteration was reached, the produced image presented a recognizable chest X-ray image. Afterwards, the picture gradually appeared throughout the iteration. Finally, in iterations > 2000, the generated image revealed an appealing result, with transparent portions of the lung area, rib bone, and even back spine bone observed, as shown in Figure 6.

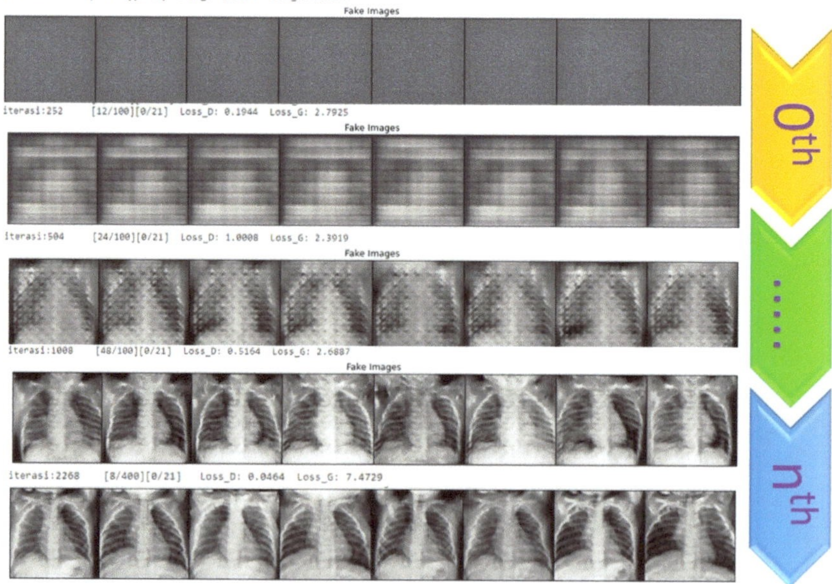

Figure 6. DCGAN iteration for image generation.

In Figure 7, during the image generation process, the generator *(G)* loss rate is relatively high initially and then gradually decreases relative to the iterations. In comparison, the discriminator has fluctuated with respect to its value since the beginning of the iterations. However, when it reaches the 1000th plus iteration, the discriminator loss is flat and reaches zero.

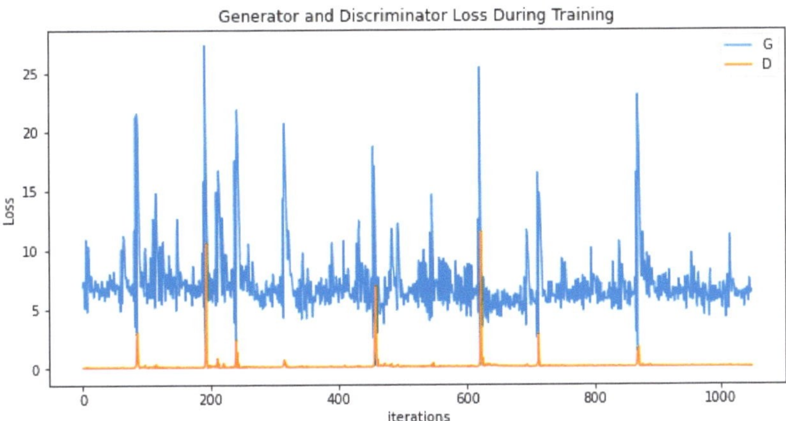

Figure 7. Generator and discriminator loss during DCGAN training.

4.2. Classification Results

We have used two primary sources of datasets (Cohen, J.P. [36] and Tabik, R. [41]) and one dataset generated from the DCGAN process, as explained in the previous subsection. During the image dataset generation process utilizing the DCGAN, the GPU's limitations introduce restrictions, resulting in the production of only 300 chest X-ray pictures in a single compilation. Subsequently, following the repetition of the process for a total of ten iterations, we successfully amassed a dataset of 3000 chest X-ray pictures that were generated. The generated data will be integrated with actual images from primary sources for the purpose of training. According to the analysis of generated images, it has been determined that a portion of the DCGAN data accounts for around 39% of the entire dataset. The generated dataset has 1000 artificially created photos, encompassing each instance of normal lung conditions, pneumonia, and COVID-19 cases.

We have trained our proposed model with 7600 datasets with three different classes: normal, pneumonia, and COVID-19. Later, we carried out a test for 89 images of normal cases, 67 images for pneumonia, and 45 images for COVID-19. We trained our model using Inception Learning V3 with 100 epochs and a batch size of 128. The result shows that the proposed model's best training and validation accuracy are 98.86% and 98.49, respectively. The validation loss is 0.015. Refer to the graph depicted in Figure 8. The graph exhibits fluctuations during the initial stages of training, which can be attributed to the limited availability of data. However, to address this issue and ensure stability during the training process, additional data were subsequently introduced. The occurrence of this volatility is also prevalent in the majority of TensorFlow training processes.

The prediction result is fascinating and can classify the tested image accordingly compared to the original dataset. Figure 9 depicts the impact of the prediction of our proposed system. Figure 9A,B,D illustrate the prediction result where the original image is pneumonia, and it is correctly predicted as pneumonia. In contrast, Figure 9C was originally normal and is indicated as a normal case.

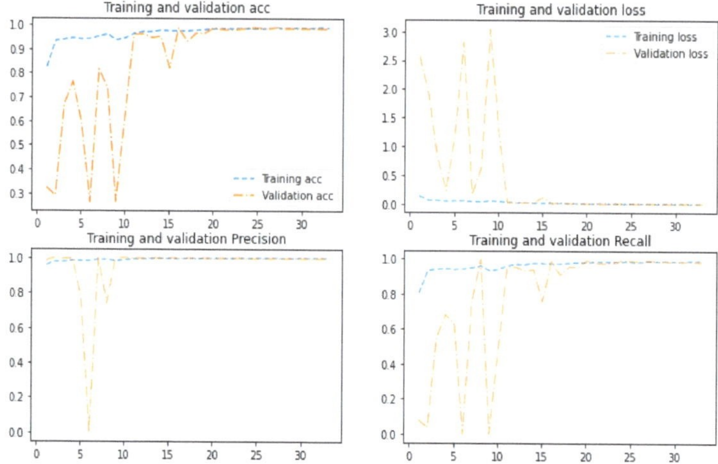

Figure 8. Training and validation accuracy.

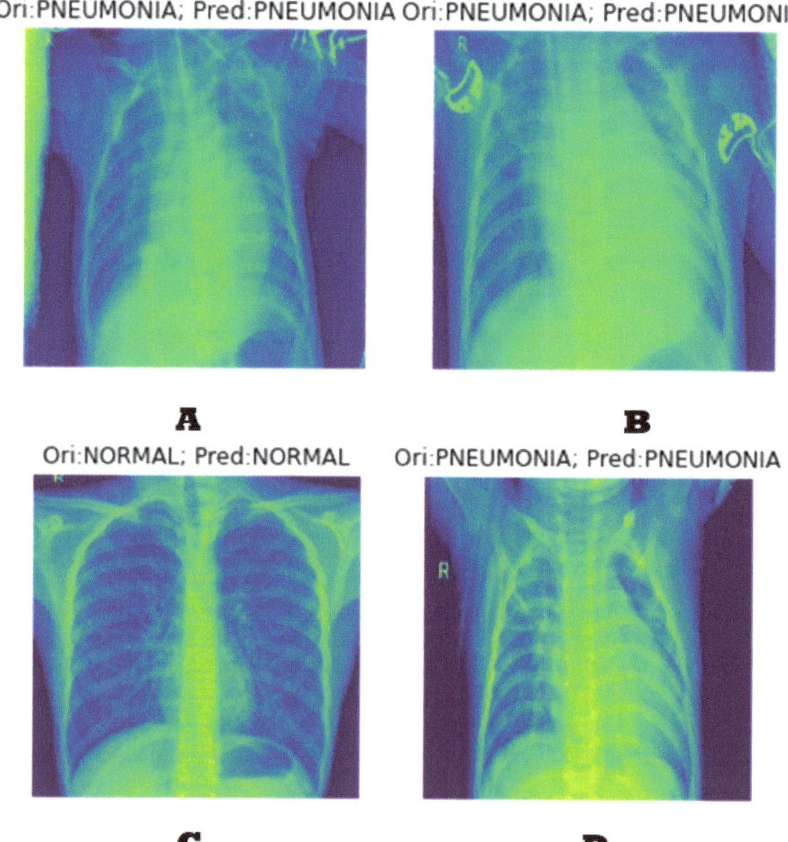

Figure 9. Prediction result samples.

4.3. Performance Evaluation

The confusion matrix also showed promising results, with 192 observed as normal and around 372 pneumonia cases (refer to Figure 10). The ROC graph also shows a positive result with a score of 96.70%, as shown in Figure 11.

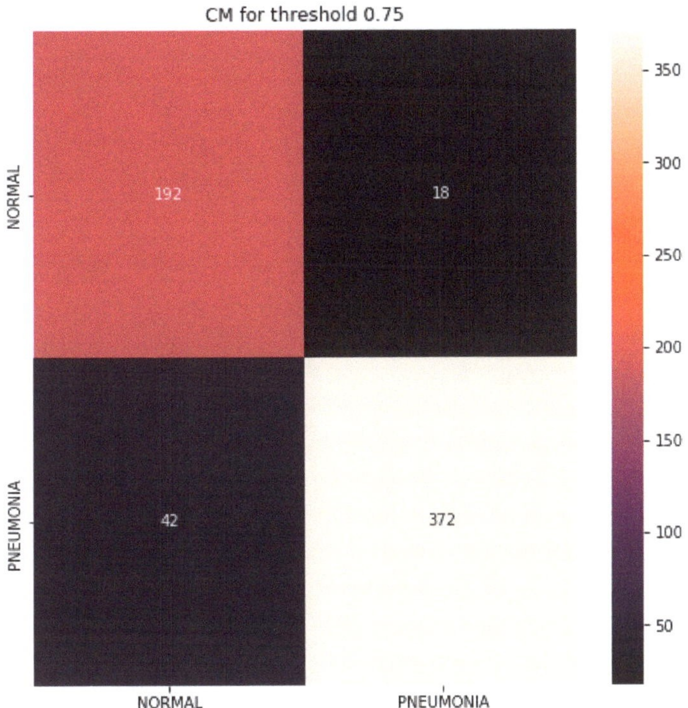

Figure 10. Confusion matrix for the proposed approach.

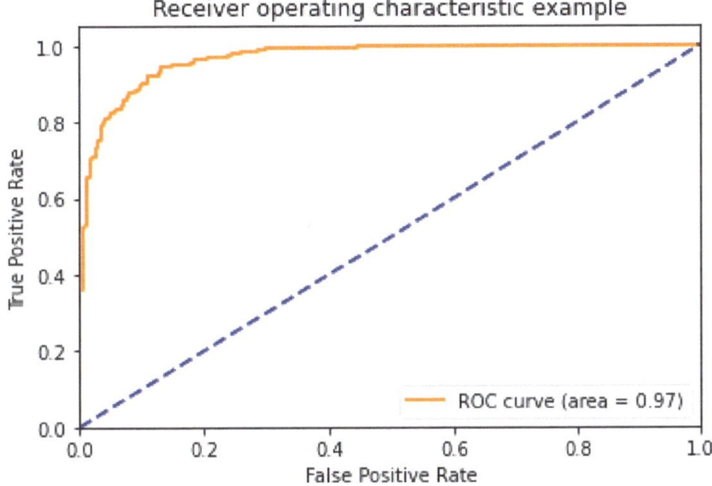

Figure 11. ROC curve.

Table 3 presents the performance analysis of three methods: DCGAN, DCGAN+ inceptionV3, and DCGAN+XGradientBoost. Based on their accuracy, precision, recall, and F1 Score, the result shows that DCGAN+XgradientBoost outperformed the other two methods by gaining an accuracy score of 98.88%, precision score of 99.1%, recall rate of 98.70%, and F1 Score of 99.3%.

Table 3. Performance analysis with various approaches.

Approach	Accuracy	Precision	Recall	F1 Score
DCGAN	96.53	95.78	94.57	95.4
DCGAN+Inception V3	97.86	97.6	97.43	97.2
DCGAN+XGradientBoost	98.88	99.1	98.7	99.3

Figure 12 compares the proposed approach with the other two techniques. It is observed visually that the overall value of DCGAN+Xboost is exceedingly high compared to other techniques with significant scores.

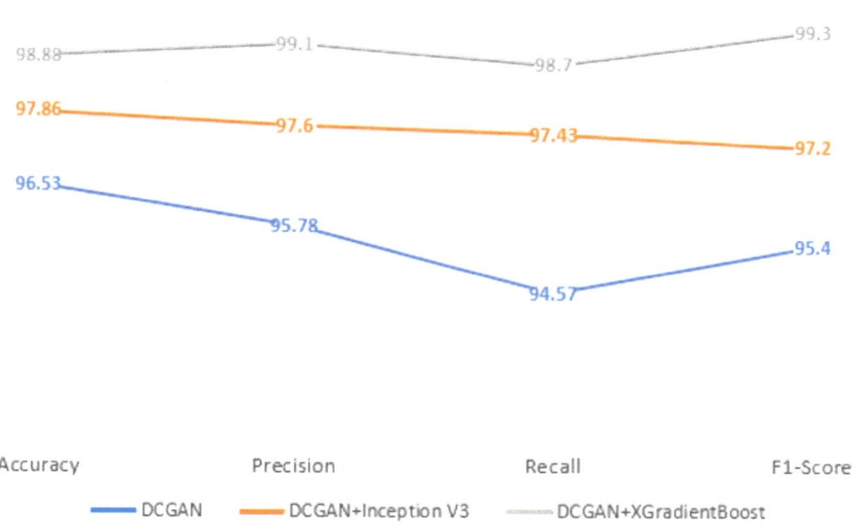

Figure 12. Performance metric illustration of DCGAN-XGradientBoost with DCGAN.

4.4. Analysis and Comparison

In this subsection, the remarks made during training are expressed. The comparison with three previous studies that trained with different strategies is exposed. The comparison of the performance of the proposed model is shown in Figure 13. It is clearly observed that the augmented image using the DCGAN approach with the Inception V3 model outperforms the other approach. Moreover, the comparison performance with the latest work is presented in Table 4.

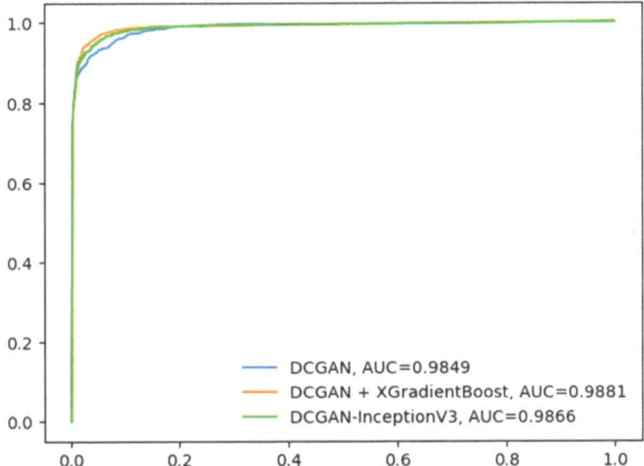

Figure 13. Accuracy rating comparison between DCGAN-XGradientBoost, DCGAN, and DCGAN + Inception V3.

Table 4. Comparison to other methods for chest X-ray classification.

Related Work	Dataset	Method	Accuracy	Precision	Recall	F1
[5]	Chest X-ray	Covinet System + CNN	98.62	95.77	93.66	93.69
[42]	Chest X-ray + Wasserstein GAN	Wasserstein GAN	95.34	99.1	-	-
[43]	Chest X-ray + PGGAN	PGGAN + SMANet	96.28	-	-	-
[44]	Chest X-ray + GAN	DenseNet121 + GAN	80.1	72.7	83.4	79.3
[45]	Chest X-ray + IAGAN	Inception + IAGAN	82	84	69	-
[46]	Chest X-ray	Transfer Learning	98.51	94.1	98.46	98.46
[47]	Chest X-ray	ResNet34 & HRNets	97.02	95.6	98.41	96.98
[48]	Chest X-ray	AlexNet mode	94.18	93.4	89.1	98.9
Proposed Method	Chest X-ray and Augmented Chest X-ray with DCGAN + XGradientBoost	Inception V3 + DCGAN	98.88	99.1	98.70	98.60

According to the findings presented in Table 4, a researcher employed the transfer learning methodology to categorize the chest X-ray pictures, resulting in an accuracy rate of 98.62%. This level of accuracy is comparable to the results reached by another researcher who utilized Covinet and CNN techniques. The AlexNet model has achieved an accuracy rate of 94.18% and a precision rate of around 93.4%. When comparing the suggested strategy to comparable methods, it is evident that PGGAN + SMANet achieves the highest accuracy (96.28%), with Wasserstein GAN coming in second (95.34%). The approach we have provided demonstrates compelling outcomes, achieving an accuracy rate of 98.88%, a precision score of 99.1%, a recall rate of 98.7%, and an F1 score of 98.60.

The outcome is influenced by the enhancement of chest X-ray image quality using the DCGAN technique. The integration of the original and produced DCGAN data via InceptionV3 during training achieves a seamless mix. Additionally, XGradientBoost contributes to the process of selecting feature mappings. According to the data presented in Figure 14, our proposed solution has demonstrated superior performance compared to the other four methods.

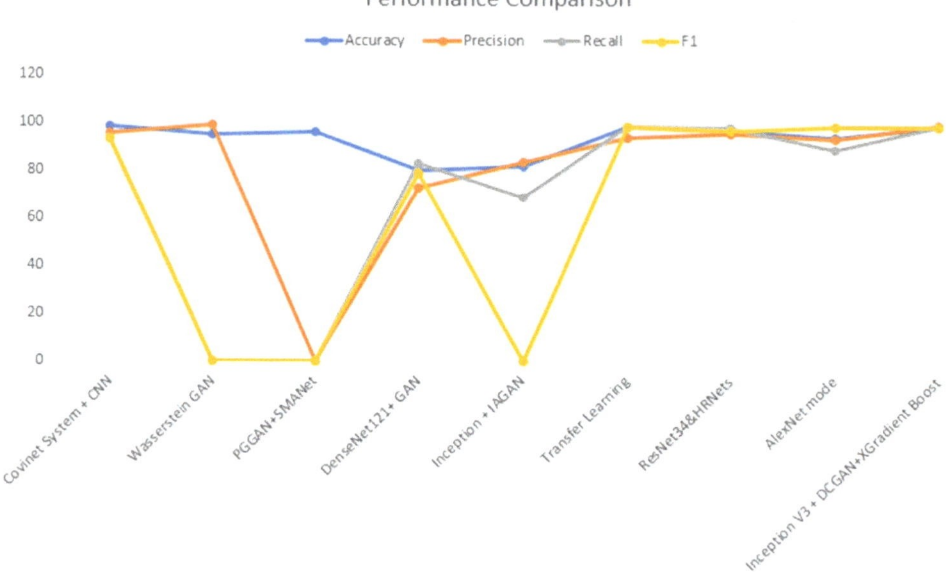

Figure 14. Performance comparison with relevant works.

4.5. Observations about the Experiment

○ By employing a fusion of DCGAN data, the Inception V3 learning model, and XGradient boost for feature mapping, it is possible to enhance both the recognition rate and quality of chest X-ray data.

○ Figure 13 illustrates the ROC curve of the suggested approach, which demonstrates the superior performance of our DCGAN, DCGAN + Inception V3, and DCGAN + Inception V3 + XGradient Boost models compared to previous research findings.

○ Table 3 shows a comparison between the proposed approach and various techniques of data augmentation, such as the DCGAN. The improved suggested method has given a promising improvement to the results.

○ The proposed methodology has demonstrated enhancements in accuracy, precision, recall, and F1 score ranging from 1.1% to 2.1% in a standard experimental setting. The observed enhancement can be attributed to the utilization of a hybrid approach involving the combination of the DCGAN and inception V3, together with the incorporation of XGradient Boost. If a one-to-one comparison is conducted, certain accomplishments may exhibit a 4.6% enhancement.

5. Conclusions

In light of the conclusion of the COVID-19 pandemic, it is currently recognized that the virus has transitioned into an endemic phase. These illnesses have become a significant cause of distress for individuals across various demographic groups. Consequently, a significant cohort of researchers is continuously progressing their investigations in the field of computer vision and artificial intelligence in order to address a diverse array of medical intricacies. The application of chest X-ray data in the prediction of COVID-19 has garnered significant interest among the academic community. The proposed methodology prioritizes the employment of artificially generated images via the implementation of the DCGAN (deep convolutional generative adversarial network) to enhance the overall quality of datasets. The synthetic and actual datasets were combined at a ratio of 39:61. The enhancement of chest X-ray image quality can be achieved via the utilization of the rotation, inspiration, and penetration (RIP) technique. Consequently, a hybrid approach was employed, combining the DCGAN framework, the inception V3 learning model, and XGradient Boost to enhance the accuracy of the prediction. The project utilizes three distinct methodologies: the deep convolutional generative adversarial network (DCGAN), DCGAN integrated with Inception V3, and DCGAN combined with Inception V3 with Xtreme Gradient Boost. The evaluation of a model's performance is greatly influenced by the metrics of accuracy, precision, recall, and F1 score. The examination produced the following results: an accuracy rate of 98.88%, a sensitivity rate of 99.1%, a recall rate of 98.70%, and an F1 score of 98.60%. In order to realize better future performance on real-world case data, potential improvements may involve adjusting the proportion of mixed datasets and supplementing the artificially generated image dataset.

Author Contributions: Conceptualization, A.H.B.; formal analysis, A.H.B. and S.J.M.; methodology, A.H.B. and S.J.M.; writing—original draft, A.H.B.; writing—review and editing, A.H.B., S.J.M. and S.A. All authors have read and agreed to the published version of the manuscript.

Funding: The authors extend their appreciation to the Deputyship for Research and Innovation, Ministry of Education, in Saudi Arabia for funding this research work through project number "IFPRC-216-830-2020" and King Abdulaziz University, DSR, Jeddah, Saudi Arabia.

Institutional Review Board Statement: Not applicable.

Informed Consent Statement: Not applicable.

Data Availability Statement: The data presented in this study are available upon request from the corresponding author. The data are not publicly available as the present dataset necessitates compilation with additional datasets prior to its prospective online publication.

Conflicts of Interest: The authors declare no conflict of interest.

References

1. Health, M.O. COVID-19 Command and Control Center CCC, The National Health Emergency Operation Center NHEOC. 2021. Available online: https://covid19.moh.gov.sa (accessed on 23 October 2021).
2. Worldmeter. Available online: https://www.worldometers.info/coronavirus/country/saudi-arabia/ (accessed on 23 October 2021).
3. Wang, S.; Sun, J.; Mehmood, I.; Pan, C.; Chen, Y.; Zhang, Y. Cerebral micro-bleeding identification based on a nine-layer convolutional neural network with stochastic pooling. *Concurr. Comput. Pract. Exp.* **2019**, *32*, e5130. [CrossRef]
4. Wang, S.; Tang, C.; Sun, J.; Zhang, Y. Cerebral Micro-Bleeding Detection Based on Densely Connected Neural Network. *Front. Neurosci.* **2019**, *13*, 422. [CrossRef] [PubMed]
5. Lafraxo, S.; Ansari, M.E. CoviNet: Automated COVID-19 Detection from X-rays using Deep Learning Techniques. In Proceedings of the 2020 6th IEEE Congress on Information Science and Technology (CiSt), Agadir, Morocco, 5–12 June 2021.
6. Akram, T.; Attique, M.; Gul, S.; Shahzad, A.; Altaf, M.; Naqvi, S.S.R.; Damaševičius, R.; Maskeliūnas, R. A novel framework for rapid diagnosis of COVID-19 on computed tomography scans. *Pattern Anal. Appl.* **2021**, *24*, 951–964. [CrossRef] [PubMed]
7. Jangam, E.; Barreto, A.A.D.; Annavarapu, C.S.R. Automatic detection of COVID-19 from chest CT scan and chest X-rays images using deep learning, transfer learning and stacking. *Appl. Intell.* **2021**, *52*, 2243–2259. [CrossRef] [PubMed]
8. Akyol, K.; Şen, B. Automatic Detection of COVID-19 with Bidirectional LSTM Network Using Deep Features Extracted from Chest X-ray Images. *Interdiscip. Sci. Comput. Life Sci.* **2021**, *14*, 89–100. [CrossRef] [PubMed]
9. Autee, P.; Bagwe, S.; Shah, V.; Srivastava, K. StackNet-DenVIS: A multi-layer perceptron stacked ensembling approach for COVID-19 detection using X-ray images. *Phys. Eng. Sci. Med.* **2020**, *43*, 1399–1414. [CrossRef] [PubMed]
10. Alqudaihi, K.S.; Aslam, N.; Khan, I.U.; Almuhaideb, A.M.; Alsunaidi, S.J.; Ibrahim, N.M.A.R.; Alhaidari, F.A.; Shaikh, F.S.; Alsenbel, Y.M.; Alalharith, D.M.; et al. Cough Sound Detection and Diagnosis Using Artificial Intelligence Techniques: Challenges and Opportunities. *IEEE Access* **2021**, *9*, 102327–102344. [CrossRef]
11. Zhang, R.; Guo, Z.; Sun, Y.; Lu, Q.; Xu, Z.; Yao, Z.; Duan, M.; Liu, S.; Ren, Y.; Huang, L.; et al. COVID19XrayNet: A Two-Step Transfer Learning Model for the COVID-19 Detecting Problem Based on a Limited Number of Chest X-ray Images. *Interdiscip. Sci. Comput. Life Sci.* **2020**, *12*, 555–565. [CrossRef]
12. Kumar, N.; Gupta, M.; Gupta, D.; Tiwari, S. Novel deep transfer learning model for COVID-19 patient detection using X-ray chest images. *J. Ambient. Intell. Humaniz. Comput.* **2021**, *14*, 469–478. [CrossRef]
13. Narin, A. Detection of COVID-19 Patients with Convolutional Neural Network Based Features on Multi-class X-ray Chest Images. In Proceedings of the 2020 Medical Technologies Congress (TIPTEKNO), Antalya, Turkey, 19–20 November 2020.
14. Saif, A.F.M.; Imtiaz, T.; Rifat, S.; Shahnaz, C.; Zhu, W.P.; Ahmad, M.O. CapsCovNet: A Modified Capsule Network to Diagnose COVID-19 from Multimodal Medical Imaging. *IEEE Trans. Artif. Intell.* **2021**, *2*, 608–617. [CrossRef]

15. Fang, Y.; Zhang, H.; Xie, J.; Lin, M.; Ying, L.; Pang, P.; Ji, W. Sensitivity of Chest CT for COVID-19: Comparison to RT-PCR. *Radiology* **2020**, *296*, E115–E117. [CrossRef] [PubMed]
16. Xie, X.; Zhong, Z.; Zhao, W.; Zheng, C.; Wang, F.; Liu, J. Chest CT for Typical Coronavirus Disease 2019 (COVID-19) Pneumonia: Relationship to Negative RT-PCR Testing. *Radiology* **2020**, *296*, E41–E45. [CrossRef] [PubMed]
17. Zhao, D.; Zhu, D.; Lu, J.; Luo, Y.; Zhang, G. Synthetic Medical Images Using F&BGAN for Improved Lung Nodules Classification by Multi-Scale VGG16. *Symmetry* **2018**, *10*, 519. [CrossRef]
18. Thepade, S.D.; Jadhav, K. COVID-19 Identification from Chest X-ray Images using Local Binary Patterns with assorted Machine Learning Classifiers. In Proceedings of the 2020 IEEE Bombay Section Signature Conference (IBSSC), Mumbai, India, 4–6 December 2020.
19. Thepade, S.D.; Chaudhari, P.R.; Dindorkar, M.R.; Bang, S.V. COVID-19 Identification using Machine Learning Classifiers with Histogram of Luminance Chroma Features of Chest X-ray images. In Proceedings of the 2020 IEEE Bombay Section Signature Conference (IBSSC), Mumbai, India, 4–6 December 2020.
20. Qjidaa, M.; Mechbal, Y.; Ben-Fares, A.; Amakdouf, H.; Maaroufi, M.; Alami, B.; Qjidaa, H. Early detection of COVID19 by deep learning transfer Model for populations in isolated rural areas. In Proceedings of the 2020 International Conference on Intelligent Systems and Computer Vision (ISCV), Fez, Morocco, 9–11 June 2020.
21. Darapaneni, N.; Ranjane, S.; Satya US, P.; Reddy, M.H.; Paduri, A.R.; Adhi, A.K.; Madabhushanam, V. COVID-19 Severity of Pneumonia Analysis Using Chest X-rays. In Proceedings of the 2020 IEEE 15th International Conference on Industrial and Information Systems (ICIIS), Rupnagar, India, 26–28 November 2020.
22. Qayyum, A.; Razzak, I.; Tanveer, M.; Kumar, A. Depth-wise dense neural network for automatic COVID19 infection detection and diagnosis. *Ann. Oper. Res.* **2021**, *ahead of print.* [CrossRef] [PubMed]
23. Shah, F.M.; Joy, S.K.S.; Ahmed, F.; Hossain, T.; Humaira, M.; Ami, A.S.; Paul, S.; Jim, A.R.K.; Ahmed, S. A Comprehensive Survey of COVID-19 Detection Using Medical Images. *SN Comput. Sci.* **2021**, *2*, 434. [CrossRef] [PubMed]
24. Shastri, S.; Singh, K.; Kumar, S.; Kour, P.; Mansotra, V. Deep-LSTM ensemble framework to forecast COVID-19: An insight to the global pandemic. *Int. J. Inf. Technol.* **2021**, *13*, 1291–1301. [CrossRef] [PubMed]
25. Shankar, K.; Perumal, E.; Tiwari, P.; Shorfuzzaman, M.; Gupta, D. Deep learning and evolutionary intelligence with fusion-based feature extraction for detection of COVID-19 from chest X-ray images. *Multimed. Syst.* **2021**, *28*, 1175–1187. [CrossRef]
26. Litjens, G.; Kooi, T.; Bejnordi, B.E.; Setio, A.A.A.; Ciompi, F.; Ghafoorian, M.; van der Laak, J.A.W.M.; van Ginneken, B.; Sánchez, C.I. A survey on deep learning in medical image analysis. *Med. Image Anal.* **2017**, *42*, 60–88. [CrossRef]
27. Greenspan, H.; Ginneken, B.V.; Summers, R.M. Guest Editorial Deep Learning in Medical Imaging: Overview and Future Promise of an Exciting New Technique. *IEEE Trans. Med. Imaging* **2016**, *35*, 1153–1159. [CrossRef]
28. Roth, H.R.; van Ginneken, B.; Summers, R.M. Improving Computer-Aided Detection Using Convolutional Neural Networks and Random View Aggregation. *IEEE Trans. Med. Imaging* **2016**, *35*, 1170–1181. [CrossRef]
29. Tajbakhsh, N.; Lu, L.; Liu, J.; Yao, J.; Seff, A.; Cherry, K.; Kim, L.; Summers, R.M. Convolutional Neural Networks for Medical Image Analysis: Full Training or Fine Tuning? *IEEE Trans. Med. Imaging* **2016**, *35*, 1299–1312. [CrossRef] [PubMed]
30. Mikołajczyk, A.; Grochowski, M. Data augmentation for improving deep learning in image classification problem. In Proceedings of the 2018 International Interdisciplinary PhD Workshop (IIPhDW), Swinoujscie, Poland, 9–12 May 2018.
31. Mahase, E. Coronavirus COVID-19 has killed more people than SARS and MERS combined, despite lower case fatality rate. *Br. Med. J.* **2020**, *368*, m641. [CrossRef]
32. Wang, W.; Xu, Y.; Gao, R.; Lu, R.; Han, K.; Wu, G.; Tan, W. Detection of SARS-CoV-2 in Different Types of Clinical Specimens. *J. Am. Med. Assoc.* **2020**, *323*, 1843–1844. [CrossRef] [PubMed]
33. Corman, V.M.; Landt, O.; Kaiser, M.; Molenkamp, R.; Meijer, A.; Chu, D.K.; Bleicker, T.; Brünink, S.; Schneider, J.; Schmidt, M.L.; et al. Detection of 2019 novel coronavirus (2019-nCoV) by real-time RT-PCR. *Eurosurveillance* **2020**, *25*, 2000045. [CrossRef] [PubMed]
34. Lal, A.; Mishra, A.K.; Sahu, K.K. CT chest findings in coronavirus disease-19 (COVID-19). *J. Formos. Med. Assoc. Taiwan Yi Zhi* **2020**, *119*, 1000–1001. [CrossRef] [PubMed]
35. Chest X-ray Pneumonia. Available online: https://github.com/ieee8023/covid-chestxray-dataset (accessed on 25 January 2022).
36. Cohen, J.P.; Morrison, P.; Dao, L. COVID-19 image data collection. *arXiv* **2020**, arXiv:2003.11597. Available online: https://github.com/ieee8023/covid-chestxray-dataset (accessed on 25 January 2022).
37. Hitawala, S. Comparative study on generative adversarial networks. *arXiv* **2018**, arXiv:1801.04271.
38. Ahmadinejad, M.; Ahmadinejad, I.; Soltanian, A.; Mardasi, K.G.; Taherzade, N. Using new technicque in sigmoid volvulus surgery in patients affected by COVID-19. *Ann. Med. Surg.* **2021**, *70*, 102789. [CrossRef]
39. Szegedy, C.; Vanhoucke, V.; Ioffe, S.; Shlens, J.; Wojna, Z. Rethinking the Inception Architecture for Computer Vision. In Proceedings of the Computer Vision and Pattern Recognition, Las Vegas, NV, USA, 27–30 June 2016. Available online: https://arxiv.org/abs/1512.00567v3 (accessed on 4 August 2022).
40. Chen, T.; Guestrin, C. XGBoost: A Scalable Tree Boosting System. Available online: https://arxiv.org/pdf/1603.02754.pdf (accessed on 30 September 2023).
41. Tabik, S.; Gómez-Ríos, A.; Martín-Rodríguez, J.L.; Sevillano-García, I.; Rey-Area, M.; Charte, D.; Guirado, E.; Suárez, J.L.; Luengo, J.; Valero-González, M.A.; et al. COVIDGR Dataset and COVID-SDNet Methodology for Predicting COVID-19 Based on Chest X-ray Images. *IEEE J. Biomed. Health Inform.* **2020**, *24*, 3595–3605. [CrossRef]

42. Hussain, B.Z.; Andleeb, I.; Ansari, M.S.; Joshi, A.M.; Kanwal, N. Wasserstein GAN based Chest X-ray Dataset Augmentation for Deep Learning Models: COVID-19 Detection Use-Case. In Proceedings of the 2022 44th Annual International Conference of the IEEE Engineering in Medicine & Biology Society (EMBC), Glasgow, UK, 11–15 July 2022.
43. Ciano, G.; Andreini, P.; Mazzierli, T.; Bianchini, M.; Scarselli, F. A Multi-Stage GAN for Multi-Organ Chest X-ray Image Generation and Segmentation. *Mathematics* **2021**, *9*, 2896. [CrossRef]
44. Sundaram, S.; Hulkund, N. GAN-based Data Augmentation for Chest X-ray Classification. In Proceedings of the KDD DSHealth, Singapore, 14–18 August 2021.
45. Motamed, S.; Rogalla, P.; Khalvati, F. Data augmentation using Generative Adversarial Networks (GANs) for GAN-based detection of Pneumonia and COVID-19 in chest X-ray images. *Inform. Med. Unlocked* **2021**, *27*, 100779. [CrossRef] [PubMed]
46. Ohata, E.F.; Bezerra, G.M.; das Chagas, J.V.S.; Neto, A.V.L.; Albuquerque, A.B.; de Albuquerque, V.H.C.; Filho, P.P.R. Automatic detection of COVID-19 infection using chest X-ray images through transfer learning. *IEEE/CAA J. Autom. Sin.* **2021**, *8*, 239–248. [CrossRef]
47. Al-Waisy, A.S.; Al-Fahdawi, S.; Mohammed, M.A.; Abdulkareem, K.H.; Mostafa, S.A.; Maashi, M.S.; Arif, M.; Garcia-Zapirain, B. COVID-CheXNet: Hybrid deep learning framework for identifying COVID-19 virus in chest X-rays images. *Soft Comput.* **2020**, *27*, 2657–2672. [CrossRef] [PubMed]
48. Ibrahim, A.U.; Ozsoz, M.; Serte, S.; Al-Turjman, F.; Yakoi, P.S. Pneumonia Classification Using Deep Learning from Chest X-ray Images During COVID-19. *Cogn. Comput.* **2021**, *ahead of print*. [CrossRef]

applied sciences

MDPI

Article

Connection-Aware Heuristics for Scheduling and Distributing Jobs under Dynamic Dew Computing Environments

Pablo Sanabria [1,2,*], Sebastián Montoya [1,2], Andrés Neyem [1,2,*], Rodrigo Toro Icarte [1,2], Matías Hirsch [3] and Cristian Mateos [3]

1   Computer Science Department, Pontificia Universidad Católica de Chile, Macul 7820436, Región Metropolitana, Chile; simontoya@uc.cl (S.M.); rntoro@uc.cl (R.T.I.)
2   Centro Nacional de Inteligencia Artificial CENIA, Macul 7820436, Región Metropolitana, Chile
3   ISISTAN-UNCPBA-CONICET, Tandil CP 7000, Buenos Aires, Argentina; matias.hirsch@isistan.unicen.edu.ar (M.H.); cristian.mateos@isistan.unicen.edu.ar (C.M.)
*   Correspondence: psanabria@uc.cl (P.S.); aneyem@uc.cl (A.N.)

Abstract: Due to the widespread use of mobile and IoT devices, coupled with their continually expanding processing capabilities, dew computing environments have become a significant focus for researchers. These environments enable resource-constrained devices to contribute computing power to a local network. One major challenge within these environments revolves around task scheduling, specifically determining the optimal distribution of jobs across the available devices in the network. This challenge becomes particularly pronounced in dynamic environments where network conditions constantly change. This work proposes integrating the "reliability" concept into cutting-edge human-design job distribution heuristics named ReleSEAS and RelBPA as a means of adapting to dynamic and ever-changing network conditions caused by nodes' mobility. Additionally, we introduce a reinforcement learning (RL) approach, embedding both the notion of reliability and real-time network status into the RL agent. Our research rigorously contrasts our proposed algorithms' throughput and job completion rates with their predecessors. Simulated results reveal a marked improvement in overall throughput, with our algorithms potentially boosting the environment's performance. They also show a significant enhancement in job completion within dynamic environments compared to baseline findings. Moreover, when RL is applied, it surpasses the job completion rate of human-designed heuristics. Our study emphasizes the advantages of embedding inherent network characteristics into job distribution algorithms for dew computing. Such incorporation gives them a profound understanding of the network's diverse resources. Consequently, this insight enables the algorithms to manage resources more adeptly and effectively.

Keywords: dew computing; reinforcement learning; connection-aware scheduling; mobility models; heuristics; transfer learning; simulation

Citation: Sanabria, P.; Montoya, S.; Neyem, A.; Toro Icarte, R.; Hirsch, M.; Mateos, C. Connection-Aware Heuristics for Scheduling and Distributing Jobs under Dynamic Dew Computing Environments. *Appl. Sci.* **2024**, *14*, 3206. https://doi.org/10.3390/app14083206

Academic Editors: Yirui Wu and Shaohua Wan

Received: 6 March 2024
Revised: 4 April 2024
Accepted: 5 April 2024
Published: 11 April 2024

1. Introduction

The significant surge in computation-intensive tasks within mobile applications has placed substantial computational burdens on devices with limited resources in recent years. Despite advancements in hardware for devices like smartphones and IoT devices, they frequently struggle to meet the demands of these resource-intensive tasks. In response to this issue, dew computing has emerged as a solution, aiming to alleviate the situation by offloading computation-intensive tasks to more potent devices, typically those nearby. In this context, "nearby devices" are defined as devices connected to the same local network [1–5]. The central premise is to transfer tasks, for example, from a smartphone to a laptop, leveraging the increased processing power of the latter. This approach not only results in faster task execution but also prevents excessive depletion of the smartphone's battery resources. While dew computing holds promising potential, its practical implementation hinges on addressing a critical challenge: the efficient job distribution across nearby devices.

This study focuses on resolving the job distribution challenge within dew environments. Dew environments comprise a collection of devices interconnected within a local network. These devices exhibit variations in their attributes and capabilities, encompassing factors such as power source, battery capacity, processor/CPU core count, storage capacity, and sensory capabilities. Furthermore, users may interact with selected devices at diverse points in time. To ensure the efficient distribution of tasks within a dew environment, it becomes imperative to consider all these factors comprehensively.

Current methods for distributing jobs in dew environments follow human-designed heuristics. These heuristics balance the workload among devices by following predefined rules. Some of these rules combine devices' features such as computing capability, the remaining battery level, current queued jobs, and job requirement information to select the most appropriate device to be assigned with a job. Some examples include the *Enhanced Simple Energy-Aware Scheduler (E-SEAS)* [6] and the *Batch Processing Algorithm (BPA)* [7]. *Round Robin (RR)* [8] is also considered for distributing jobs, quite frequently as a baseline method for new proposals. Unfortunately, the performance of these methods remains unknown when the network topology dynamically changes over time due to network conditions. Consequently, they lead to suboptimal decision making and the wastage of valuable resources.

Exploring the benefits of considering network-related parameters as part of job distribution algorithms, we propose a reliability score that accounts for the time a device—also called a node—is connected to a dew environment. The score is integrated into previously proposed human-designed heuristics and thoroughly evaluated. The results obtained encourage us to incorporate the reliability score into the learning process of a reinforcement learning (RL) agent.

Prior research [9] highlights that RL effectively accommodates dew computing environments [10]. RL is a subfield of artificial intelligence that studies how to develop agents that can learn optimal behavior by interacting with an environment. Every interaction with the environment delivers a reward signal that the agent seeks to maximize. The agent improves its current policy (mapping observations to actions) by learning from past experiences to accomplish this. Powered by deep learning, RL agents have been used to solve complex decision-making problems across different research areas, from robotics [11] to conversational agents [12] to molecule discovery [13]. Here, we propose letting an RL agent learn how to distribute jobs effectively in dew environments with nodes' mobility presence.

The main contributions of our work are, then, as follows. First, we extend EdgeDewSim [7] with all necessary features to support dynamic changes in the network conditions as a consequence of nodes' mobility. The new features include the capability to generate reusable connection–disconnection traces per node. Trace reusability is the key to experiment reproducibility. Second, we proposed an easy-to-implement "reliability" score that gives E-SEAS and BPA network condition awareness and boosts the performance achieved in dew environments with nodes' mobility. Such integration derives into new heuristics that we call ReleSEAS and RelBPA. Third, we propose an RL approach with awareness of network conditions, and we demonstrate that using this approach can outperform human-designed heuristics.

Finally, as a summary of our empirical findings, we discovered that giving awareness of node connection activity to the scheduler gives better endurance and adaptability when the network conditions are dynamic. Also, we show that providing awareness of the policies improves the job completion rate by up to 95% without giving up performance.

This paper is organized as follows: In Section 2, we introduce the necessary concepts of dew computing in the context of our work; in Section 3, we make a review of the state of the art related to scheduling algorithms. In Section 4, we describe the necessary changes to extend the simulation capabilities to represent dynamic network-related features as an effect of nodes' mobility. Section 5 describes the mobility models used in this work. In Section 6, we show the definitions of the proposed algorithms for this job. In Section 7, we describe the experimentation methodology and the metrics used to compare the performance of our defined algorithms, followed by our results with their respective discussion. Finally, in Section 8, we show our findings and propose the next steps toward improving this field of study.

2. Edge and Dew Computing

Mobile edge computing is a paradigm that seeks to solve latency and network traffic problems found in mobile cloud computing environments [14]. Ref. [15] defines edge computing as *"a model that allows a cloud-based computing capacity providing services making use of the infrastructure that is on the edge of the network"*. In this way, mobile edge computing allows servers or workstations within a computer network, thus ensuring low latency while enabling the efficient processing of information, which permits the deployment of more robust applications. Furthermore, this paradigm lets edge servers work with other nodes in proximity or collaborate with cloud services, thus deploying much larger and more efficient applications [16,17]. However, while edge computing helps to reduce the network's problems, it still depends on the network backbone that may not be available or reachable in certain situations, like working with IoT devices in mines and on ships, in deserts, or on moving vehicles.

Dew computing is a new paradigm where connected devices offload jobs to nearby devices in the same network. This paradigm proposes an architecture that reduces network latency, the energy cost of remote data communication, and the costs inherent to cloud infrastructure usage [18]. Through this, dew computing optimizes the usage of mobile and IoT devices in two manners. First, it treats mobile devices as clients in the network infrastructure to offload their work to other devices in the same network [19]. Second, dew computing considers mobile and IoT devices as resources to increase the available computational power from an existing system. In this approach, one device can offload its work onto another available device in the network (including other mobile and IoT devices) [20,21].

We note that a network topology is needed to use mobile and IoT devices as resources in a local network [3]. The *Smart Cluster at the Edge (SCE)* is a network topology commonly used for that purpose in dew computing. Figure 1 shows how the devices are organized in this type of network. This topology can be established wherever an access point and a group of mobile and IoT devices coexist. The topology's main feature is a central scheduler coordinating the task assignment among the network's available resources. This central scheduler can be any capable device in the network [22]. In this work, we address the problem of distributing jobs in a dew environment, assuming that the network topology is an SCE.

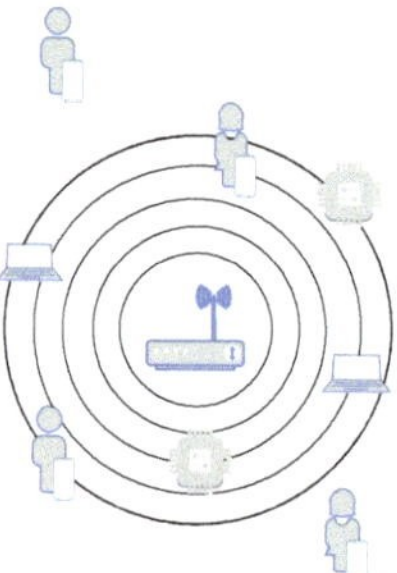

Figure 1. Example of SCE architecture: In this example, six devices are connected to the local network. The devices include two smartphones, two Raspberry Pis, two laptops, and one scheduler inside the network. Outside of the network are two smartphones.

3. Related Work

Distributing jobs in dew environments *optimally* is a challenging combinatorial problem [6,23]. Many factors must be taken into account to assign jobs to devices correctly. Those factors range from a device's CPU speed to a device's availability within a time window, from how often jobs arrive at the scheduler to how urgent jobs must be completed. Many previous works have proposed heuristic methods to deal with this complex problem. A heuristic method is a practical approach to obtaining approximated yet satisfactory solutions to problems where finding the optimal solution is computationally intractable, in this case, the job assignment problem. This paper seeks human-designed heuristics that consider features from the SCE and are aware of network changes through time. Also, we explore an alternative approach, which consists of *learning* a policy to distribute jobs using RL. To provide a coherent view of the related work, we divided this discussion into two parts: We first discuss existing heuristic methods to distribute jobs in an SCE and then review previous works at the intersection of RL with cloud and edge computing.

In the context of SCEs, different algorithms have been proposed to optimize the system's utility and job execution time, using the mobile devices' remaining battery as a formal constraint of the resource allocation problem formulation [24–28]. These algorithms, however, assume complete and accurate information regarding the energy spent and the execution time for every candidate node, making it challenging to apply in real-life scenarios.

In Refs. [29,30], they addressed the previous limitation by proposing algorithms that do not rely on complete information. This approach seeks to exploit the nodes' proximity and cost-effectiveness of node transferring capabilities. However, their studies focused on analyzing the effect of nodes' mobility rather than balancing the load to efficiently utilize the battery and processing power of the available resources.

Refs. [6,22] presented other types of heuristics, which distribute jobs by considering the mobile device's battery level and computing scores obtained from benchmarks. These methods outperformed traditional scheduling algorithms (such as round robin) but have the limitation of only considering battery-dependent devices. This problem was recently addressed by Ref. [7]. In Ref. [7], the authors studied hybrid SCEs, which combine both battery-dependent and non-battery-dependent devices, and proposed heuristic methods that consider the device's battery level, computing score, and current workload to distribute jobs in dew environments.

In contrast to those works, where the primary assumption was that the network topology remained static, meaning that the structure and connections within the network did not change over time, this work takes a different approach. Our research recognizes the limitations of static topologies, particularly in real-world scenarios where network environments are often dynamic and subject to change. By "dynamic environments", we specifically refer to situations where mobile devices can enter or exit the boundary of an SCE over time, leading to fluctuations in the network's composition and connection throughput. To address these challenges, we propose new heuristic algorithms specifically designed to adapt to network dynamic conditions. To achieve this, we improved the simulation support offered by EdgeDewSim [7], enhancing its capability to support the development of new connection-aware job distribution as the topology evolves, configure nodes' mobility in an easy-to-reproduce manner, and represent network changes as an effect of nodes' mobility. By incorporating these advancements, our approach aims to provide a more robust and flexible solution for job distribution in dynamic network environments.

RL Methods in Edge and Dew Computing

In cloud computing, task scheduling focuses on deciding which resources (servers) should process an incoming task. In this problem setup, previous works have shown that RL agents can reduce the execution time in distributed systems [31] and avoid overloading (and deadlocking) cloud servers [32]. Furthermore, Ref. [33] showed that deep RL can help schedule tasks in large-scale cloud service providers, surpassing traditional methods in terms of energy cost and reject rates.

As for edge computing, the task scheduling problem is similar. The edge server receives tasks and has to decide whether to send those tasks to a nearby server or the cloud. Several works have explored how to distribute jobs in edge computing using deep learning techniques like convolutional neural networks or reinforcement learning according to different performance metrics. These metrics include reducing computing times [34,35], energy consumption [36], latency [37,38], task failure rate [39], or a combination of the previous [40–42]. We note that in edge computing, the action space is usually limited. For instance, sometimes the agent can only decide whether to send (or not) a job to the cloud [41]. In contrast, the action space in a dew environment typically ranges from tens to hundreds (i.e., one action per device), making the decision-making problem considerably harder.

RL has been increasingly utilized for job scheduling in environments that are closely related to one another, such as cloud computing [31–33] and edge computing [34,36–42]. With emerging studies in dew computing, RL has been applied in cloud and edge computing for job scheduling. For instance, in Ref. [43], they propose a vehicular dew computing architecture using RL for content delivery optimization. Similarly, in Ref. [44], they suggest a dew computing-based vehicular edge caching architecture to enhance vehicle stability and high-definition map acquisition. In Ref. [45], they introduce dew quantum machine learning (DewQML) to improve decision making and reduce latency in dew computing environments. In Ref. [46], they present a dew computing-based microservice execution (DoME) scheme using RL to minimize service delay and optimize costs in mobile edge computing.

Additionally, research shows deep RL agents can effectively offload jobs in dew computing environments. Finally, in the context of job offloading in dew computing environments, prior research demonstrates that deep RL agents can offload jobs more effectively than traditional state-of-the-art heuristic methods, even when faced with previously unseen scenarios. The study conducted by [9] empirically proves that the agent learns to generalize in network environments that lack dynamic components when continuously exposed to new situations. This means that the agent can appropriately distribute sequences of jobs that arrive in patterns and sizes not encountered during training. Furthermore, the agent can learn to effectively distribute jobs in fixed dew environments, significantly outperforming state-of-the-art heuristics regarding the number of instructions executed per second. This highlights the potential of RL in enhancing the efficiency and adaptability of job scheduling in dew computing environments, paving the way for more responsive and robust computing systems [9].

In the same line of Ref. [9], there are three main differences between our work and the existing work on RL with cloud and edge computing. We address a different problem. Distributing jobs in a dew environment has challenges that do not typically arise in cloud or edge computing. For instance, in dew computing, some devices might run out of battery, or users might start interacting with them. When the network conditions are dynamic, the availability of the devices cannot be assured. Taking these elements into account is a nontrivial task. Second, we thoroughly examine the transfer learning capabilities exhibited by the policies acquired by the reinforcement learning agent. This is a crucial step toward applying RL methods in natural systems since, in practice, it is unlikely that the agent would encounter scenarios that it had previously seen during the training. Finally, we use an RL agent better suited for generalization than the agents used in previous works [47,48].

4. EdgeDewSim Extended Simulator

The EdgeDewSim Simulator [7] was built as an extended version of DewSim [49], in which new features were added. It incorporates the notion of worker nodes that do not depend on batteries to cooperate in job execution, i.e., edge nodes without energy constraints. In the present work, we extended EdgeDewSim with the functionality to model the effect of node mobility on connectivity status. We will refer to the latter extension, i.e., EdgeDewSim, which has mobility functionality, as the EdgeDewSim Extended Simulator.

The previous version of the EdgeDewSim Simulator [7,9] did not have support for allowing devices to connect and disconnect from the network; because of this, the first task of this work was to extend the EdgeDewSim Simulator to include this new feature. The extended simulator version allows researchers to provide a connection file, in the form of a trace, for each device containing all the connection and disconnection events the device will go through during the simulation. The connection file provided to the simulator uses the following syntax:

```
# Connection Event; Event Time (ms)
ENTER_NETWORK;1000
LEAVE_NETWORK;2000
```

where the field before the semicolon indicates the device status for the SCE, and the field after the semicolon refers to the time (expressed in milliseconds) when the status change event will occur. Such a time is always relative to the time zero of the simulation. More explicitly, a device configured with the connection file given in the example will join the SCE a second after the simulation starts and leave the SCE two seconds after the simulation begins, providing a permanence time of one second within the SCE.

In the EdgeDewSim Simulator [7], devices were configured to connect to the network as soon as the simulation started, and there were no events related to device disconnection due to mobility. Instead, devices would only disconnect from the cluster when their battery was depleted. This behavior was implemented through a series of events, including device joining, device leaving, and state of battery updates, to which a scheduler entity would subscribe to maintain an up-to-date list of candidate devices for executing jobs.

Two main features must be addressed to accommodate device connection and disconnection due to mobility. Firstly, devices should be able to discharge their battery even when they are not part of the cluster or network. This adds a layer of realism to the simulation, reflecting the natural battery consumption of mobile devices in real-world scenarios. Secondly, the schedulers require a memory mechanism to keep track of potential devices that can execute a job every time they connect to the network. This memory should also be updated to remove devices from the list whenever they disconnect from the network. This feature is crucial for efficiently managing the dynamic nature of mobile device clusters, as it ensures that the scheduler has an accurate pool of devices to choose from for job execution.

The EdgeDewSim Extended Simulator incorporates the above-mentioned features and offers backward compatibility for conducting experiments before implementing these features. This is achieved by using connection files in which devices are set to connect at the start of the simulation and remain connected throughout without any disconnection events caused by mobility. This simplifies the simulation environment and allows for testing other aspects of the system without the added complexity of the device mobility feature. However, incorporating the above-mentioned features is essential for a more comprehensive and realistic simulation of mobile device clusters.

4.1. Device Connection Score

The scheduler must determine which device is more suitable for executing a job. To start considering the connection and disconnection, it is necessary to add logic to the device to report a connection score as a heartbeat. This concept will be explained in more detail in Section 6; nevertheless, it is relevant to know that other authors [50] have used WiFi to predict the length of a connection to an access point, geo-located tags [51] to indicate the following location of the device, and other sources of data from the device such as battery state, WiFi signal strength, and GPS traces, among others.

4.2. Device Connection Event

To address this issue, certain events were modified, and new ones were created. The event type *DEVICE_START* was modified to solely initiate CPU usage and battery discharge of the device upon joining an SCE. Before the modification, such an event causes the device to be added to the scheduler's list, and the device's state is reported to the scheduler. With the modification, two new event types were introduced. The $DEVICE\_CONNECT$ event now manages the actions previously performed with a $DEVICE\_START$ event and updates the connection score from Section 4.1.

4.3. Device Disconnection Event

The event $DEVICE\_DISCONNECT$ removes the device from the scheduler's list of possible devices that can execute jobs. Also, the device stops sending the scheduler updates of its battery state of charge. The $DEVICE\_DISCONNECT$ also needs to handle what happens to the jobs being executed by the device at the moment of disconnection and the ones that are queued for future execution; in this implementation, these jobs are canceled when the device disconnects from the network. This could be improved or changed in future work to simulate that jobs are executed in the background and results are cached and sent once the connection is reestablished; it all depends on the scenario being tested. Regarding the network model, this had to be modified to accept the disconnection of devices and be able to cancel the transfer of messages related to the device disconnecting from the network.

5. Human Mobility Modeling

Humans constantly move to our jobs, universities, schools, or any other destination daily. However, this behavior is nothing new; throughout history, humans have migrated between different territories countless times, as mentioned in Ref. [52]. A significant change between a few decades ago and now is the high usage of smart devices within the population; this growth in usage is mainly due to the high adoption of smartphones in our lives [53,54].

This characteristic of the current state in the adoption of smartphones has attracted the interest of scientists in researching human mobility, mainly because now we have the tools to collect accurate data that can help to create models that can mimic our mobility patterns. Nowadays, there are multiple ways to obtain data from human mobility. GPS traces are one of these sources for mobility patterns. Also, CDRs (called detail records) have been used to learn from human mobility, and, finally, from some datasets of connection to an access point like a WiFi [55,56] point. In this field of research, although there has been significant progress in terms of data sources, the use of synthetic data from mobility models continues to be prevalent. This is because synthetic data remain a convenient method for studying and proposing new mobility models.

As mentioned in Ref. [57], we can divide human mobility models into nonsocial, social, and hybrid categories. Others [53] divide them into movement-based, linked-based, and network-based. Individual or movement-based mobility models are generated from individual traces of either GPS, CDR, or any other data source that can describe how a human moves without the influence of other humans over his or her patterns. On the other hand, social, network-based, or linked-based models try to model human mobility patterns, considering the different interactions we, as humans, have and how these social interactions can affect our mobility patterns [53]. In this work, we will focus on individual or movement-based mobility models, notably random walk and random waypoint, to generate the connection and disconnection events from an access point. We mainly chose these two models to generate synthetic data because they represent the most dynamic scenarios, since nodes tend to move randomly. This way, we could expose our proposed human-designed heuristics (RelBPA, ReleSEAS) and the RL agent to hostile environments. To develop these data, we used PyMobility [58] as a base for the mobility algorithms and added a layer of code to check the connections and disconnections from the devices. Then, these data were used as input for the EdgeDewSim Extended Simulator.

5.1. Random Walk

Random walk is the most simple mobility algorithm [57], where an individual's movements are simulated by a statistical model, which sets an individual in a particular scenario defined by its minimum and maximum displacement velocity and the rotation angle. This algorithm randomly chooses the individual's velocity at each step and the direction, which can be in the range of $(0, 2\pi)$; this process repeats at every simulation step. This model is the most basic one because it is far from representing realistic human movements that are not random at all.

5.2. Random Waypoint

Random waypoint is a variation of the previous model [57,59], with the difference that now an individual moves to a waypoint where it waits a certain amount of time until it can choose randomly its speed and direction.

6. Connection-Aware Scheduling Heuristics

This paper proposes two new scheduling heuristics, ReleSEAS and RelBPA, based on previously developed heuristics, E-SEAS and BPA.

6.1. Reliability Score

To ensure that the stability of the devices in the network is considered, we suggest introducing a new score that the heuristics should consider. This score, similar to the battery level, aims to influence the decision-making process used by the heuristic. We refer to this score as the "reliability score", denoted by *Reliability*. It is calculated as follows:

$$Reliability = \begin{cases} curr\_execution\_time & \text{if } not\,first\_disconnection \\ avg\_connection\_time & \text{otherwise} \end{cases} \tag{1}$$

where $first\_disconnection$ is a variable that each device has to express if it has disconnected from the network for the first time; the variable $curr\_execution\_time$ represents the amount of time in milliseconds executed since the start of the execution to the time the *Reliability* score is calculated; and finally, $avg\_connection\_time$ corresponds to the average time that the device has been connected to the network, taking into consideration these connection events can happen multiple times over a simulation and that the metric is calculated only for the devices connected to the network.

The following example demonstrates the calculation of the reliability score. In a hypothetical case, we have two devices. The first one has the subsequent connection and disconnection events: $t = 1$ (connect), $t = 2$ (disconnect), $t = 5$ (connect), $t = 10$ (disconnect). The second device connects at $t = 0$ and disconnects at $t = 10$. Assuming the simulation clock is at $t = 11$, the reliability score for the first device will be $3 = (1 + 5)/2$ and 10 for the second device. In this example, we can see what the reliability score aims to incorporate as new information about how stable a device is for the time it plays the role of a worker node within a cluster. The proxy calculates the reliability score when a job arrives and needs to be distributed; in this way, it can obtain the ranking for each device. Also, every time a device disconnects, its last connection duration is stored to be used for the following calculation of the reliability score.

6.2. ReleSEAS

The main feature of E-SEAS [6] is that it is easy to implement in real-life environments because it requires easy-to-obtain information, such as battery level, nodes' computing capability measured in FLOPs (float point operations per second), and remaining jobs as nJobs to build a rank of devices joined to the cluster. Then, the E-SEAS ranking formula is defined as follows:

$$E - SEAS = \frac{flops \cdot SOC}{nJobs} \tag{2}$$

A scheduler applying E-SEAS criteria assigns jobs using this ranking. Every incoming job triggers a recalculation of the ranking, which uses fresh information about jobs' and nodes' statuses, such as job queue size and battery level at each node. Job counters and node status records are maintained by a centralized component called the proxy, from where the scheduler operates. This component keeps track of all information necessary to recalculate the ranking. In this paper, we propose incorporating a new score (reliability score) into the E-SEAS ranking formula to boost the ranking of devices connected to the network for extended periods. This modification derives into what we call ReleSEAS:

$$ReleSEAS = \frac{flops \cdot SOC \cdot Reliability}{nJobs} \tag{3}$$

where $flops$ is the device capabilities (processor speed) measured in float point operations per second, SOC is the last report of the state of charge (battery level), and the reliability is calculated using Equation (1).

6.3. RelBPA

BPA [7] is a scheduling algorithm that considers the jobs loaded onto devices and the present hardware capabilities to select the best device to execute an incoming task. The ranking formula for BPA is defined as follows:

$$BPA = \frac{\Sigma OP\_jobs}{flops \cdot Battery} \tag{4}$$

where OP_jobs is the current job load in a device's queued jobs in terms of how many operations are needed to finish those jobs; $flops$ is the device capability (processor speed) measured in floating-point operations per second; and $Battery$ is the remaining battery of the device expressed in values between the range of $0 < Battery \leq 1$ [7].

The reliability batch processing algorithm (RelBPA) is an adaptation of BPA that incorporates the reliability score introduced in Section 6.1. In the same way as ReleSEAS, the purpose of adding the reliability score is to boost devices that have been connected to the network for more extended periods; the reliability score is added to the denominator to make the metric as small as possible. The ranking formula for RelBPA is defined as follows:

$$RelBPA = \frac{\Sigma OP\_jobs}{flops \cdot Battery \cdot Reliability} \tag{5}$$

where the variables used in the formula have the same meaning as in BPA, and the reliability score is calculated using Equation (1).

6.4. Connection-Aware Reinforcement Learning Agent

In this section, we discuss how we integrate dew computing with RL. First, we formally define the problem of job scheduling in a dew environment. Then, we describe how to solve such a problem using RL. Finally, we discuss the software architecture behind our proposed solution, combining the EdgeDewSim Extended Simulator with the RL framework OpenAI Gym v0.21.0 [60]. But first, we briefly discuss the notation that we use in this section.

6.4.1. Notation

Below, we use the following notation. We use uppercase letters to refer to sets of elements and lowercase letters to refer to individual elements in those sets. For instance, we use J to denote the possible jobs that arrive in the dew environment and $j \in J$ to indicate one particular job in J. In addition, $|J|$ denotes the number of elements in J. Elements in a set have different features. To refer to the value of a feature, we use $x.feature$. For instance, j.ops refers to the number of giga-operations of job $j \in J$.

6.4.2. Problem Definition: Job Scheduling in Dew Computing

We tackle the problem of distributing jobs in a dew environment. The dew environment consists of devices connected to a local network. Some of these devices can be *IoT devices* that have an unlimited power supply (e.g., Raspberry Pis and personal computers), and others are *mobile devices* with a limited battery (e.g., tablets and smartphones). In addition, one node in the local network is assigned as the *scheduler*. The scheduler's purpose is to receive and distribute jobs among the devices.

Once a job is assigned to a device, the time it takes to complete (and how much battery it consumes) depends on the job's features and the device's current state. For instance, since some devices have limited battery life, a job might not be completed if assigned to a device with a low battery level. Users might also interact with the devices, and the local network might have congestion issues. The scheduler must consider all these factors to distribute jobs effectively.

In more detail, assigning a job to a device adds that job to the device's job queue. Then, the device will keep running the jobs in its queue, one by one, until they are completed or running out of battery. Once a device runs out of battery, all the jobs in its current queue are discarded. If a job is assigned to a device that has no battery, that job is also discarded. In our experiments, a job is not reassigned to a different device when the job is discarded. This forces the scheduler to be extra careful when assigning a job to a battery-dependent device.

To evaluate the effectiveness of a given job distribution, we use the *giga-instructions per second (GIPS)* that are completed in the dew environment:

$$\text{GIPS}(t_0, t) = \frac{\sum_{j \in J_c} j.\text{ops}}{t - t_0}, \tag{6}$$

where $J_c \subseteq J$ is the subset of jobs that were completed within the time interval $[t_0, t]$ and $j.\text{ops}$ is the number of giga-operations of job $j \in J_c$. Intuitively, GIPS measures how many operations are completed in a unit of time. The scheduler aims to distribute the jobs to maximize the GIPS in the dew environment.

In this study, we postulate that the dew computing environment is subject to a limitation regarding the maximum number of devices that can be integrated into the network at various intervals. This constraint results from the restrictions imposed by the size of the RL model, which is crucial for optimizing network performance and decision making and cannot accommodate an undefined number of devices. Understanding the implications of this restriction is essential for managing the network's scalability and ensuring that the dew environment can efficiently handle the dynamic addition of devices over time.

6.4.3. Environment Definition: States, Actions, and Rewards

The first step in applying RL in dew computing is to define the environment in which the agent will interact adequately, that is, to define the states S of the environment, the actions A that the agent can perform, and the reward signal $r(s, a, s')$ that the agent will optimize for. We also have to define the transition probabilities $p(s'|s, a)$ if the agent does not interact with a real system (or with a predefined simulator). Whether the agent succeeds or fails at distributing jobs in a dew environment partially depends on how we define those four elements: S, A, r, and p.

To define a dew environment in terms of S, A, r, and p, our starting point was EdgeDewSim Extended Simulator. As presented in Section 4, EdgeDewSim Extended Simulator inherits the simulation features of DewSim and EdgeDewSim, including the energy consumption of each device, battery-dependent and non-battery-dependent devices as contributors of an SCE, the job executions, and user behaviors, and, in this work, it incorporates features to represents network status changes due to nodes' mobility. Thus, the transition probabilities p in the environment are modeled by EdgeDewSim Extended Simulator.

To define the state space S, we considered two critical features of RL. First, most RL agents do not have memory. They learn a policy $\pi(a_t|s_t)$ that makes decisions purely based on the information available in the current state $s_t \in S$. Thus, s_t must include all the relevant information the agent needs to assign a job to the correct device. Second, we note that any information identical for all the states $s \in S$ is useless for the agent. The reason is that the agent has to discriminate whether action a is a good action in state s. If a subset of information $x \subset s$ is identical for all states, then x provides no discriminatory information to the agent.

We then define the state space as follows: $S = D \times J \times C$, where $(d, j, c) \in S$. Specifically, $d \in D$ contains information on the current state of each device. This information includes, for each device, its CPU usage percentage, its remaining battery, the network strength, the connection status, and its current job queue. Then, $j \in J$ provides information about the job the scheduler must assign next. This information includes the job's ops, input size, and output size. Finally, $c \in C$ contains general statistics about the previously completed jobs. These statistics include the number of jobs that have arrived, the number of jobs that have been completed, the sum of the jobs' ops that have been completed, and the elapsed time since the beginning of the simulation (i.e., $t - t_0$). We note that $s \in S$ comprises all the necessary information to correctly assign job j to a device in a fixed dew environment. If the environment is not fixed, we might want to add some additional features, such as the number of instructions per second a device can execute (or its battery capacity). We discuss this further in Section 8.

Our definition of the action space A is as expected. We define one possible action per device in the dew environment. Then, whenever the agent executes action $a_i \in A$ given the current state $(d, j, c) \in S$, the job j is assigned to the device associated with action a_i.

Finally, the reward function is equal to the number of successful jobs executed in the dew environment, as defined in Equation (6). Formally,

$$r(s, a, s') = \begin{cases} \frac{|J_c|}{|J|} & \text{if } s' \text{ is terminal} \\ 0 & \text{otherwise} \end{cases} \tag{7}$$

Note that this reward function only rewards the agent in terminal states. After the agent finishes distributing the whole sequence of jobs, the agent receives a reward equivalent to the successfully executed jobs in the dew environment. As a result, any optimal policy $\pi^*(a|s)$ will optimally distribute jobs according to our problem definition from Section 6.4.2.

As a summary of this section, Figure 2 illustrates the training loop and how a learning agent interacts with our proposed environment. The agent assigns the current job to a particular device using its actions. In response, the dew environment returns the next state s_t and a reward r_t. The state includes information about the current job to be assigned and the state of the devices. The reward r_t will be zero unless the agent has just distributed the last job. If that is the case, r_t will be equivalent to the successfully executed jobs. Since the agent tries to maximize the reward received from the environment, it will learn to distribute jobs to improve the performance of the dew environment.

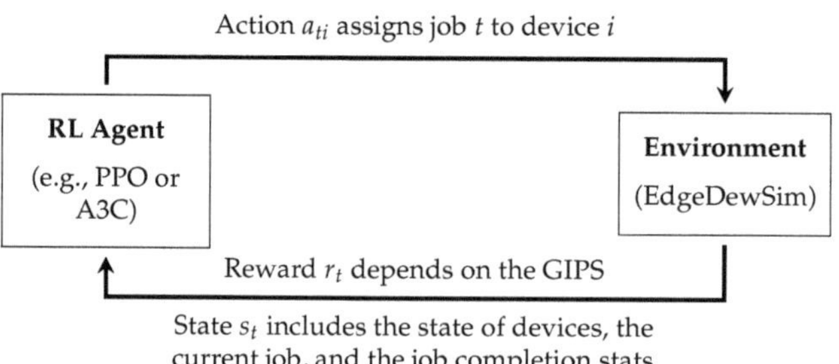

Figure 2. Training loop of the EdgeDewSim Extended Simulator environment.

6.4.4. Implementation Details

For the dew environment simulator, we used EdgeDewSim Extended Simulator. This version can handle both battery-dependent and non-battery-dependent devices. To simplify the process of training and testing deep RL methods, we also developed an OpenAI Gym environment [60]. This allowed us to quickly and easily test different RL agents using existing implementations made for the OpenAI Gym library.

Since EdgeDewSim Extended Simulator is written in java and OpenAI Gym is written in python, we developed an interface that allowed us to control EdgeDewSim Extended Simulator from OpenAI Gym. To do so, we added logic on top of EdgeDewSim Extended Simulator to delegate the scheduling decision to an external agent so that every time a job arrives, the simulator broadcasts the general state of the simulation and then listens for which device has to be selected, as shown in Figure 3. Then, we developed a new RL environment, following the guidelines from OpenAI Gym [60], that undertakes three main things: (i) it establishes the connection with the simulator, (ii) it sends actions to the simulator, and (iii) it resets the environment to restart from an initial state. Further details about our implementation can be found in Section 4.

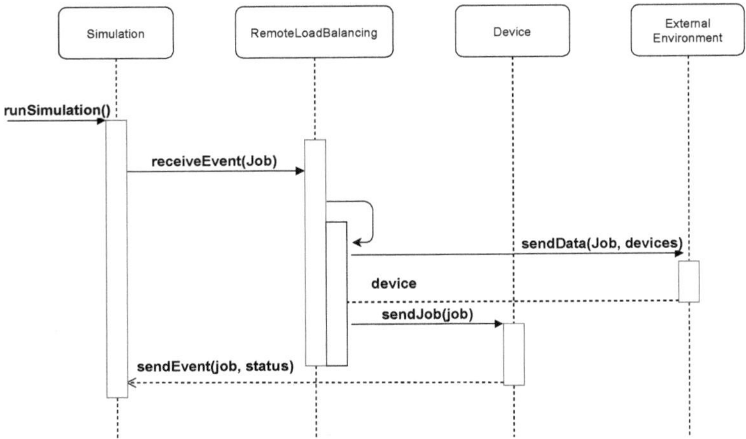

Figure 3. Sequence diagram of one job scheduling with an external environment.

7. Methodology and Experimentation

7.1. Methodology

The experimental design utilized in this study can be categorized into two main groups. The initial group focused solely on human-designed heuristics. It aimed to evaluate the effectiveness of the proposed heuristics, specifically ReleSEAS and RelBPA, in comparison to random device selection and previously examined heuristics, namely, eSEAS and BPA, as elaborated in Section 7.2. The second group was devoted to substantiating the capacity of a reinforcement learning agent to emulate the performance of human-designed heuristics. Furthermore, it sought to demonstrate the agent's adaptability to novel conditions by utilizing transfer learning techniques; this will be explained in depth in Section 7.3.

7.2. Human-Designed Heuristics

The method employed for the assessment of human-designed heuristics comprised two sequential stages. The initial step involved generating connection and disconnection events for individual devices, while the subsequent stage entailed the execution of simulations utilizing the EdgeDewSim Extended Simulator.

As mentioned in Section 5, we employed random walk and random waypoint models to generate synthetic data for our experimental work. For each distinct human mobility scenario, we introduced variations generated randomly using a variation of generation of some parameters of these models (like speed, initial position, connection, and disconnection events). This approach yielded a range of scenarios representing diverse conditions for device mobility within the network context.

We performed 20 simulation runs for each model with unique parameter settings that affect human movement patterns. Despite the variations, certain fundamental elements remained unchanged in both models. These constants encompassed a 60 m by 60 m grid defining device movement, a consistent use of 25 devices, and a centrally located Wi-Fi access point covering a 30 m radius. Given the reliance of these models on random variables, we conducted 10 sets of 20 simulations to obtain an average value for job completion and performance. The objective behind using this number of sets is to have multiple cases using the mobility models to generate data by adjusting the parameters so the heuristics and the RL agent can interact with multiple different scenarios.

7.3. Reinforcement Learning Agent

The training method used for the RL agent was divided into two phases. The first was to train an agent using a set randomly selected from 20 random walk environments. A different selection was made for each iteration in the training phase. The environments were the same used in Section 5, each composed of 25 devices, a 60 m by 60 m grid with centrally located Wi-Fi access covering a 30 m radius. The Job J set contains jobs with [15–25] Kb data input and [1–2] Kb data output and computing requirements in the range of [80–125] mega float-point operations.

For the RL agent, we used a convolutional neural network (CNN) with five layers of 512 neurons, based on previous experience [9] where this architecture demonstrated a good balance between computational efficiency and learning capacity. The learning rate was set to 1×10^{-5}, chosen to ensure a stable and gradual learning progress. We utilized 32 Gym environments to parallel and speed up the training process. The lambda and gamma functions were both set to 1.0 to give equal importance to all states in the value function estimation. The model was trained for 5×10^7 steps with a batch size of 64, which were determined through experimental tuning to optimize the trade-off between training time and model performance.

The other phase was related to the transfer learning phase due to the high probability of overfitting in RL [61,62]. We trained a new environment using the previous model as a base, and with fewer steps, we achieved convergence in the new environment. We repeated these steps with different environments and collected the data. The new random waypoint environments with 25 devices and 1500 JOBs were new. We trained the latest models in 1×10^6 steps and a batch size of 64.

7.4. Experimentation

This section aims to summarize the results obtained from the experiments defined in the previous section, which will be separated into two subsections. In conformity with [7], the most relevant metrics to review are the job completion rate and the executed operations at the SCE level, i.e., considering that an SCE, as a virtual processing node, renders execution services to an external entity through the devices integrating it. As mentioned in Section 7.2, we have 200 instances by model and set of jobs, with 800 experimentation scenarios. The experiment results are presented in the following subsection, and in Section 7.6 they will be further analyzed. Due to resource computing limitations, we could not fully simulate a video stream on the RL agent. We made a sample of 1500 jobs to train the agent to have representative results. On the other hand, the human-designed heuristics do not require heavy computation resources, so we ran experiments with the 1500 jobs sample and the complete set of 36,000 jobs.

7.4.1. Job Completion

This section is divided into two parts: the results regarding human-designed heuristics and those obtained by the RL agent. In Figure 4, we present the average from 20 simulations when using random walk along with the 1500 jobs set; analogously, the same results are given when using random waypoint in Figure 5. In Figures 6 and 7, we present the same 20 simulations but using the entire dataset of jobs.

As mentioned previously, we performed experiments on two different sets of jobs. The primary goal of the comprehensive dataset of jobs is to highlight that when the simulation runs longer and thus encompasses more events, the proposed human-designed heuristics demonstrate improved performance compared to their original counterparts. The results shown in Table 1 are proof of this since in the set with 1500 jobs, the original human-designed heuristics do not deviate too much from the new variation we present in this paper. For instance, the difference in job completion between BPA and RelBPA is 8.21%, whereas between eSEAS and ReleSEAS it is 10.58%. On the other hand, when using the complete set with 36,000 jobs, the difference between BPA and RelBPA increases to 16.49%, then again for eSEAS and ReleSEAS, the difference rises to 18.83%. In both cases, the difference in job completion between the original human-crafted policy and the new version almost doubles when using the complete set of jobs over the small set. This same behavior can be seen in Table 2, in which the usage of Random Waypoint makes this difference smaller than the results obtained using Random Walk. An additional Random scheduler is added to these results to use as a baseline, which distributes the jobs randomly every time the given event is received.

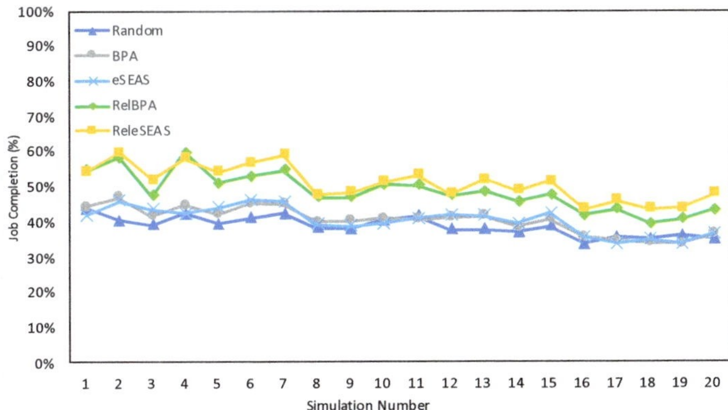

Figure 4. Average job completion for each simulation using random walk to generate connection events and the 1500 jobs set.

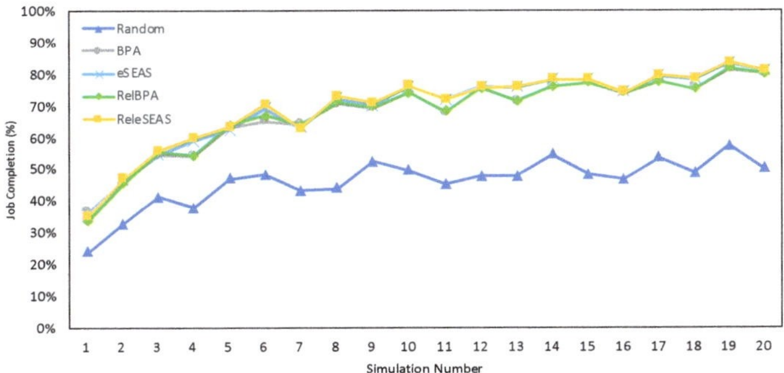

Figure 5. Average job completion for each simulation using random waypoint to generate connection events and the 1500 jobs set.

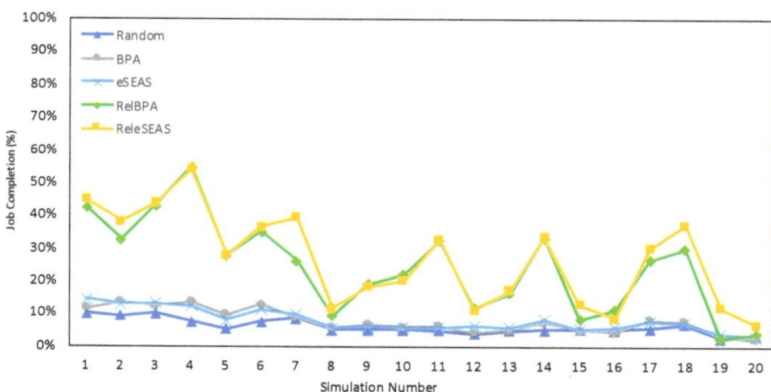

Figure 6. Average job completion for each simulation using random walk to generate connection events and the full set of jobs.

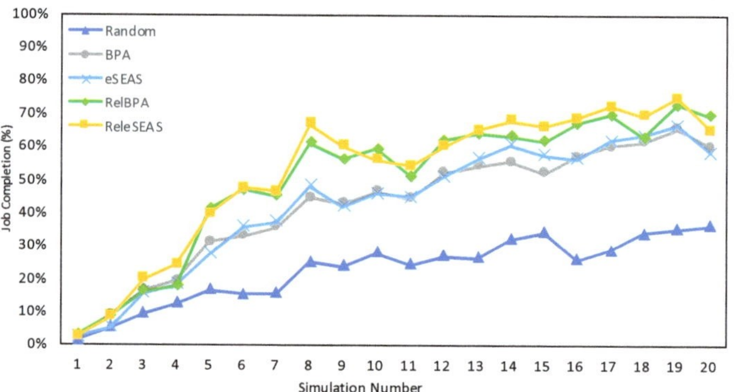

Figure 7. Average job completion for each simulation using random waypoint to generate connection events and the full set of jobs.

Table 1. Average job completion throughout all simulations using random walk.

Job Set	**Random**	**BPA**	**eSEAS**	**RelBPA**	**ReleSEAS**
1500	38.69%	40.38%	40.40%	48.59%	50.98%
36,000 (full set)	5.75%	7.56%	7.63%	24.05%	26.39%

Table 2. Average job completion throughout all simulations using random waypoint.

Job Set	**Random**	**BPA**	**eSEAS**	**RelBPA**	**ReleSEAS**
1500	45.97%	67.58%	69.17%	67.75%	69.51%
36,000 (full set)	22.74%	42.21%	42.60%	49.73%	51.52%

7.4.2. Performance

As in Section 7.4.1, the current section is also divided into two subsections, one regarding human-designed heuristics results and the other for the RL agent. In Figures 8 and 9, we present the performance measured in GIPS for each human-designed heuristic using random walk as the model to generate connection events for 20 simulations. As seen in Figure 8, the values obtained for each simulation across all heuristics are not spread out but are instead close, which translates to small differences between the proposed heuristics with the reliability score and those that do not have the score. On the other hand, Figure 9 shows a different scenario where ReleSEAS and RelBPA can separate from the competitors in a significant way in most cases. In Figures 10 and 11, which present the performance when using random waypoint, we can see the same behavior but with less strength.

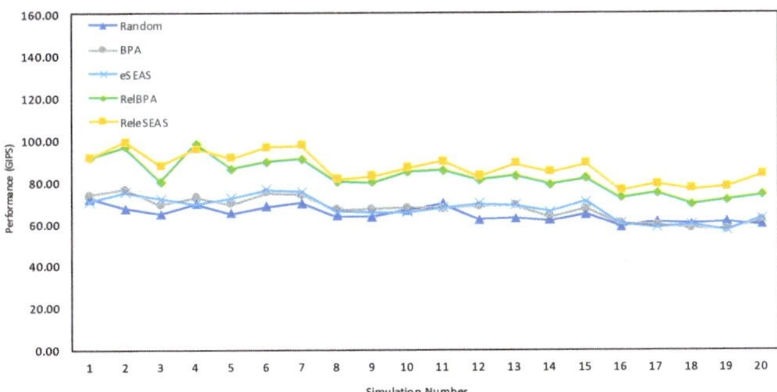

Figure 8. Average performance in GIPS for each simulation using random walk to generate connection events and the 1500 jobs set.

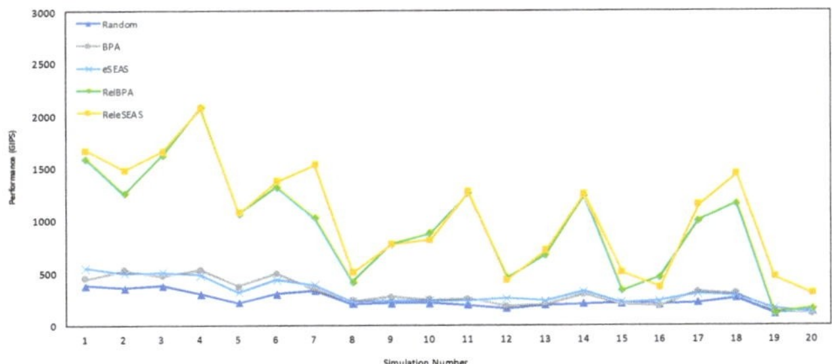

Figure 9. Average performance in GIPS for each simulation using random walk to generate connection events and the full set of jobs.

In this metric, we use the complete and small datasets to explore whether the same relation is in job completion. Even though both metrics are highly related, having more completed jobs does not necessarily mean that the performance in GIPS will be equally high. This is because modeled jobs can—as in real life—be simple or complex; we might want to process a simple math calculation or run an image through an object detection model. Nevertheless, the exact relationship between the original human-designed heuristics and the proposed versions remains for both metrics.

In Figure 8, the values obtained for each simulation across all heuristics are not scattered but are instead close, which translates to slight differences between ReleSEAS and RelBPA with regard to the original versions. On the other hand, Figure 9 shows a different scenario where ReleSEAS and RelBPA can significantly separate from the rest of the policies when using the complete set of jobs. In regard to the performance metric, we can see in Table 3 a difference of 14.98 GIPS for BPA and RelBPA, whereas for eSEAS and ReleSEAS, the difference is 19.07 GIPS. On the other hand, when using the complete set of jobs, the difference between BPA and RelBPA rises up to 631.9 GIPS, and then again, for eSEAS and ReleSEAS, the difference is 724.46. This also holds when using random waypoint, as can be seen in Table 4 and in Figures 10 and 11. Still, on a smaller scale, since this algorithm is more stable than the random walk, the new versions of human-designed heuristics presented in this paper thus do not outperform the classic versions in the same ways as in the random walk scenarios, which are more dynamic or unstable.

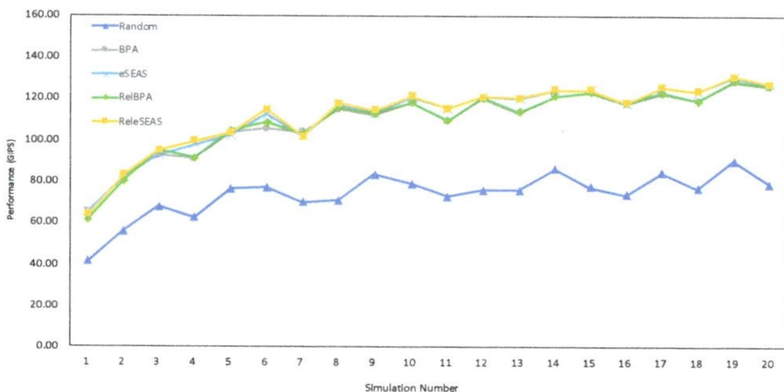

Figure 10. Average performance in GIPS for each simulation using random waypoint to generate connection events and the 1500 jobs set.

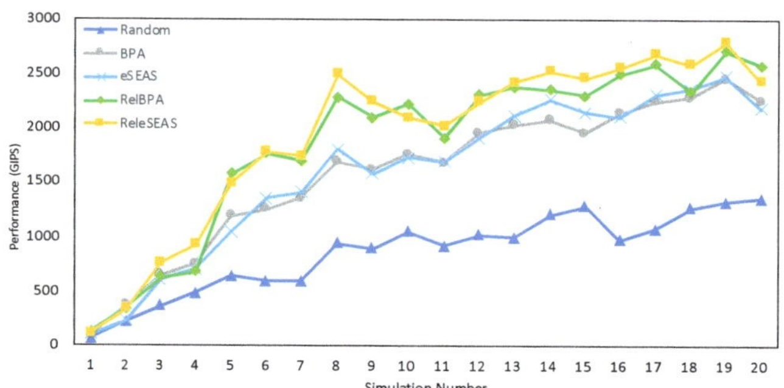

Figure 11. Average performance in GIPS for each simulation using random waypoint to generate connection events and the full set of jobs.

Table 3. Average performance in GIPS across all simulations using random walk.

Job Set	Random	BPA	eSEAS	RelBPA	ReleSEAS
1500	64.36	67.05	67.18	82.03	86.25
36,000 (full set)	231.69	303.27	305.30	935.17	1029.76

Table 4. Average performance in GIPS across all simulations using random waypoint.

Job Set	Random	BPA	eSEAS	RelBPA	ReleSEAS
1500	73.68	109.30	111.60	109.65	112.23
36,000 (full set)	860.52	1588.68	1604.73	1868.67	1934.31

7.5. Reinforcement Learning Agent Results

After the training process described in Section 7.3, we compared the job completion rate with other algorithms using the same simulation environment. The results, as shown in Figure 12, demonstrate that a trained model can quickly adapt to environmental variations and surpass the performance of human-designed job completion heuristics. Notably, around iteration 50, the model in the environment mentioned in Section 7.3 outperforms the best human-designed heuristic regarding job completion, and it continues to improve in this metric in subsequent iterations. One of the challenges in training a reinforcement learning agent for this use case is that the model needs to learn from multiple scenarios to generalize effectively, which is both time-consuming and resource-intensive. However, our objective in this paper is to provide evidence that, despite being trained in different environments, the model can quickly adapt to new ones and excel in the metrics mentioned.

Using the obtained model, we trained new and specific environments using the weights obtained in the first training phase, transferring the learned weights to the new environment. The results show that this new RL agent converges on the job completion rate for each new simulation. Figure 13 details how the agent converges to a solution in each training iteration. The new environments, as said in Section 7.3, were new random waypoint environments, with 1500 JOBS and 25 devices.

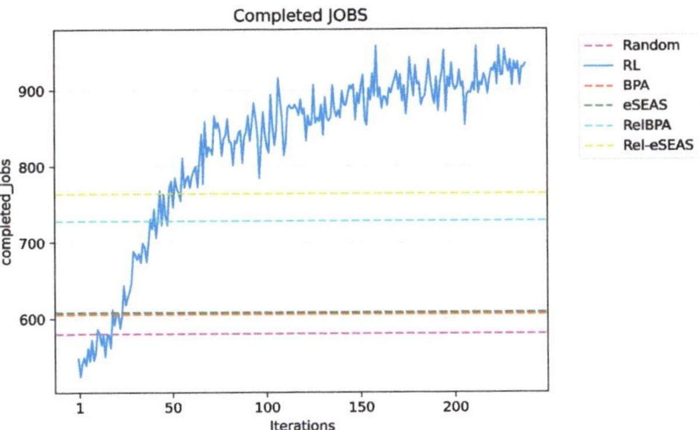

Figure 12. RL convergence for job completion for each simulation using random walk compared to other algorithms in the 1500 jobs set.

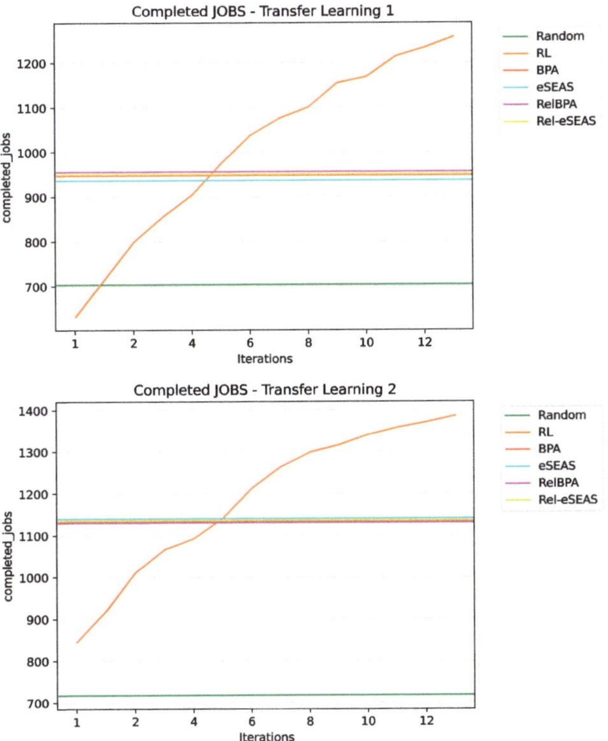

Figure 13. RL Convergence for job completion for each simulation using transfer learning in a random waypoint environment compared to other algorithms in the 1500 jobs set.

7.6. Discussion

We could extract critical results and insights from the experiments for future work. First, the proposed reliability score, which gives a scheduler awareness of node mobility, improves the job completeness metric with regard to human-designed heuristics, not considering such aspect in all run experiments. Moreover, in all experiments, it can be observed that human-crafted policies performed better with random waypoint than with random walk. This result agrees with what the authors in Ref. [53] state: random walk is the simplest model, and humans rarely move randomly.

We also showed that RL is viable when the RL agent is adapted to connection and disconnection events. With sufficient training time, RL can devise a solution and even surpass human-crafted policies in terms of job completion rate. Regarding the overfitting problem when the agent needs to be applied in different environments, we suggest training the model with a set of varying environment events data to reduce the probability of overfitting and using the base model as a starting point to specialize one agent version to a specific environment.

8. Conclusions and Future Work

This work presents the EdgeDewSim Extended Simulator, which supports the connection and disconnection of devices into the cluster and its network. With this capability, we were able to experiment with two simple mobility models that tend to simulate a human's movements and incorporate a new reliability score into two very well-studied algorithms, BPA and E-SEAS [7,22]. The reliability score offers improvements in job completeness and operations executed, which vary depending on the scenario. For future work, it is necessary to extend the mobility models used to generate the connection and disconnection events to reflect real-world human movements more accurately and explore the usage of human-generated movement data to validate the heuristics' effectiveness. Also, as mentioned in Ref. [9], using RL techniques makes it possible to surpass the human-designed heuristics used in this type of environment. However, it is necessary to explore new RL techniques to handle a broader range of dynamic jobs and explore generalization techniques to improve the RL model's adaptability to diverse environments.

A significant branch of future work involves the practical implementation and testing of the proposed scheduling algorithms in real use-case scenarios. This includes deploying the algorithms in dew computing environments, such as smart city infrastructure or industrial IoT setups, to assess their performance and robustness in real-world conditions. By evaluating the algorithms in diverse applications, we can identify potential challenges, refine them based on real-world feedback, and ensure their readiness for deployment in practical settings. These efforts will bridge the gap between theoretical advancements and practical applicability, paving the way for more reliable and effective device management solutions in dynamic dew computing environments.

Author Contributions: Conceptualization, P.S., S.M., A.N., R.T.I., M.H. and C.M.; methodology, P.S., S.M., A.N., R.T.I., M.H. and C.M.; software, P.S., S.M., A.N., R.T.I., M.H. and C.M.; validation, P.S., S.M., A.N., R.T.I., M.H. and C.M.; formal analysis, P.S., S.M., A.N., R.T.I., M.H. and C.M.; investigation, P.S., S.M., A.N., R.T.I., M.H. and C.M.; resources, P.S., S.M., A.N., R.T.I., M.H. and C.M.; data curation, P.S., S.M., A.N., R.T.I., M.H. and C.M.; writing—original draft preparation, P.S., S.M., A.N., R.T.I., M.H. and C.M.; writing—review and editing, P.S., S.M., A.N., R.T.I., M.H. and C.M.; visualization, P.S., S.M., A.N., R.T.I., M.H. and C.M.; supervision, P.S., S.M., A.N., R.T.I., M.H. and C.M.; project administration, P.S., S.M., A.N., R.T.I., M.H. and C.M.; funding acquisition, P.S., S.M., A.N., R.T.I., M.H. and C.M. All authors have read and agreed to the published version of the manuscript.

Funding: This work was funded by the National Agency for Research and Development (ANID)/ Scholarship Program/DOCTORADO NACIONAL/2020-21200979. We also gratefully acknowledge funding from the National Center for Artificial Intelligence CENIA FB210017, Basal ANID. Additionally, we thank the funding provided by CONICET grant number PIBAA-28720210101298CO and grant number PIP-11220210100138CO.

Institutional Review Board Statement: Not applicable.

Informed Consent Statement: Not applicable.

Data Availability Statement: Source code and datasets are available at: https://github.com/psanabr iaUC/gym-EdgeDewSim, accessed on 27 November 2023.

Conflicts of Interest: The authors declare no conflicts of interest.

References

1. Wang, Y. Definition and categorization of dew computing. *Open J. Cloud Comput. (OJCC)* **2016**, *3*, 1–7.
2. Ray, P.P. An introduction to dew computing: Definition, concept and implications. *IEEE Access* **2017**, *6*, 723–737. [CrossRef]
3. Hirsch, M.; Mateos, C.; Zunino, A. Augmenting computing capabilities at the edge by jointly exploiting mobile devices: A survey. *Future Gener. Comput. Syst.* **2018**, *88*, 644–662. [CrossRef]
4. Khalid, M.N.B. Deep Learning-Based Dew Computing with Novel Offloading Strategy. In Proceedings of the International Conference on Security, Privacy and Anonymity in Computation, Communication and Storage, Nanjing, China, 18–20 December 2020; Springer: Berlin/Heidelberg, Germany, 2020; pp. 444–453.
5. Nanakkal, A. A Brief Survey of Future Computing Technologies in Cloud Environment. *Ir. Interdiscip. J. Sci. Res. (IIJSR)* **2021**, *4*, 63–70. [CrossRef]
6. Hirsch, M.; Rodriguez, J.M.; Zunino, A.; Mateos, C. Battery-aware centralized schedulers for CPU-bound jobs in mobile Grids. *Pervasive Mob. Comput.* **2016**, *29*, 73–94. [CrossRef]
7. Sanabria, P.; Tapia, T.F.; Neyem, A.; Benedetto, J.I.; Hirsch, M.; Mateos, C.; Zunino, A. New Heuristics for Scheduling and Distributing Jobs under Hybrid Dew Computing Environments. *Wirel. Commun. Mob. Comput.* **2021**, *2021*, 8899660. [CrossRef]
8. Samal, P.; Mishra, P. Analysis of variants in Round Robin Algorithms for load balancing in Cloud Computing. *Int. J. Comput. Sci. Inf. Technol.* **2013**, *4*, 416–419.
9. Sanabria, P.; Tapia, T.F.; Toro Icarte, R.; Neyem, A. Solving Task Scheduling Problems in Dew Computing via Deep Reinforcement Learning. *Appl. Sci.* **2022**, *12*, 7137. [CrossRef]
10. Sutton, R.S.; Barto, A.G. *Reinforcement Learning: An Introduction*; MIT Press: Cambridge, MA, USA, 2018.
11. Akkaya, I.; Andrychowicz, M.; Chociej, M.; Litwin, M.; McGrew, B.; Petron, A.; Paino, A.; Plappert, M.; Powell, G.; Ribas, R.; et al. Solving rubik's cube with a robot hand. *arXiv* **2019**, arXiv:1910.07113.
12. Li, J.; Monroe, W.; Ritter, A.; Galley, M.; Gao, J.; Jurafsky, D. Deep reinforcement learning for dialogue generation. *arXiv* **2016**, arXiv:1606.01541.
13. Popova, M.; Isayev, O.; Tropsha, A. Deep reinforcement learning for de novo drug design. *Sci. Adv.* **2018**, *4*, eaap7885. [CrossRef] [PubMed]
14. Mao, Y.; You, C.; Zhang, J.; Huang, K.; Letaief, K.B. A survey on mobile edge computing: The communication perspective. *IEEE Commun. Surv. Tutorials* **2017**, *19*, 2322–2358. [CrossRef]
15. Khan, W.Z.; Ahmed, E.; Hakak, S.; Yaqoob, I.; Ahmed, A. Edge computing: A survey. *Future Gener. Comput. Syst.* **2019**, *97*, 219–235. [CrossRef]
16. Drolia, U.; Martins, R.; Tan, J.; Chheda, A.; Sanghavi, M.; Gandhi, R.; Narasimhan, P. The case for mobile edge-clouds. In Proceedings of the 2013 IEEE 10th International Conference on Ubiquitous Intelligence and Computing and 2013 IEEE 10th International Conference on Autonomic and Trusted Computing, Vietri sul Mare, Italy, 18–21 December 2013; IEEE: Vietri sul Mare, Italy, 2013; pp. 209–215.
17. Benedetto, J.I.; González, L.A.; Sanabria, P.; Neyem, A.; Navón, J. Towards a practical framework for code offloading in the Internet of Things. *Future Gener. Comput. Syst.* **2019**, *92*, 424–437. [CrossRef]
18. Shi, W.; Cao, J.; Zhang, Q.; Li, Y.; Xu, L. Edge computing: Vision and challenges. *IEEE Internet Things J.* **2016**, *3*, 637–646. [CrossRef]
19. Yu, W.; Liang, F.; He, X.; Hatcher, W.G.; Lu, C.; Lin, J.; Yang, X. A survey on the edge computing for the Internet of Things. *IEEE Access* **2017**, *6*, 6900–6919. [CrossRef]
20. Olaniyan, R.; Fadahunsi, O.; Maheswaran, M.; Zhani, M.F. Opportunistic edge computing: Concepts, opportunities and research challenges. *Future Gener. Comput. Syst.* **2018**, *89*, 633–645. [CrossRef]
21. Aslam, S.; Michaelides, M.P.; Herodotou, H. Internet of ships: A survey on architectures, emerging applications, and challenges. *IEEE Internet Things J.* **2020**, *7*, 9714–9727. [CrossRef]
22. Hirsch, M.; Rodriguez, J.M.; Mateos, C.; Alejandro, Z. A Two-Phase Energy-Aware Scheduling Approach for CPU-Intensive Jobs in Mobile Grids. *J. Grid Comput.* **2017**, *15*, 55–80. [CrossRef]
23. Hirsch, M.; Mateos, C.; Rodriguez, J.M.; Zunino, A.; Garí, Y.; Monge, D.A. A performance comparison of data-aware heuristics for scheduling jobs in mobile grids. In Proceedings of the 2017 XLIII Latin American Computer Conference (CLEI), Córdoba, Argentina, 4–8 September 2017; IEEE: Córdoba, Argentina, 2017; pp. 1–8.
24. Chen, X.; Pu, L.; Gao, L.; Wu, W.; Wu, D. Exploiting Massive D2D Collaboration for Energy-Efficient Mobile Edge Computing. *IEEE Wirel. Commun.* **2017**, *24*, 64–71. [CrossRef]

25. Mtibaa, A.; Fahim, A.; Harras, K.A.; Ammar, M.H. Towards resource sharing in mobile device clouds. *ACM SIGCOMM Comput. Commun. Rev.* **2013**, *43*, 51–56. [CrossRef]
26. Li, B.; Pei, Y.; Wu, H.; Shen, B. Heuristics to allocate high-performance cloudlets for computation offloading in mobile ad hoc clouds. *J. Supercomput.* **2015**, *71*, 3009–3036. [CrossRef]
27. Chunlin, L.; Layuan, L. Exploiting composition of mobile devices for maximizing user QoS under energy constraints in mobile grid. *Inf. Sci.* **2014**, *279*, 654–670. [CrossRef]
28. Birje, M.N.; Manvi, S.S.; Das, S.K. Reliable resources brokering scheme in wireless grids based on non-cooperative bargaining game. *J. Netw. Comput. Appl.* **2014**, *39*, 266–279. [CrossRef]
29. Loke, S.W.; Napier, K.; Alali, A.; Fernando, N.; Rahayu, W. Mobile Computations with Surrounding Devices. *ACM Trans. Embed. Comput. Syst.* **2015**, *14*, 1–25. [CrossRef]
30. Shah, S.C. Energy efficient and robust allocation of interdependent tasks on mobile ad hoc computational grid. *Concurr. Comput. Pract. Exp.* **2015**, *27*, 1226–1254. [CrossRef]
31. Orhean, A.I.; Pop, F.; Raicu, I. New scheduling approach using reinforcement learning for heterogeneous distributed systems. *J. Parallel Distrib. Comput.* **2018**, *117*, 292–302. [CrossRef]
32. Kaur, P. DRLCOA: Deep Reinforcement Learning Computation Offloading Algorithm in Mobile Cloud Computing. *SSRN Electron. J.* **2019**, *12*, 238–246. [CrossRef]
33. Cheng, M.; Li, J.; Nazarian, S. DRL-cloud: Deep reinforcement learning-based resource provisioning and task scheduling for cloud service providers. In Proceedings of the 2018 23rd Asia and South Pacific Design Automation Conference (ASP-DAC), Jeju, Republic of Korea, 22–25 January 2018; IEEE: Jeju, Republic of Korea, 2018; pp. 129–134. [CrossRef]
34. Huang, L.; Bi, S.; Zhang, Y.J.A. Deep Reinforcement Learning for Online Computation Offloading in Wireless Powered Mobile-Edge Computing Networks. *IEEE Trans. Mob. Comput.* **2020**, *19*, 2581–2593. [CrossRef]
35. Ha, S.; Choi, E.; Ko, D.; Kang, S.; Lee, S. Efficient Resource Augmentation of Resource Constrained UAVs Through EdgeCPS. In Proceedings of the 38th ACM/SIGAPP Symposium on Applied Computing, Tallinn, Estonia, 27–31 March 2023; pp. 679–682.
36. Ren, J.; Xu, S. DDPG Based Computation Offloading and Resource Allocation for MEC Systems with Energy Harvesting. In Proceedings of the 2021 IEEE 93rd Vehicular Technology Conference (VTC2021-Spring), Virtual, 25 April–19 May 2021; IEEE: Helsinki, Finland, 2021; pp. 1–5.
37. Zhao, R.; Wang, X.; Xia, J.; Fan, L. Deep reinforcement learning based mobile edge computing for intelligent Internet of Things. *Phys. Commun.* **2020**, *43*, 101184. [CrossRef]
38. Tefera, G.; She, K.; Shelke, M.; Ahmed, A. Decentralized adaptive resource-aware computation offloading & caching for multi-access edge computing networks. *Sustain. Comput. Inform. Syst.* **2021**, *30*, 100555. [CrossRef]
39. Baek, J.; Kaddoum, G. Heterogeneous Task Offloading and Resource Allocations via Deep Recurrent Reinforcement Learning in Partial Observable Multi-Fog Networks. *IEEE Internet Things J.* **2020**, *8*, 1041–1056. [CrossRef]
40. Lu, H.; Gu, C.; Luo, F.; Ding, W.; Liu, X. Optimization of lightweight task offloading strategy for mobile edge computing based on deep reinforcement learning. *Future Gener. Comput. Syst.* **2020**, *102*, 847–861. [CrossRef]
41. Li, J.; Gao, H.; Lv, T.; Lu, Y. Deep reinforcement learning based computation offloading and resource allocation for MEC. In Proceedings of the 2018 IEEE Wireless Communications and Networking Conference (WCNC), Barcelona, Spain, 15–18 April 2018; IEEE: Barcelona, Spain, 2018; pp. 1–6. [CrossRef]
42. Alfakih, T.; Hassan, M.M.; Gumaei, A.; Savaglio, C.; Fortino, G. Task Offloading and Resource Allocation for Mobile Edge Computing by Deep Reinforcement Learning Based on SARSA. *IEEE Access* **2020**, *8*, 54074–54084. [CrossRef]
43. Zhao, L.; Li, H.; Zhang, E.; Hawbani, A.; Lin, M.; Wan, S.; Guizani, M. Intelligent Caching for Vehicular Dew Computing in Poor Network Connectivity Environments. *ACM Trans. Embed. Comput. Syst.* **2024**, *23*, 1–24. [CrossRef]
44. Khatua, S.; Manerba, D.; Maity, S.; De, D. Dew Computing-Based Sustainable Internet of Vehicular Things. In *Dew Computing: The Sustainable IoT Perspectives*; Springer: Berlin/Heidelberg, Germany, 2023; pp. 181–205.
45. Pal, M.N.; Sengupta, D.; Tran, T.A.; De, D. Machine Learning-Based Sustainable Dew Computing: Classical to Quantum. In *Dew Computing: The Sustainable IoT Perspectives*; Springer: Berlin/Heidelberg, Germany, 2023; pp. 149–177.
46. Chakraborty, S.; De, D.; Mazumdar, K. DoME: Dew computing based microservice execution in mobile edge using Q-learning. *Appl. Intell.* **2023**, *53*, 10917–10936. [CrossRef]
47. Cobbe, K.; Klimov, O.; Hesse, C.; Kim, T.; Schulman, J. Quantifying generalization in reinforcement learning. In Proceedings of the 36th International Conference on Machine Learning (ICML), Long Beach, CA, USA, 9–15 June 2019; pp. 1282–1289.
48. Cobbe, K.; Hesse, C.; Hilton, J.; Schulman, J. Leveraging procedural generation to benchmark reinforcement learning. In Proceedings of the 37th International Conference on Machine Learning (ICML), Virtual Event, 13–18 July 2020; pp. 2048–2056.
49. Hirsch, M.; Mateos, C.; Rodriguez, J.M.; Zunino, A. DewSim: A trace-driven toolkit for simulating mobile device clusters in Dew computing environments. *Softw. Pract. Exp.* **2020**, *50*, 688–718. [CrossRef]
50. Manweiler, J.; Santhapuri, N.; Choudhury, R.R.; Nelakuditi, S. Predicting length of stay at wifi hotspots. In Proceedings of the 2013 IEEE INFOCOM, Turin, Italy, 14–19 April 2013; pp. 3102–3110.
51. Blanford, J.I.; Huang, Z.; Savelyev, A.; MacEachren, A.M. Geo-located tweets. Enhancing mobility maps and capturing cross-border movement. *PLoS ONE* **2015**, *10*, e0129202. [CrossRef]
52. Barbosa, H.; Barthelemy, M.; Ghoshal, G.; James, C.R.; Lenormand, M.; Louail, T.; Menezes, R.; Ramasco, J.J.; Simini, F.; Tomasini, M. Human mobility: Models and applications. *Phys. Rep.* **2018**, *734*, 1–74. [CrossRef]

Appl. Sci. **2024**, *14*, 3206

53. Solmaz, G.; Turgut, D. A Survey of Human Mobility Models. *IEEE Access* **2019**, *7*, 125711–125731. [CrossRef]
54. Falaki, H.; Mahajan, R.; Kandula, S.; Lymberopoulos, D.; Govindan, R.; Estrin, D. Diversity in Smartphone Usage. In Proceedings of the 8th International Conference on Mobile Systems, Applications, and Services (MobiSys'10), New York, NY, USA, 15–18 June 2010; pp. 179–194. [CrossRef]
55. Keramat Jahromi, K.; Zignani, M.; Gaito, S.; Rossi, G.P. Simulating human mobility patterns in urban areas. *Simul. Model. Pract. Theory* **2016**, *62*, 137–156. [CrossRef]
56. Henderson, T.; Kotz, D.; Abyzov, I. The changing usage of a mature campus-wide wireless network. *Comput. Netw.* **2008**, *52*, 2690–2712. [CrossRef]
57. Gorawski, M.; Grochla, K. Review of mobility models for performance evaluation of wireless networks. In *Man-Machine Interactions 3*; Springer: Berlin/Heidelberg, Germany, 2014; pp. 567–577.
58. Panisson, A. Pymobility v0.1—Python Implementation of Mobility Models. 2015. Available online: https://zenodo.org/records/9873 (accessed on 27 November 2023).
59. Zhao, K.; Tarkoma, S.; Liu, S.; Vo, H. Urban human mobility data mining: An overview. In Proceedings of the 2016 IEEE International Conference on Big Data (Big Data), Washington, DC, USA, 5–8 December 2016; pp. 1911–1920.
60. Brockman, G.; Cheung, V.; Pettersson, L.; Schneider, J.; Schulman, J.; Tang, J.; Zaremba, W. Openai gym. *arXiv* **2016**, arXiv:1606.01540.
61. Zhang, A.; Ballas, N.; Pineau, J. A dissection of overfitting and generalization in continuous reinforcement learning. *arXiv* **2018**, arXiv:1806.07937.
62. Zhang, C.; Vinyals, O.; Munos, R.; Bengio, S. A study on overfitting in deep reinforcement learning. *arXiv* **2018**, arXiv:1804.06893.

Article

Energy-Efficient Joint Partitioning and Offloading for Delay-Sensitive CNN Inference in Edge Computing

Zhiyong Zha [1], Yifei Yang [2], Yongjun Xia [3], Zhaoyi Wang [2], Bin Luo [3], Kaihong Li [4], Chenkai Ye [2], Bo Xu [2,5] and Kai Peng [2,*]

1   State Grid Information Telecommunication Co., Ltd., Wuhan 430048, China; hustboyzzy@163.com
2   Hubei Key Laboratory of Smart Internet Technology, School of Electronic Information and Communications, Huazhong University of Science and Technology, Wuhan 430074, China; u202113885@hust.edu.cn (Y.Y.); u202113908@hust.edu.cn (Z.W.); u202113953@hust.edu.cn (C.Y.); xubosite@hust.edu.cn (B.X.)
3   Hubei Huazhong Electric Power Technology Development Co., Ltd., Wuhan 430079, China; xiayj6@hb.sgcc.com.cn (Y.X.); luobin1425@163.com (B.L.)
4   School of Electronic Information Science and Technology, Wuhan University, Wuhan 430072, China; likaihong1245@163.com
5   Hubei ChuTianYun Co., Ltd., Wuhan 430076, China
*   Correspondence: pkhust@hust.edu.cn

Abstract: With the development of deep learning foundation model technology, the types of computing tasks have become more complex, and the computing resources and memory required for these tasks have also become more substantial. Since it has long been revealed that task offloading in cloud servers has many drawbacks, such as high communication delay and low security, task offloading is mostly carried out in the edge servers of the Internet of Things (IoT) network. However, edge servers in IoT networks are characterized by tight resource constraints and often the dynamic nature of data sources. Therefore, the question of how to perform task offloading of deep learning foundation model services on edge servers has become a new research topic. However, the existing task offloading methods either can not meet the requirements of massive CNN architecture or require a lot of communication overhead, leading to significant delays and energy consumption. In this paper, we propose a parallel partitioning method based on matrix convolution to partition foundation model inference tasks, which partitions large CNN inference tasks into subtasks that can be executed in parallel to meet the constraints of edge devices with limited hardware resources. Then, we model and mathematically express the problem of task offloading. In a multi-edge-server, multi-user, and multi-task edge-end system, we propose a task-offloading method that balances the tradeoff between delay and energy consumption. It adopts a greedy algorithm to optimize task-offloading decisions and terminal device transmission power to maximize the benefits of task offloading. Finally, extensive experiments verify the significant and extensive effectiveness of our algorithm.

Keywords: edge computing; task offloading; CNN inference partitioning; greedy algorithm

Citation: Zha, Z.; Yang, Y.; Xia, Y.; Wang, Z.; Luo, B.; Li, K.; Ye, C.; Xu, B.; Peng, K. Energy-Efficient Joint Partitioning and Offloading for Delay-Sensitive CNN Inference in Edge Computing. *Appl. Sci.* **2024**, *14*, 8656. https://doi.org/10.3390/app14198656

Academic Editor: Stefan Fischer

Received: 25 August 2024
Revised: 14 September 2024
Accepted: 19 September 2024
Published: 25 September 2024

1. Introduction

Recent years have witnessed the explosive growth of IoT devices and the emergence of MaaS (Model as a Service) in IoT application types [1], and the lack of memory and computing resources of edge computing [2,3] is insufficient to satisfy the inference of the DNN or CNN [4]. However, placing it on traditional cloud server computing will not only lead to a huge delay caused by frequent communication but also expose private and sensitive information in the process of data transmission and remote processing [5,6]. The rapid explosion and the scale of data collection will make the cloud-based centralized data processing infeasible in the near future [7].

In addition, fog computing [8] or edge computing has also been proposed to utilize the computing resources closer to the data collection endpoints, such as gateways and edge node devices. However, the question of how to effectively partition, distribute, and schedule CNN inference in a locally connected but severely resource-constrained cluster of IoT edge devices is a challenging problem that has not been fully resolved. The existing hierarchical partitioning methods for deep neural network inference applications lead to a significant amount of local processing of intermediate-feature map data. This brings about significant memory occupancy that exceeds the capabilities of typical IoT devices. Furthermore, existing distributed DNN/CNN [9,10] edge-processing solutions based on static partitioning and distribution schemes cannot optimally explore the availability of dynamically changing data sources and computing resources in typical IoT edge clusters [11].

Existing research on service deployment [12,13] and task offloading [14,15] often focuses on how to offload some ordinary, computationally lightweight tasks to servers for processing. However, the above methods are not applicable to the offloading of large model inference tasks in the large-scale edge computing network studied in this paper because most of them are limited to task offloading and do not fully consider task splitting, let alone the computing characteristics of large model tasks. Therefore, this paper innovatively combines distributed computing for large model application tasks with partitioning the large model before task offloading so that edge devices can also complete DNN/CNN inference [16] with limited memory and computing resources. In addition, most of the existing task offloading methods also adopt batch processing, but in reality, most tasks arrive in real-time. We innovatively propose a task offloading method for real-time task arrival.

Inspired by [17], we decided to adopt a method based on matrix convolution for the division of DNN/CNN inference. In this context, how to offload these partition tasks to various servers for execution has become a major problem. This paper proposes a solution to this problem based on this premise.

In this article, we propose a horizontal partitioning method for DNN/CNN and design a real-time greedy algorithm to seek the optimal real-time offloading solution. This method includes the following main innovations and contributions:

(1)   We formulate the problem of partitioning and offloading DNN/CNN inference tasks into a real-time integer linear programming program and consider the dependency relationship between different partitioning granularity and offloading schemes.
(2)   We propose a real-time Greedy Partitioning and Offloading Algorithm (GPOA) for effectively finding the optimal granularity partitioning and unloading scheme that can achieve minimum latency and energy consumption. We achieve our goals and achieve good performance by utilizing the coupling between particle size partitioning and unloading schemes.
(3)   We focus on finding the optimal partitioning and offloading solution in real-time when tasks arrive and evaluate the performance of the GPOA based on data. We compare three baseline algorithms, and the results showed that our algorithm is superior in terms of latency, energy consumption, and overall performance.

The remainder of this paper is organized as follows. In Section 2, we briefly review related work. Section 3 describes the system model and problem formulation. In Section 4, we present the greedy partitioning and task offloading algorithm. In Section 5, we evaluate the performance of our algorithms through extensive simulations, followed by presenting the conclusions of our work in Section 6.

2. Related Work

We aimed to offload the tasks that are disassembled from the foundation model inference to edge servers with limited computing resources and memory, thus reducing latency and improving security. Therefore, we have conducted investigations from the following aspects.

2.1. Foundation Model Inference Partitioning

Due to the mismatch between the computing power required for foundation model inference and the computational resources of edge servers, how to decompose foundation model inference has become a crucial issue that urgently needs to be addressed. To solve this problem, several related solutions have been proposed.

In response to the above issues, many people have put forward the idea of cloud-edge collaboration. Li et al. [18] proposed the JALAD method, a joint accuracy and latency-aware execution framework. By decoupling the structure of deep neural networks, JALAD enables part of the model to be executed on edge devices and the rest to run in the cloud, minimizing data transmission. This method aims to reduce latency while maintaining model accuracy, improving the performance and user experience of deep learning services. Li et al. [19] proposed a real-time DNN inference acceleration method based on edge computing. The core idea of this method is to utilize the powerful computing capabilities of edge servers to offload part or all of the DNN inference tasks to the edge servers for processing, thereby reducing the computational burden of mobile devices and improving inference speed and accuracy.

However, numerous researchers and practitioners also believe that the effect of direct dismantling large model inference is better. Hu et al. [20] proposed the EdgeFlow method. EdgeFlow is a distributed inference system that provides natural support for DAG (Directed Acyclic Graph) structural models. The main design goal of this system is to divide and distribute the model to different devices while maintaining complex layer dependencies to ensure correct inference results. Mao et al. [21] proposed MoDNN—A locally distributed mobile computing system for deep neural network applications. MoDNN can divide trained DNN models into multiple mobile devices to accelerate DNN calculations by reducing device-level computing costs and memory usage. Two model partitioning schemes are designed to minimize non-parallel data transmission time, including wake-up time and transmission time. Zhang et al. [22] introduced DeepSlicing, a collaborative and adaptive inference system that adapts to various CNNs and supports customized flexible fine-grained scheduling. By dividing the model and data, they also designed an efficient scheduler—Proportional Synchronous Scheduler (PSS), which realizes the trade-off between computation and synchronization. Zhao et al. [17] proposed DeepThings, a framework for adaptive distributed execution of CNN-based inference applications on resource-constrained IoT edge clusters. DeepThings used fused tile partitioning to minimize memory occupancy while exposing parallelism.

2.2. Task Offloading

Current task offloading algorithms include three types: minimum energy consumption, minimum latency, and the trade-off between energy consumption and latency. We will compare these algorithms and analyze their advantages and disadvantages.

For algorithms that optimize execution time, we have found the following two options: Liu et al. [23] adopt a Markov decision process framework to tackle the given problem, wherein the scheduling of computational tasks is dictated by the queuing status of the task buffer, the operational state of the local processing unit, and the status of the transmission unit. Furthermore, they introduce an efficient one-dimensional search algorithm to identify the optimal task scheduling policy. The advantage of this algorithm is the greatly reduced execution time compared to the case of not using computing offloading. The disadvantage is that it is very complex. Mao et al. [24] proposed a low-complexity online algorithm, namely the dynamic computation offloading algorithm based on Lyapunov optimization, which jointly determines the offloading decision, the CPU cycle frequency for mobile execution, and the transmit power for computation offloading. The unique advantage of this algorithm is that the decision-making only relies on the current system state, without the need to calculate the distribution information of task requests, wireless channels, and EH processes. The disadvantage is that it has weak applicability because the data are completely offloaded.

According to the Computing and Offloading Algorithm Based on Minimization of Energy Consumption, we found that Shi et al. [25] introduced a computing offloading algorithm for distributed mobile cloud computing environments. This algorithm examined the node movement patterns and developed a network access prediction method based on subsequence tail matching. One of its primary advantages lies in its energy-saving capabilities, rendering it suitable for various scenarios. However, it fell short in overlooking the crucial aspects of time delay implementation and data processing efficiency, as well as the instability perceived by mobile devices.

Lastly, we investigated the model and algorithm for the tradeoff between energy consumption and latency. The algorithm proposed by Chen et al. [26] comprehensively addresses the factors that impact the algorithm's efficiency, incorporating the system's actual conditions to effectively offload task data. One significant advantage of this algorithm lies in its ability to conserve at least 50% of the energy compared to solutions that do not employ partial offloading techniques. However, a potential drawback is that it fails to take into account the energy consumption and time expenditure associated with data transmission. Lu et al. [27] introduced a computing offloading algorithm that leverages specific information within the Internet of Things (IoT) environment to optimize the number of feedback transmissions from nodes to the Mobile Edge Computing (MEC) server. Employing Lyapunov optimization, the algorithm ensures energy equilibrium and stability in the overall system data throughput. This approach transformed the balancing challenge of energy consumption, data backlog, and time consumption into a knapsack problem. A key advantage of this algorithm is its ability to restrict the number of nodes requiring feedback transmission in a wireless sensor network comprising 5000 IoT devices to less than 60. However, a limitation lies in the fact that the solution to the knapsack problem does not guarantee a global optimal solution, potentially rendering the offloading decisions of this algorithm suboptimal in terms of efficiency.

We can observe that decision-making for task offloading can be achieved from various perspectives, but in order to consider all aspects comprehensively, we chose to create our own energy consumption and execution time trade-off model for handling the offloading of large model inference tasks.

3. Problem Formulation

3.1. System Model

In the edge network, we divide the large model inference tasks, which require a huge amount of computation, into many subtasks based on the model mentioned in Section 3.2 to match the computing power of our edge servers, and the system model described in this chapter is used to offload these subtasks to the servers. In order to clearly demonstrate our partitioning and offloading methods, the following discussion will use CNN inference tasks as examples.

As shown in Figure 1, we demonstrate a multi-user multi-edge server-edge computing system. These edge servers are devices installed on wireless access points, with computing and storage capabilities similar to the cloud. In this edge computing system, we have multiple CNN inference tasks that need to be offloaded to the server for processing, denoted as: $U = \{U_1, U_2, \ldots U_n\}$, and an edge network: $S = \{S_1, S_2, \ldots S_m\}$ connected by multiple edge servers.

On the edge network S, there are edge servers S_m, and each edge server can be expressed as $S_i = \{P_i, M_i, N_i, P_{i,com}\}$, where p_i represents the computation frequency of the i-th server's core, m_i represents the storage resources of the i-th server, n_i represents the number of cores of the i-th server, and P_{com} represents the working power of this server when performing inference tasks.

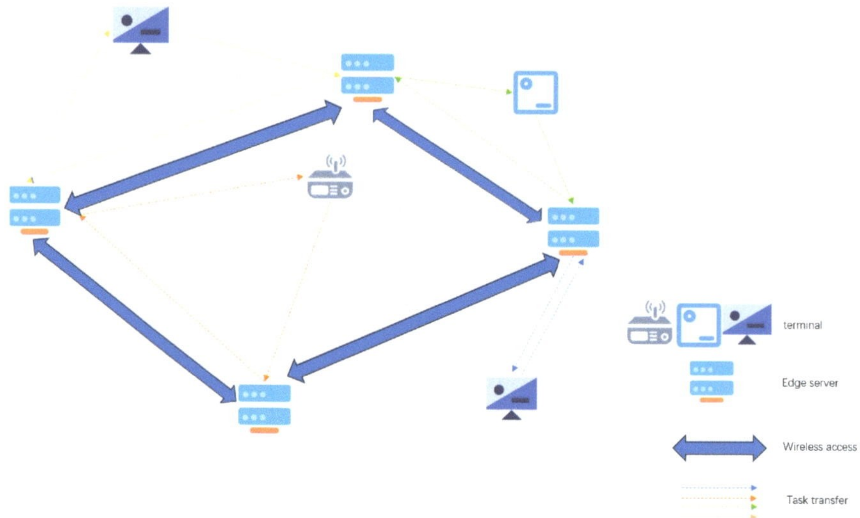

Figure 1. A multi-user multi-edge server-edge computing system.

In this scenario, the CNN inference tasks arrive in real-time, and we use t_{aj} to represent the arrival time of each large task. Additionally, we divide each large task into smaller subtasks for easier processing, denoted as: $U_j = \left\{ t_j^1, t_j^2, \ldots t_j^z \right\}$, where t_j^K represents the k-th task (this task is a divided subtask) of the j-th user. These tasks can also be expressed as $t_j^k = \left\{ C_j^k, d_j^k, X_j^k \right\}$, where C_j^k represents the computing resources required for this task, d_j^k represents the data volume of this task, and X_j^k is an m-dimensional decision variable, $X_j^k = \left\{ x_{j,k}^1, x_{j,k}^2, \ldots x_{j,k}^m \right\}$. When the value is 1, it indicates that the k-th task of the j-th user is offloaded to the i-th server, and when the value is 0, it is the opposite. In addition, we have also created a y_j variable to represent the server closest to the j-th user in order to find the shortest propagation delay.

3.2. Partitioning Model

Firstly, it should be pointed out that our partitioning method is not only designed for specific CNN architectures but is applicable to all neural networks with convolutional neural network layer structures, such as RNN and Transformer. For complex and numerous convolutional neural network architectures, our method is also applicable, as long as the granularity of partitioning is designed to be smaller, or large CNN networks can be first partitioned into vertical layers and then parallelized through our method.

In DNN architectures, multiple convolutional and pooling layers are usually stacked together to gradually extract hidden features from input data. In a CNN with L layers, for layer $l = 1 \ldots L$, the input feature map dimensions for each layer are $W_{l-1} * H_{l-1}$, and each layer has corresponding learnable filters with dimensions $F_l * F_l * D_{l-1}$. The filters slide across D_{l-1} input feature maps with a stride of S_l to perform convolution calculations and then generate D_l output feature maps with dimensions $W_l * H_l$, which, in turn, form the input maps for layer $l + 1$.

Note that in such convolution chains of CNN, due to the large dimensions of the input and intermediate feature maps, the data and computational complexity of each layer are very large. This results in extremely large memory footprints. Ordinary edge devices are unable to meet such memory and computing power requirements.

However, we can observe that each output data element only depends on a local region in the input feature maps. Thus, each original convolutional layer can be partitioned into multiple parallel execution tasks with smaller input and output regions and hence a reduced memory footprint for each layer.

Inspired by [17], we decided to adopt a method based on matrix convolution for the division of DNN/CNN inference to parallelize the convolutional operation and reduce the memory footprint. In our method, the original CNN is divided into tiled stacks of convolution and pooling operations. The feature maps of each layer are divided into small tiles in a grid fashion, where corresponding feature map tiles and operations across layers are vertically fused together to constitute an execution partition and stack. Figure 2 shows the DAG of the partitioned CNN process. As shown in Figure 3, the original set of layers is thus partitioned into M × M independent execution stacks. By dividing the fusion stack along the grid, the size of the intermediate feature maps associated with each partition can be reduced to any desired footprint based on the grid granularity. Then, multiple partitions can be iteratively executed within one device, reducing memory requirements to a maximum of one partition at a time instead of the entire convolutional feature map. In order to distribute the inference operations and input data, each partition's intermediate feature tiles' size and input region's size need to be correctly calculated based on the output partition's size.

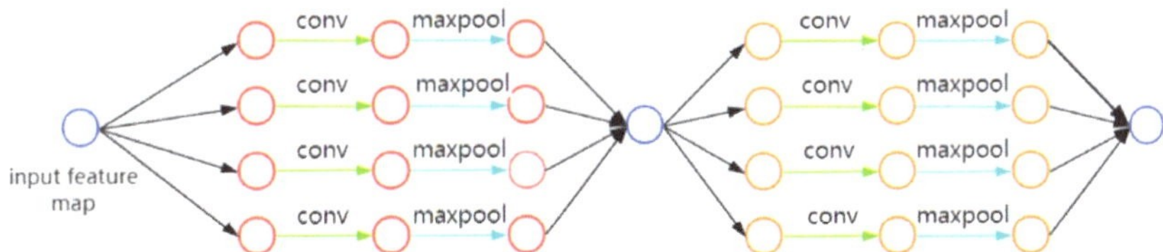

Figure 2. DAG for the partitioning process of the general CNN network framework.

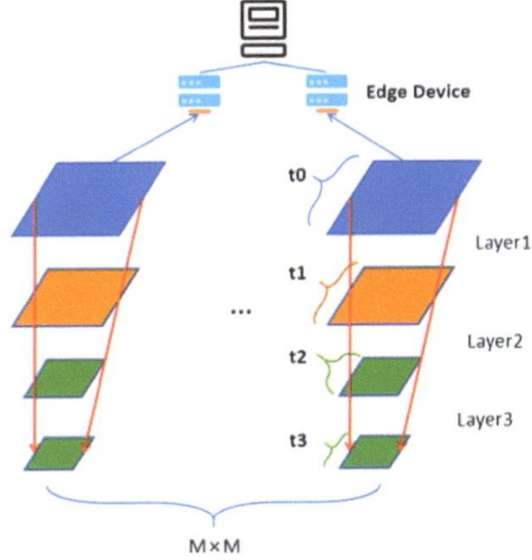

Figure 3. Partitioning method for CNN.

In our method, the region's size for a tile in the output map of layer l is represented by $t_l * t_l$. During the process, the output data are first partitioned equally into nonoverlapping grid tiles with a given grid dimension. Then, with the output offset parameters, a recursive backward traversal is performed for each partition to calculate the required tile region's size in each layer as follows:

$$t_{l-1} = (t_l - 1) * S_l + F_l \tag{1}$$

For a CNN with L layers, the final input offsets for each partition in the CNN input map at $l = 1$ can be obtained by applying (1) recursively starting from partition offsets in the output map at $l = L$ with an initial size of $t_L = W_L / M$. For the l layer, the partitioned CNN convolution task's calculation times C_l can be calculated by the following formulas:

$$C_l = (F_l^2 + F_l^2 - 1) * t_l^2 \tag{2}$$

By using the two Formulas (1) and (2), we can obtain the amount of data and calculation times required by each convolution layer in the partitioned CNN framework, which can be used as evaluation parameters for unloading CNN tasks.

3.3. Offloading Model

In this section, we first define the notations to be used in this paper. Subsequently, we briefly introduce the system model, describe the decision variables, and formulate the problem.

3.3.1. Execution Time of Task Offloading

Since this model is an optimization model based on latency and energy consumption balancing, we inevitably need to establish a model to measure latency. We know that propagation delay is negligible compared to the total latency, so we only need to consider three components: transmission delay, computational processing delay, and queuing delay arising from the real-time arrival of tasks.

Regarding the computational processing delay, we can represent it based on the ratio of the computational resource required by the task c_j^k to total computation frequency of server cores $p_i * core$:

$$t_{com} = c_j^k / (p_i * core) \tag{3}$$

In this context, "core" represents the number of server cores.

Considering the transmission delay of each task between different servers and users, we can express it by the ratio of the amount of data required for transmission, d_j^k, to the transmission rate R_j or $R_{j,i}$, where R_j represents the link transmission rate for tasks offloaded to the server closest to the user, and $R_{j,i}$ represents the link transmission rate for tasks offloaded to other servers. Both rates can be calculated using the Shannon formula:

$$R_j = B * log_2(1 + P * h_{i,j} / N_0) \tag{4}$$

After determining the transmission rate, we next discuss the transmission delay and obtain the total delay by adding it to the computational processing delay and queuing delay t' :

(1)   When the user offloads the task to the local server, i.e., $i = y$, the total delay for a single task is calculated as follows:

$$t_{j,k}^{i,1} = c_j^k / (p_i * core) + d_j^k / R_j + t' \tag{5}$$

(2)   When the local server closest to the user is busy and the task needs to be offloaded to other edge servers, i.e., $i \neq y$, the total delay for a single task is calculated as follows:

$$t_{j,k}^{i,2} = t_{j,k}^{i,1} + d_j^k / R_{y,i} \tag{6}$$

In summary, the total delay for a single task can be summarized as follows:

$$t_{j,k}^i = \begin{cases} t_{j,k'}^{i,1} & \text{if } i = y \\ t_{j,k'}^{i,2} & \text{if } i \neq y \end{cases} \tag{7}$$

From this, we can write the total delay for all tasks of a single user:

$$t_{j,k} = \sum_{i=1}^{m} (t_{j,k}^i * X_{j,k}^i) \tag{8}$$

However, since this paper discusses the offloading and processing of multiple batches of tasks on multiple edge servers, we can further take the maximum value of the offloading and processing delays of all single-user tasks as the total delay for all users and all tasks:

$$t_j = max\left\{t_{j,k}\right\} \tag{9}$$

3.3.2. Energy Consumption

Since this paper discusses the task offloading method balancing delay and energy consumption, establishing a reasonable energy consumption model is crucial. We can also divide energy consumption into two parts as the delay model: the energy consumption of users sending tasks and the energy consumption required for task offloading computation.

Regarding the energy consumption for tasks sent by users, we can also follow a similar discussion pattern as that for latency, addressing two scenarios:

Firstly, when $i = y$, which means the user offloads the task to the local server, we can sum up the energy consumption required for transmitting each task of the users:

$$e_{j,tran1} = \sum_{k=1}^{z} d_j^k * p_{tran} / R_j \tag{10}$$

However, when i is not equal to y, indicating that the local server is busy and the task needs to be sent to another server, we must take into account the energy consumption for transmission between servers:

$$e_{j,tran2} = \sum_{k=1}^{z} d_j^k * p_{tran} / R_j + d_j^k * p_{tran} / R_{y,i} \tag{11}$$

In summary, our total transmission energy consumption can be summarized as:

$$e_{tran} = \begin{cases} e_{tran1}, & \text{if } i = y \\ e_{tran2}, & \text{if } i \neq y \end{cases} \tag{12}$$

Similarly, for the computation energy consumption of task offloading, we can also perform a summation:

$$e_{j,com} = \sum_{k=1}^{z} \sum_{i=1}^{m} C_j^k * p_{i,com} * (X_{j,k}^i) / p_i \tag{13}$$

In summary, our energy consumption optimization model can combine the two parts of energy consumption to represent the total energy consumption of each user processing tasks:

$$e_j = e_{j,tran} + e_{j,com} \tag{14}$$

3.4. Optimization Objective

We formulate the offloading problem of large-model inference subtasks as a linear programming problem to improve the overall performance in a stable state. The first goal is to minimize latency, i.e., maximize the quality of service, under the premise of successful process task offloading and computing. The second optimization goal is to minimize energy consumption. The specific task offloading model is as follows:

$$\min \quad \alpha t_j + \beta e_j \tag{15}$$

$$\text{subject to} \quad \alpha + \beta = 1 \qquad \alpha, \beta \in (0, 1) \tag{16}$$

$$\sum_{i=p} d_j^k \le m_p \tag{17}$$

$$\sum_p X_{i,j}^p = 1 \qquad X_{i,j}^p \in (0, 1) \tag{18}$$

$$\sum_{j,p} X_{i,j}^p \le n_i \tag{19}$$

where α and β are linear weighting coefficients that can control whether the overall optimization goal is to achieve the lowest latency or minimize total energy consumption. By adjusting the values assigned to these coefficients, the optimization process can be tuned to prioritize either low latency or energy efficiency. Given the actual deployment scenario and mathematical reasoning, the deployment strategy is subject to some key constraints. Equation (16) represents the constraint on the weighting coefficients. Equation (17) indicates that the total amount of data required for the task should not exceed the server's storage capacity. Equation (18) imposes constraints on the 0, 1 matrix. Finally, Equation (19) states that the number of offloaded tasks should not exceed the number of cores on the server. Equations (17) and (19) are the constraints related to storage resources.

4. Algorithm

As described in the third section, we establish separate partitioning and offloading models to model the problem. In order to solve the decision-making problem of real-time task partitioning and offloading, we propose Greedy Partitioning and Offloading Algorithms (GPOA). The algorithm process is shown in Figure 4. This algorithm mainly has two decision-making parts—the granularity of the model's partitioning and the unloading scheme of the partition task. Different model partitioning schemes can lead to different optimal offloading schemes, and similarly, changes in offloading schemes can also affect the optimal partitioning scheme, and the two are coupled. Therefore, our proposed GPOA simultaneously makes decisions for both. At the same time, due to the real-time arrival of tasks, it is necessary to estimate the queuing time when deciding on the unloading plan for each task. Therefore, we need to maintain and update the core usage and queuing queue in real-time for each edge server. We have designed a Core Refreshing Algorithm (CRA) to address this issue.

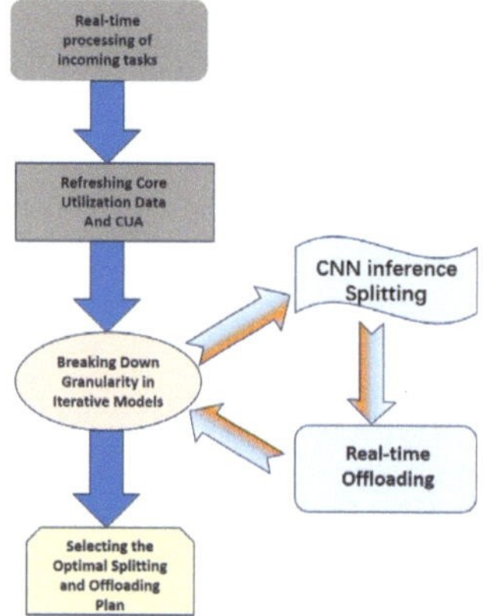

Figure 4. Flowchart of GPOA.

4.1. Core Refreshing Algorithm

As mentioned in Algorithm 1. When it is necessary to update the core state of the edge server, the core that has already completed the task is first released based on the current time and the usage of the core maintained by the system (line 3). Then, iterate through the subtasks that are currently in queue. If the idle core can meet the requirements of the subtask in the queue, allocate the core to this task and check whether the remaining cores continue to meet the requirements of the next task (line 6). If it cannot meet the requirements of queuing tasks, then this part of the core remains idle. Through this algorithm, we maintain a series of tables that play an important role in estimating the queue duration of tasks.

Algorithm 1 CRA

Input: Time t, number of servers n, the usage of the server's core before updating *core*, the queuing situation of the core server before updating *queue*;
Output: The usage of the updated server's core *core*, queuing situation of the updated server's core *queue*
1: **for** $i = 0$ to n **do**
2:     **while** The completion time of core tasks is less than the current time **do**
3:         Release the core of the earliest completed task
4:         **while** The idle core can meet the needs of the first task in the *queue* **do**
5:             Assign the core to this task
6:             Add task to *core*
7:             Delete the first task in *queue*
8:         **end while**
9:     **end while**
10: **end for**

4.2. Greedy Partitioning and Offloading Algorithms

The specific explanation for Algorithm 2 is as follows: When a large CNN inference task arrives at the system, the first step is to partition the inference task according to a certain granularity (line 4). For different partition granularities, we will obtain different numbers of subtasks. We need to find a solution that balances latency and power requirements to offload these subtasks to this edge network. In this algorithm, we first sort these subtasks by computational complexity (line 5). Starting from the task with the highest computational complexity, calculate the latency and energy consumption that would occur if subtasks are offloaded to a server (line 6). Then, offload this task to the server with the lowest latency and energy consumption. Finally, we obtain an uninstallation scheme with specific partition granularity and corresponding latency and energy consumption (line 9). Then, we compare the energy consumption and latency of different partition granularities and select the optimal partition scheme (line 11).

Algorithm 2 GPOA

Input: Inference model M, arrival time t;
Output: Optimal unloading plan x;
1: Update the status of each core of the server
2: Update queue
3: **while** each partitioning method **do**
4:     partitioning the inference task
5:     Sort the partition subtasks by computational complexity
6:     Calculate the goal of offloading each subtask to different servers
7:     Sort different servers according to goal
8:     Select the optimal offloading method
9:     Storage offloading methods and goal
10: **end while**
11: Select the optimal partitioning method based on the goal

4.3. Complexity Analysis

CRA: The outermost loop is to traverse all servers, with a time complexity of $O(|S|)$. The inner loop may execute $O(|L|)$ times in the worst-case scenario, as each iteration may involve the deletion and addition of one or more tasks. Each iteration's internal operation is $O(|L|)$. Thus, the time complexity of this algorithm is $O(|S| * |L|^2)$, where $|S|$ is the number of servers and $|L|$ is the length of the usage sequence of the server's cores.

GPOA: As mentioned above, the time complexity of updating the state is $O(|S| * |L|^2)$. For each subtask, the estimated queuing delay function will traverse the usage sequence of the kernel, with a time complexity of $O(|L|)$. At the same time, each subtask will traverse each server, so the time complexity is $O(|S| * |L|)$. Assuming the number of subtasks is $|M|$, the total time complexity is $O(|S| * |L|^2 + |M| * |S| * |L|)$

5. Performance Evaluation

In this section, we evaluate the performance of our partitioning model and task offloading algorithm through simulation experiments. We consider the following influencing factors: data size, number of Edge Node CPU Cores, and request arrival rate. Moreover, in our simulation experiments, we compare our model with three baseline algorithms and use the method of controlling variables to evaluate the impact of influencing factors on our model and algorithm.

5.1. Experiment Settings

We build a simulation environment for CNN task partitioning and offloading on the Python platform. All simulations are performed on a PC with an Intel (R) Core (TM) i7-11370H CPU @ 3.30GHz, NVIDIA GeForce RTX 3050 Ti GPU, 8GB RAM and Windows 10. The detailed experimental data settings are as follows:

Edge computing network: Our edge network consists of 10 nodes, each with distinct regional characteristics, thus forming a collaborative network. Each node is assigned a specific service area. The data of our edge network nodes and structures are abstracted from the edge computing dataset [28] of Shanghai Telecom. The dataset shows the distribution of 3233 base stations in the Shanghai area, which constitute the infrastructure of mobile communication networks. The density and distribution of base stations reflect the coverage capability and service quality of mobile communication networks in Shanghai. The CPU cores of each edge node are evenly distributed within the range of [8, 16], the core computing frequency is 1.12 GHz and the data transmission rates between nodes are randomly distributed within [5, 20] Gbps. To account for the heterogeneity in propagation paths within the edge system, we utilized a 10 × 10 bandwidth matrix to represent the diverse bandwidth conditions among the edge nodes.

Inference workload: Our partitioning model divides the linear chain structure into a nonlinear DAG structure, enabling parallel execution of the partitioned subtasks. This method can be applied to most large model inference tasks with convolutional layer structures, such as Transformer and RNN. Here, our simulation experiment takes the classic CNN model, VGG-16 [29], as an example to discuss the partitioning of inference tasks. VGG-16 consists of a total of 16 layers, but during the partitioning process, the workload of each subtask increases uniformly as the number of layers increases. Therefore, to demonstrate the performance of our partitioning model and task offloading algorithm, we can select one of the layers for simulation of partitioning and offloading. For VGG-16, the piece of input data is a 224 × 224 image with 3 channels.

The banded distribution figures, constructed from gathered data across repeated experiments, provide a comprehensive visualization that underscores the robust consistency and validity of the derived conclusions. This approach ensures a heightened level of scientific rigor and credibility in the interpretation of the experimental outcomes.

5.2. Baseline Algorithm and Metrics

In order to qualitatively and effectively evaluate the performance of our algorithm, we use the following baseline algorithms for comparison.

(1)   Randomized algorithm. The random algorithm randomly selects edge network nodes for microservice deployment in response to each application request. If the initially selected node lacks the necessary resources, the algorithm repeats the selection process until a suitable node is found. However, the random algorithm here is not simply random but also involves the process of selecting the optimal solution.

(2)   Genetic algorithm. Generate a set of initial solutions that satisfy the constraints using a random algorithm. Subsequently, the optimization objective is used as a fitness function to evaluate the quality of each solution. Perform iterative crossover, mutation, and selection operations to search for the best task offloading solution. The genetic algorithm here includes selection strategies.

(3)   Fixed partition granularity. Using a fixed granularity to partition all CNN tasks without considering the characteristics of the task and the condition of the edge server.

The following metrics are used to evaluate the performance of our partitioning and offloading algorithm.

(1)   Average delay: The average delay of all tasks from request to completion, which is defined in Equation (9).

(2)   Energy consumption: The energy consumed during transmission and computation processes, which is defined in Equation (14).

(3)   Combined metric: Combined metric based on delay and energy consumption, which is defined in Equation (15).

5.3. Experimental Results

5.3.1. The Influence of the Number of Edge Node CPU Cores

We conducted an in-depth study on the impact of the number of available cores on edge servers on the algorithm results. The number of cores on each edge server is investigated within the range of 5 to 30, with an interval of 1, as shown in Figure 5. The results indicate that as the number of server cores increases, the latency of all three algorithms gradually decreases. This is because the number of available idle cores per server increases, enabling each algorithm to better allocate corresponding cores for incoming tasks, enhancing task efficiency and parallelism, and subsequently reducing computation time and waiting time. Notably, regardless of the variation in the number of server cores, our offloading algorithm consistently exhibits the lowest latency compared to the random algorithm and genetic algorithm. Moreover, Figure 5 demonstrates that as the number of server cores increases, our algorithm maintains a declining trend in average energy consumption, consistently outperforming the other two baseline algorithms. This characteristic underscores the capability of our algorithm to minimize energy usage. Figure 5 also illustrates that our offloading algorithm surpasses the other two baseline algorithms in terms of comprehensive metrics encompassing latency and energy consumption. It can also be seen intuitively from Figure 5 that the strip distribution of multiple experimental results shows that our algorithm always outperforms the baseline algorithm, indicating the consistency of our experimental results. The specific experimental data on latency and energy consumption as the number of Edge Node CPU cores increases are presented in Table 1. The superiority of our algorithm can be specifically seen from the table. From Figure 5, we can see that as the number of cores increases, the energy consumption of our algorithm and baseline algorithm remains basically unchanged. This is because an increase in the number of cores will only reduce the queuing probability and latency of reaching subtasks. In our energy consumption calculation method, only transmission energy consumption and calculation energy consumption are included, and the reduction in queuing delay does not affect the calculation of energy consumption, so the energy consumption remains basically unchanged. For fixed partition granularity, energy consumption decreases as the number of cores increases. This is because the granularity of the split remains constant, which means that as the number of cores increases, the method cannot achieve optimal partitioning. Figure 5 further compares the performance of our method against a fixed partitioning granularity model for task partitioning. As evident from the figure, when running the same task offloading algorithm, the latency, energy consumption, and comprehensive metrics associated with fixed partitioning granularity are significantly inferior to those achieved by our method. This disparity stems from the fact that our method exhaustively explores various partitioning granularities and ultimately selects the one that optimizes the comprehensive metrics for peak performance. The rationale behind this approach is that excessively large partitioning granularities result in oversized subtasks that fail to fully capitalize on the parallelism benefits of task partitioning, while excessively small granularities increase data redundancy and subsequently expand the task workload. Our method addresses these issues by striking a balance between the two extremes to achieve optimal performance.

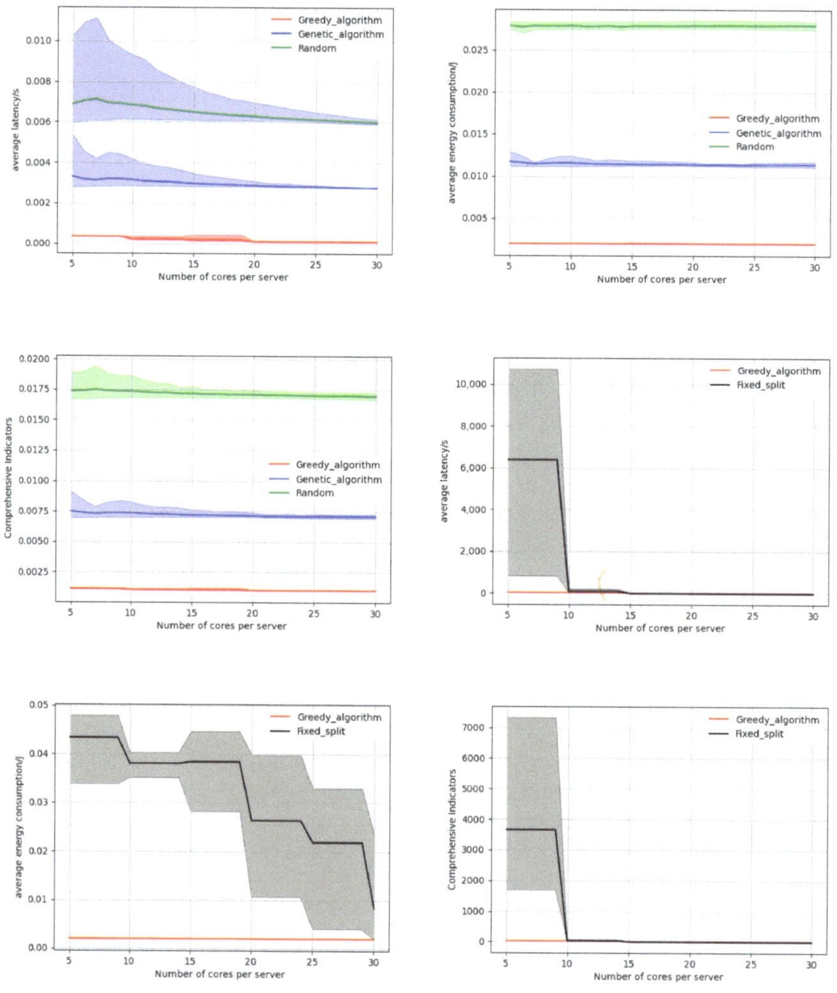

Figure 5. The influence of the number of Edge Node CPU Cores.

Table 1. The influence of the number of Edge Node CPU Cores.

Number of Cores	Delay/ms				Energy Consumption/J			
	Greedy	Genetic	Random	Fixed	Greedy	Genetic	Random	Fixed
5	0.33	3.28	6.86	7,331,934	0.00187	0.0116	0.0277	0.0434
10	0.19	3.13	6.81	71,607	0.00189	0.0115	0.0278	0.0381
15	0.17	2.96	6.48	4925	0.00190	0.0113	0.0278	0.0385
20	0.071	2.86	6.27	252	0.00189	0.0113	0.0279	0.0264
25	0.071	2.80	6.10	26	0.00189	0.0113	0.0278	0.0219
30	0.071	2.75	5.98	1.27	0.00189	0.0113	0.0279	0.0084

5.3.2. The Influence of the Data Size

Subsequently, we delve into the performance of the three algorithms under varying task data sizes. We change the size of the input data from 224 × 224 to 504 × 504 with a step length of 56. When the size of the data is large, it is equivalent to simulating the performance of our algorithm under high task loads. As the data size increases, the time and power consumption required for servers to complete tasks escalate, leading to a higher likelihood and duration of task queuing, which in turn elevates the average latency and energy consumption across all tasks. This investigation is conducted using 10 edge servers, each equipped with 8 to 15 cores. As depicted in Figure 6, the experimental results reveal a consistent trend in the performance metrics of all three algorithms: as the task data size grows, the average latency, average energy consumption, and the composite indicator all increase. However, our algorithm consistently outperforms the two baseline algorithms. As shown in the strip distribution figure, the superiority of our algorithm is consistent. This superiority can be attributed to our greedy algorithm, which dynamically determines the optimal queuing sequence and assigns the most suitable server for each task upon the arrival of multiple tasks. Such adaptability enables our algorithm to alleviate latency and energy consumption issues, ultimately enhancing overall performance. Furthermore, Figure 6 compares the performance of our method against that of fixed partition granularity. It is evident that our method surpasses the fixed partition granularity approach in terms of latency, energy consumption, and the composite indicator. This underscores the advantage of dynamically selecting the optimal partition granularity for each task. The specific experimental data on latency and energy consumption as the data size increases are presented in Table 2.

Table 2. The influence of the data size.

Size of Data	Delay/ms				Energy Consumption/J			
	Greedy	Genetic	Random	Fixed	Greedy	Genetic	Random	Fixed
224 * 224	0.20	6.72	11.61	259,839	0.00191	0.0207	0.0509	0.0364
280 * 280	0.32	6.99	11.81	494,168	0.00297	0.0209	0.0510	0.0564
336 * 336	0.48	7.04	12.05	671,661	0.00427	0.0215	0.0528	0.0752
392 * 392	0.82	7.68	12.77	1,009,451	0.00581	0.0230	0.0558	0.1030
448 * 448	2.79	8.26	13.70	1,395,492	0.00757	0.0246	0.0600	0.1299

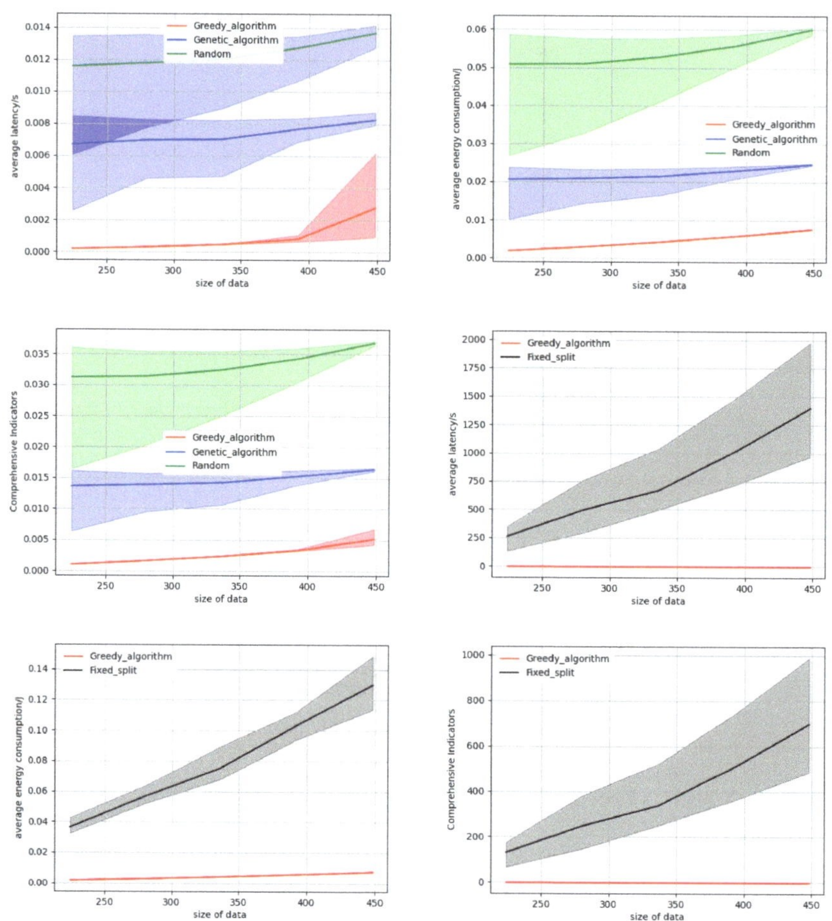

Figure 6. The influence of the data size.

5.3.3. The Influence of the Request Arrival Rate

Finally, we investigated the performance of three algorithms when the time interval between adjacent inference tasks changed. We study the performance of our method under high task loads by increasing the arrival rate of the request. As the time interval increases, the number of idle kernels for the next arriving task also increases, resulting in a decrease in the likelihood and duration of task queuing, thereby reducing the average latency of all tasks. This survey is conducted on 10 edge servers, each equipped with a specific number of cores. As shown in Figure 7, the experimental results reveal the performance indicator trends of three algorithms: for the average delay, as the time interval increases, the greedy algorithm always searches for the optimal solution, the possibility and duration of task queuing decrease, and the average delay continues to decrease. And the band distribution graph in Figure 7 demonstrates the consistency of our algorithm's performance over other baseline algorithms. For average energy consumption, increasing the time interval can only change the queue delay, so the impact on average energy consumption is minimal. Overall, compared to the two baseline algorithms, our algorithm consistently maintains excellent performance and consistent results. In Figure 7, we compare the performance of our method with that of fixed partition granularity. It is evident that our algorithm has significant advantages over fixed partitioning in terms of latency, energy consumption, and overall metrics, highlighting the advantage of our dynamic partitioning in selecting the optimal partition granularity for each task. Table 3 presents specific experimental data on latency and energy consumption as the request arrival rate increases.

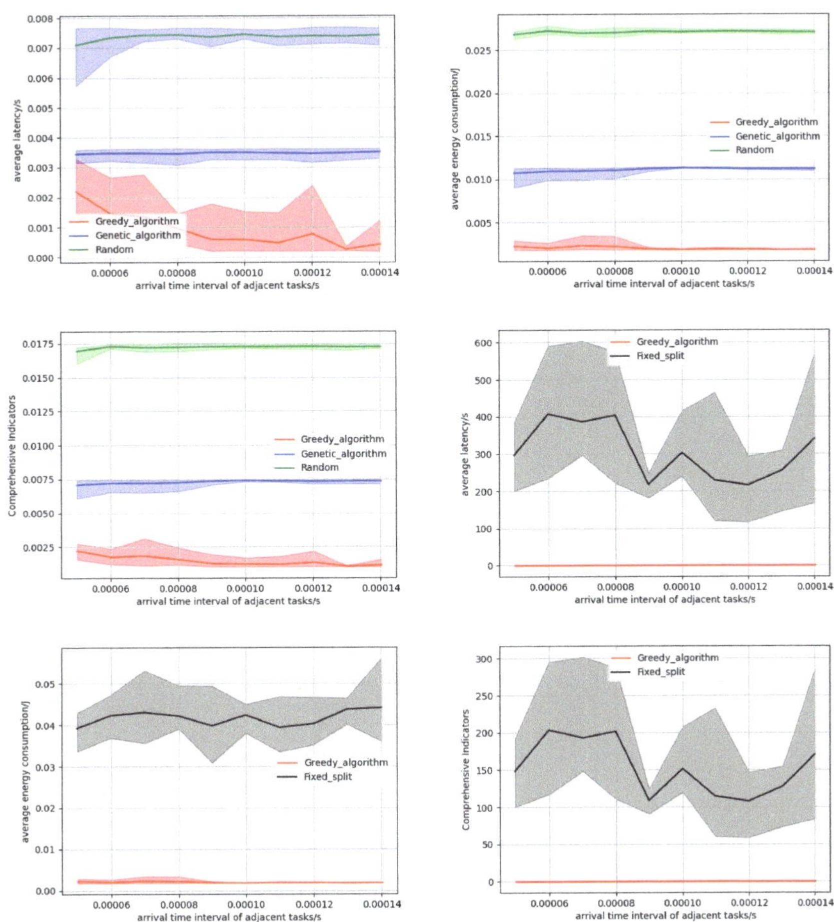

Figure 7. The influence of the request arrival rate.

Table 3. The influence of the request arrival rate.

Time Interval/ms	Delay/ms				Energy Consumption/J			
	Greedy	Genetic	Random	Fixed	Greedy	Genetic	Random	Fixed
0.06	1.47	3.48	7.34	406,554	0.00204	0.0109	0.0272	0.0423
0.08	0.96	3.46	7.44	402,913	0.00220	0.0112	0.0270	0.0422
0.10	0.59	3.50	7.46	302,445	0.00187	0.0113	0.0271	0.0424
0.12	0.77	3.46	7.41	215,295	0.00191	0.0112	0.0271	0.0402
0.14	0.43	3.51	7.43	340,176	0.00188	0.0112	0.0271	0.0441

6. Conclusions

This paper has investigated the issues of CNN task partitioning and task offloading in edge computing networks to enable the deployment of large CNN tasks on edge servers with limited hardware resources while minimizing task latency and energy consumption.

Our approach has comprised two main components: task partitioning and offloading. Instead of the conventional vertical partitioning method, we have opted for horizontal partitioning during partitioning, leveraging the characteristic that convolutional results are only relevant to part of the input data. This approach has divided the CNN task into multiple subtasks that can be executed in parallel. In modeling task offloading, we have accounted for the random arrival of tasks and employed a greedy algorithm to dynamically allocate servers in real-time, aiming to minimize task latency and energy consumption. Finally, we have designed simulation experiments to compare our algorithm with two baseline algorithms, and we have also compared scenarios with fixed partitioning granularity. The experimental results indicate that our method consistently identifies the optimal partitioning granularity for each task, achieving optimal performance when considering both latency and energy consumption.

Lastly, it should be noted that our partitioning method is not designed for a specific CNN architecture but is applicable to all neural networks with a convolutional neural network layer structure, such as RNN and Transformer. This is because our resolution method determines the corresponding resolution scale according to the size of the characteristic graph output of the inference task and the hardware conditions of the device, while considering the coupling of the resolution and offloading process, and then determines the specific partition granularity based on the comprehensive index of delay and energy consumption, which is suitable for most existing CNN models. Furthermore, our partitioning method is designed based on the data locality of convolution computation. The result of the multi-layer convolution operation in CNN is only related to a specific region of the input feature graph data, and the size of this specific region can be calculated by the calculation formula given by us. For convolutional neural network architectures with complex structures and large numbers, our method is also applicable, just the granularity of the partition needs to be designed to be smaller, or large CNN networks can be divided into vertical hierarchies first, and then partitioned in parallel by our method.

Author Contributions: Conceptualization, Y.Y. and K.P.; methodology, Z.Z.; software, Y.Y.; validation, Z.W., Y.Y. and Y.X.; formal analysis, B.L.; investigation, Y.X.; resources, K.L.; data curation, C.Y.; writing—original draft preparation, Z.Z.; writing—review and editing, Z.W.; visualization, B.X.; supervision, K.P.; project administration, Z.Z.; funding acquisition, K.P. All authors have read and agreed to the published version of the manuscript.

Funding: This research was funded in part by the Key Research and Development Program of Hubei Province under grant 2022BAA038, in part by the Key Research and Development Program of Hubei Province under grant 2023BAB074, in part by the special fund for Wuhan Artificial Intelligence Innovation under grant 2022010702040061.

Institutional Review Board Statement: Not applicable.

Informed Consent Statement: Not applicable.

Data Availability Statement: The original contributions presented in the study are included in the article, further inquiries can be directed to the corresponding author.

Conflicts of Interest: Author Zhiyong Zha was employed by the company State Grid Information Telecommunication Co., Ltd. Authors Yongjun Xia, Bin Luo were employed by the company Hubei Huazhong Electric Power Technology Development Co., Ltd. Author Bo Xu was employed by the company Hubei ChuTianYun Co., Ltd. The remaining authors declare that the research was conducted in the absence of any commercial or financial relationships that could be construed as a potential conflict of interest

Nomenclatures

The following abbreviations are used in this manuscript:

Notation	Definition
S	A set of edge servers
P_i	The computing frequency of a single kernel on the i-th edge server
M_i	The storage capacity of the i-th edge server
N_i	The number of cores in the i-th edge server
U	A group of users who need to use large model inference
y_j	The nearest edge server for the j-th user
t_j^k	The k-th task of the j-th user
C_j^k	The computing resources required for the k-th task of the j-th user
d_j^k	The amount of data for the k-th task of the j-th user
$x_{j,k}^i$	Binary decision variables
W_l	The width of the feature map for the l-th layer
H_l	The height of the feature map for the l-th layer
F_l	The dimension of the filter for the l-th layer
D_l	The number of output feature maps for the l-th layer
t_l	The region's size for a tile in the output map of layer l
C_l	The calculation times for the l-th layer
M	Partition the feature map into $M * M$ parts

References

1. Gubbi, J.; Buyya, R.; Marusic, S.; Palaniswami, M. Internet of Things(IoT): A vision, architectural elements, and future directions. *Future Gener. Comput. Syst.* **2013**, *29*, 1645–1660. [CrossRef]
2. Chien, S.Y.; Chan, W.K.; Tseng, Y.H.; Lee, C.; Somayazulu, V.S.; Chen, Y. Distributed computing in IoT: System-on-a-chip for smart cameras as an example. In Proceedings of the 20th Asia and South Pacific Design Automation Conference (ASP-DAC), Chiba, Japan, 19–22 January 2015; pp. 130–135.
3. Shi, W.; Cao, J.; Zhang, Q.; Li, Y.; Xu, L. Edge computing: Vision and challenges. *IEEE Internet Things J.* **2016**, *3*, 637–646.
4. Lane, N.D.; Bhattacharya, S.; Georgiev, P.; Forlivesi, C.; Kawsar, T.F. An early resource characterization of deep learning on wearables, smartphones and Internet-of-Things devices. In Proceedings of the International Workshop Internet Things Towards Appl. (IoT-A), Seoul, Republic of Korea, 26–28 October 2015; pp. 7–12.
5. Kang, A.M.; Hauswald, J.; Gao, C.; Rovinski, A.; Mudge, T.; Mars, J.; Tang, L. Neurosurgeon: Collaborative intelligence between the cloud and mobile edge. In Proceedings of the International Conference on Architectural Support for Programming Languages and Operating Systems (ASPLOS), Xi'an, China, 8–12 April 2017; pp.615–629.
6. Teerapittayanon, S.; McDanel, B.; Kung, H.-T. Distributed deep neural networks over the cloud, the edge and end devices. In Proceedings of the International Conference on Distributed Computing Systems (ICDCS), Atlanta, GA, USA, 5–8 June 2017; pp. 328–339.
7. Zhang, B.; Mor, N.; Kolb, J.; Chan, D.S.; Lutz, K.; Allman, E.; Wawrzynek, J.; Lee, E.; Kubiatowicz, J. The cloud is not enough: Saving IoT from the cloud. In Proceedings of the USENIX Workshop Hot Topics Cloud Computing (HotCloud), Santa Clara, CA, USA, 6–7 July 2015; p. 21.
8. Ribeiro, F.M.; Kamienski, C.A. Timely Anomalous Behavior Detection in Fog-IoT Systems using Unsupervised Federated Learning. In Proceedings of the IEEE 8th World Forum on Internet of Things (WF-IoT), Yokohama, Japan, 26 October–11 November 2022.
9. Krizhevsky, A.; Sutskever, I.; Hinton, G.E. ImageNet classification with deep convolutional neural networks. In Proceedings of the Advances in Neural Information Processing Systems (NIPS), Lake Tahoe, NV, USA, 3–6 December 2012; pp. 1097–1105.
10. He, K.; Zhang, X.; Ren, S.; Sun, J. Deep residual learning for image recognition. In Proceedings of the IEEE Conference on Computer Vision and Pattern Recognition, Las Vegas, NV, USA, 26 June–1 July 2016; pp.770–778.
11. Samie, F.; Tsoutsouras, V.; Xydis, S.; Bauer, L.; Soudris, D.; Henkel, J. Distributed QoS management for Internet of Things under resource constraints. In Proceedings of the 2016 International Conference on Hardware/Software Codesign and System Synthesis (CODES+ISSS), Pittsburgh, PA, USA, 2–7 October 2016; pp. 1–10.
12. Jiao, Z.; Zhang, J.; Yao, P.; Wan, L.; Ni, L. Service Deployment of C4ISR Based on Genetic Simulated Annealing Algorithm. *IEEE Access* **2020**, *8*, 65498–65512. [CrossRef]
13. Fan, W.; Cui, Q.; Li, X.; Huang, X.; Tao, X. On Credibility-Based Service Function Chain Deployment. *IEEE Open J. Comput. Soc.* **2021**, *2*, 152–163. [CrossRef]
14. Li, R.; Lim, C.S.; Rana, M.E.; Zhou, X. A Trade-Off Task-Offloading Scheme in Multi-User Multi-Task Mobile Edge Computing. *IEEE Access* **2022**, *10*, 129884–129898.
15. Mahjoubi, A.; Ramaswamy, A.; Grinnemo, K.J. An Online Simulated Annealing-Based Task Offloading Strategy for a Mobile Edge Architecture. *IEEE Access* **2024**, *12*, 70707–70718.

16. Liu, C.S.W.; Jia, Y.; Sermanet, P.; Reed, S.; Anguelov, D.; Erhan, D.; Vanhoucke, V.; Rabinovich, A. Going deeper with convolutions. In Proceedings of the 2015 IEEE Conference on Computer Vision and Pattern Recognition (CVPR), Boston, MA, USA, 7–12 June 2015; pp. 1–9.
17. Zhao, Z.; Barijough, K.M.; Gerstlauer, A. DeepThings: Distributed adaptive deep learning inference on resource-constrained IoT edge clusters. *IEEE Trans. Comput.-Aided Des. Integr. Circuits Syst.* **2018**, *37*, 2348–2359.
18. Li, H.; Hu, C.; Jiang, J.; Wang, Z.; Wen, Y.; Zhu, W. JALAD: Joint Accuracy- and Latency-Aware Deep Structure Decoupling for Edge-Cloud Execution. In Proceedings of the IEEE 24th International Conference on Parallel and Distributed Systems (ICPADS), Singapore, 11–13 December 2018; pp. 671–678.
19. Li, E.; Zeng, L.; Zhou, Z.; Chen, X. Edge AI: On-Demand Accelerating Deep Neural Network Inference via Edge Computing. *IEEE Trans. Wirel. Commun.* **2020**, *19*, 447–457. [CrossRef]
20. Hu, C.; Li, B. Distributed Inference with Deep Learning Models across Heterogeneous Edge Devices. In Proceedings of the IEEE Infocom, London, UK, 2–5 May 2022; pp. 330–339.
21. Mao, J.; Chen, X.; Nixon, K.W.; Krieger, C.; Chen, Y. MoDNN: Local distributed mobile computing system for deep neural network. In Proceedings of the IEEE Infocom, Atlanta, GA, USA, 1–4 May 2017; pp.1396–1401.
22. Zhang, S.; Zhang, S.; Qian, Z.; Wu, J.; Jin, Y.; Lu, S. DeepSlicing: Collaborative and Adaptive CNN Inference With Low Latency. *IEEE Trans. Parallel Distrib. Syst.* **2021**, *32*, 2175–2187. [CrossRef]
23. Liu, J.; Mao, Y.; Zhang, J. Delay-optimal computation task scheduling for mobile-edge computing systems. In Proceedings of the IEEE International Symposium on Information Theory, Barcelona, Spain, 10–15 July 2016; pp.1451–1455.
24. Mao, Y.; Zhang, J.; Letaief, K.B. Dynamic Computation Offloading for Mobile-Edge Computing With Energy Harvesting Devices. *IEEE J. Sel. Areas Commun.* **2016**, *34*, 3590–3605. [CrossRef]
25. Shi, Y.; Chen, S.; Xu, X. MAGA: A Mobility-Aware computation Decision for Distributed Mobile Cloud Computing. *IEEE Internet Things J.* **2018**, *5*, 164–174. [CrossRef]
26. Chen, M.H.; Liang, B.; Dong, M. A semidefinite relaxation approach to mobile cloud offloading with computing access point. In Proceedings of the IEEE International Workshop on Signal Processing Advances in Wireless Communications, Stockholm, Sweden, 28 June–1 July 2015; pp.186–190.
27. Lu, W.; Ni, W.; Tian, H.; Liu, R.P.; Wang, X.; Giannakis, G.B.; Paulraj, A. Optimal Schedule of Mobile Edge Computing for Internet of Things using Partial Information. *IEEE J. Sel. Areas Commun.* **2017**, *35*, 2606–2615. [CrossRef]
28. Wang, S.; Zhao, Y.; Xu, J.; Yuan, J.; Hsu, C.-H. Edge server placement in mobile edge computing. *J. Parallel Distrib.* **2019**, *127*, 160–168. [CrossRef]
29. Simonyan, K.; Zisserman, A. Very deep convolutional net works for large-scale image recognition. In Proceedings of the 3rd International Conference on Learning Representations, San Diego, CA, USA, 7–9 May 2015.

Article

Techniques for Detecting the Start and End Points of Sign Language Utterances to Enhance Recognition Performance in Mobile Environments

Taewan Kim and Bongjae Kim *

Department of Computer Engineering, Chungbuk National University, Cheongju 28644, Republic of Korea; taewangim05@gmail.com
* Correspondence: bjkim@chungbuk.ac.kr

Abstract: Recent AI-based technologies in mobile environments have enabled sign language recognition, allowing deaf individuals to communicate effectively with hearing individuals. However, varying computational performance across different mobile devices can result in differences in the number of image frames extracted in real time during sign language utterances. The number of extracted frames is a critical factor influencing the accuracy of sign language recognition models. If the number of extracted frames is too small, the performance of the sign language recognition model may decline. Additionally, detecting the start and end points of sign language utterances is crucial for improving recognition accuracy, as the period before the start point and after the end point often involves no action being performed. These parts do not capture the unique characteristics of each sign language. Therefore, this paper proposes a technique to dynamically adjust the sampling rate based on the number of frames extracted in real time during sign language utterances in mobile environments, with the aim of accurately detecting the start and end points of the sign language. Experiments were conducted to compare the proposed technique with the fixed sampling rate method and with the no-sampling method as a baseline. Our findings show that the proposed dynamic sampling rate adjustment method improves performance by up to 83.64% in top-5 accuracy and by up to 66.54% in top-1 accuracy compared to the fixed sampling rate method. The performance evaluation results underscore the effectiveness of our dynamic sampling rate adjustment approach in enhancing the accuracy and robustness of sign language recognition systems across different operational conditions.

Keywords: sign language recognition; mobile environments; dynamic sampling rate; artificial intelligence

Citation: Kim, T.; Kim, B. Techniques for Detecting the Start and End Points of Sign Language Utterances to Enhance Recognition Performance in Mobile Environments. *Appl. Sci.* **2024**, *14*, 9199. https://doi.org/10.3390/app14209199

Academic Editor: Andrea Prati

Received: 30 August 2024
Revised: 4 October 2024
Accepted: 9 October 2024
Published: 10 October 2024

1. Introduction

Recently, with the improvement in computing performance of mobile devices such as smartphones, there has been active research on AI-powered mobile applications utilizing artificial intelligence models. For example, various AI-based services are continuously being developed in fields such as education, real-time translation, and privacy protection [1–3]. In relation to AI-powered mobile applications, research aimed at improving communication between hearing and deaf individuals through the use of artificial intelligence has also been actively conducted [4–9]. A notable example is the use of avatars in sign language services [10–12]. In contemporary society, mobile devices play a significant role in communication, information accessibility, and service usage. In this context, recognizing and translating sign language in mobile environments is crucial not only for deaf individuals but also for effectively facilitating smooth communication with hearing individuals.

However, recognizing and translating sign language in the current mobile environment has limitations [13]. One major limitation is that mobile environments lack the computational power compared to server environments [14,15], resulting in fewer image frames being extracted within the same time period. If the number of extracted frames is too small, the performance of the sign language recognition model may decrease. Additionally, accurately detecting the start and end points of a sign language utterance is critical for improving the accuracy of sign language recognition models. As seen in automatic speech recognition (ASR), the failure to accurately detect start and end points can severely degrade the performance of the recognition system [16]. The period before the start and after the end of a sign language utterance often involves no action being performed. Such segments do not adequately capture the unique characteristics of each sign language. Therefore, including these segments in the training of a sign language recognition model can reduce its accuracy. Addressing these challenges is essential for enhancing the performance and reliability of sign language recognition on mobile devices.

To accurately detect the start and end points of a sign language utterance in a mobile environment, it is necessary to dynamically adjust the sampling rate based on the number of frames extracted, rather than relying on a fixed sampling rate, as proposed in previous research [17]. This is because when fewer frames are extracted during a sign language utterance, the movement changes between consecutive frames are greater compared to when more frames are extracted. Therefore, it is necessary to adjust the sampling rate to determine the start and end points of the sign language utterance based on the number of extracted frames. This approach can better accommodate the computational limitations of mobile environments compared to server environments.

In previous research [17], the focus was on detecting the start and end points of sign language utterances by analyzing keypoint variations in sign language videos. This research was conducted in standard computing environments, such as personal computers or server computers. However, the method proposed in that research is not as suitable for resource-constrained mobile computing environments as it is for server environments. This is because mobile environments have varying computational performance across different devices, leading to differences in the number of sign language utterance image frames that can be obtained within the same time period. In practice, the average frame rate was measured to be 17 fps when extracting keypoint images during real-time sign language utterances on a Galaxy Z Flip4 (Snapdragon 8 Gen 2). Additionally, if multiple applications are running on a mobile device, the frame rate may decrease. Similarly, high memory usage on the mobile device can also cause a drop in frame rate [18]. Consequently, applying the fixed sampling rate method proposed in the previous research may degrade the performance of sign language recognition models in mobile environments.

This paper proposes an improved dynamic sampling rate adjustment method for mobile environments based on the technique suggested in previous research. Our method dynamically adjusts the sampling rate to detect the start and end points of sign language utterances based on the number of frames extracted in real time during sign language utterances in mobile environments. By applying this dynamic sampling rate adjustment method, we can enhance the accuracy of sign language recognition models.

To evaluate the recognition accuracy of the proposed dynamic sampling rate adjustment method, we compared its performance to that of a fixed sampling rate method and a no-sampling baseline across various frame rates: 5, 10, 15, 20, and 25 fps. The experimental results indicated that the proposed method consistently outperformed the fixed sampling rate method, particularly at lower frame rates of 5 and 10 fps, demonstrating its effectiveness in computationally constrained environments. Even at higher frame rates, the dynamic method maintained superior accuracy. Overall, it improved top-5 accuracy by up to 83.64% and top-1 accuracy by up to 66.54%, underscoring its ability to enhance both the accuracy and robustness of sign language recognition systems.

The remainder of this paper is organized as follows. Section 2 reviews related research. Section 3 provides a detailed explanation of the proposed dynamic sampling rate adjustment method. Section 4 describes the performance evaluation environment and presents the results of the performance evaluation in detail. Finally, Section 5 concludes the paper with a discussion of the conclusions and future work.

2. Related Works

In this section, we will discuss related research. Various sign language recognition techniques have been studied to improve the accuracy of sign language detection. Tables 1 and 2 summarize key related works in this area.

Table 1. Overview of sign language recognition techniques and their backbone networks.

Authors	Techniques and Their Explanations	Backbone Network
Ekbote et al. [19]	The paper developed an Indian Sign Language numeral recognition system (0–9) using ANN and SVM classifiers.	SVM
Pathan et al. [20]	The paper employed two layers of image processing: the first for whole images, and the second for hand landmarks.	CNN
Katoch et al. [21]	The paper presented a system for Indian Sign Language recognition using the SURF (Speeded Up Robust Features) method for feature extraction, combined with SVM (support vector machine) and CNN (convolutional neural network) classifiers. The system utilizes the Bag of Visual Words (BOVW) model to enhance the recognition process.	CNN
Kothadiya et al. [22]	The paper employed LSTM and GRU models, combined them sequentially, used dropout for regularization, and trained on the IISL2020 dataset to enhance Indian Sign Language recognition.	LSTM and GRU
Kothadiya et al. [23]	The paper used a transformer encoder with positional embedding patches and self-attention layers, followed by a multilayer perceptron network, to enhance the recognition of fixed Indian Sign Language.	Transformer
Alharthi et al. [24]	The paper employed transfer learning with various pretrained models (MobileNet, Xception, Inception, InceptionResNet, DenseNet, BiT) and vision transformers (ViT, Swin) to recognize Arabic Sign Language, comparing their performance with CNNs trained from scratch.	ViT, Swin

Table 2. Overview of keypoint-based techniques for sign language recognition.

Authors	Techniques and Their Explanations	Backbone Network
Tripathi et al. [25]	The paper developed a sign language recognition system employing video-based, skeleton-based, and deep learning techniques, enhanced by data augmentation for robust performance across various environments.	PoseConv3D
Bird et al. [26]	The paper proposed a neuroevolution approach to enhance fingerspelling recognition in sign language by optimizing deep neural networks through evolutionary algorithms. The technique involves processing ASL fingerspelling images into normalized keypoints, which are then used as inputs for neural networks.	Neuroevolution-optimized deep neural network
Kim et al. [17]	The paper proposed a method for detecting start and end points of sign language utterances using the MediaPipe Holistic model.	R(2 + 1)D

Ekbote et al. proposed the development of an automatic recognition system specifically designed for Indian Sign Language numerals (0–9). The system was built using a self-created database of 1000 images, with 100 images dedicated to each numeral. To effectively capture the distinguishing features of these numerals, the study employed advanced feature extraction techniques, including shape descriptors, scale-invariant feature transform (SIFT), and histogram of oriented gradients (HOG). For classification, two robust machine learning models, artificial neural networks (ANN) and support vector machine (SVM), were utilized.

The proposed system achieved a remarkable accuracy of up to 99%, demonstrating its potential for accurately recognizing Indian Sign Language numerals [19].

In the field of computer vision, convolutional neural networks (CNNs) have become foundational architectures, significantly impacting various applications, including sign language recognition, facial recognition, object detection, medical image analysis, and autonomous driving. For instance, Pathan et al.'s proposed methodology employs a two-layer image processing technique. In the first layer, images are processed as a whole for training purposes, while the second layer focuses on extracting hand landmarks to refine the recognition process. This approach is implemented through a multiheaded convolutional neural network (CNN) model, which effectively manages the dual processing tasks [20].

Katoch et al. proposed a novel Indian Sign Language (ISL) recognition system that combines Speeded Up Robust Features (SURF), support vector machine (SVM), and convolutional neural network (CNN) to improve communication for the hearing and speech-impaired communities. Through extensive experiments with a well-defined ISL gesture dataset, the system achieved over 90% accuracy, with CNN playing a key role in enhancing feature extraction and classification [21].

Kothadiya et al. proposed a model that utilizes long short-term memory (LSTM) and gated recurrent unit (GRU) networks to detect and recognize gestures from isolated video frames of Indian Sign Language (ISL). The authors developed their own dataset, IISL2020, and achieved approximately 97% accuracy in recognizing 11 different signs using a combination of LSTM and GRU layers [22].

More recently, transformer models have been applied to sign language recognition, showing promising results. Transformers, known for their capability to handle long-range dependencies and parallelize training, are particularly effective for sequential data, such as sign language videos.

Kothadiya et al. proposed a novel approach to recognizing static Indian Sign Language using a transformer encoder. The proposed system, which utilizes a vision transformer to process sign language through positional embedding patches and a transformer block with self-attention layers, significantly outperforms traditional convolutional architectures. Achieving an accuracy of 99.29% with minimal training, the proposed SIGNFORMER demonstrates robustness and effectiveness in real-world applications [23].

Alharthi et al. addressed the gap in existing research by applying pretrained models and vision transformers, typically used in image classification, to Arabic Sign Language (ArSL). Utilizing a dataset of 54,049 images representing 32 different Arabic letters, the research compared transfer learning models, including MobileNet, Xception, and ResNet, with CNN architectures. The results show that transfer learning approaches, particularly ResNet and InceptionResNet, achieved high accuracy of 98%, demonstrating the potential of these methods for enhancing sign language recognition in low-resourced languages [24].

Recent research in sign language recognition leverages keypoint-based approaches, drawing on methods proposed in [25]. This study significantly improves 3D human pose estimation by achieving an 18% reduction in prediction error compared to previous unsupervised methods, demonstrating effective 3D pose prediction from 2D inputs without relying on 3D training data. The results, validated on the Human3.6M dataset, show that PoseNet3D outperforms existing techniques, delivering robust and accurate 3D pose estimations with natural and realistic transitions across frames. The success of the model is attributed to the joint fine-tuning of teacher and student networks using temporal, self-consistency, and adversarial losses, enhancing overall prediction accuracy.

Bird et al. [26] explored the use of neuroevolution techniques to optimize deep neural networks for recognizing American Sign Language (ASL) fingerspelling. By employing hyperheuristic algorithms to fine-tune network architectures and hyperparameters, the study achieved a high mean accuracy of 97.44% on a dataset of 1678 images. The research highlights the potential of these techniques in enhancing the accuracy and efficiency of sign language recognition systems, making them more effective for individuals with hearing impairments.

Together, these studies illustrate the critical role of keypoints in improving the accuracy and efficiency of computer vision systems across diverse applications, from human action and pose estimation to facial and hand gesture recognition. By leveraging keypoint data, these models can achieve high performance while addressing the specific challenges associated with each application area.

Kim et al. [17] proposed a novel approach for detecting the start and end points of sign language utterances. This method, which can be considered a preliminary study to our work, utilizes keypoint data extracted from sign language video frames to identify the precise moments when a sign language utterance begins and ends. Kim et al. employed the MediaPipe Holistic model to extract keypoints from video frames, focusing on 59 keypoints, including 17 from the Pose model and 21 from each Hand model (left and right). Their method calculates the change in position of these keypoints between frames to determine the start and end of sign language gestures. The study compared the performance of their method using two different sets of keypoints: a set of 10 keypoints (including elbows, wrists, and middle joints of thumbs, middle fingers, and little fingers) and a set of 22 keypoints (focusing on wrists and the start and end joints of all fingers). It showed that using 10 keypoints yielded better results, with an average detection error of 2.2 frames and an execution time of 3.7 ns. This suggests that considering the overall arm movement, including the elbow, is more effective in detecting the start and end points of sign language utterances than focusing solely on hand movements. This work provides a foundation for improving sign language recognition models by accurately isolating the actual utterance portions of sign language videos, potentially contributing to increased recognition accuracy in sign language translation systems. Our proposed method builds upon this work, addressing some of its limitations and extending its capabilities to propose a dynamic sampling rate adjustment method that can be effectively used in mobile environments.

3. Dynamic Sampling Rate Adjustment Method

In this section, we introduce our dynamic sampling rate adjustment method, designed to improve the performance of sign language recognition in mobile environments by adjusting the sampling rate based on the number of frames extracted in real time during sign language utterances.

Mobile environments typically have less computational power compared to server environments, resulting in greater variability in the number of frames that can be captured in real time during sign language utterances. The method suggested in previous research [17] does not account for these variations, potentially leading to suboptimal recognition accuracy in mobile environments. Our proposed dynamic sampling rate adjustment method addresses this issue by dynamically adjusting the sampling rate according to the actual frame rate obtained during sign language utterances. Table 3 provides a comprehensive list of symbols and their descriptions to explain our proposed method.

In the previous research of [17], the start and end points of sign language utterances were detected based on the amount of change in the movement of each keypoint. SR_{server} is the sampling rate for counting and measuring keypoint coordinates. The keypoint coordinates are counted and measured every SR_{server} frames of the data frame. In the experiment, SR_{server} was set to 5, and N_{change} represents the total number of keypoint position changes. Therefore, N_{change} can be calculated using Equation (1). The counted and measured keypoint information data are structured as (x, y) coordinates. C_i is the i-th keypoint change information, counted and measured by SR_{server}. C_i can be calculated using Equation (2).

$$N_{change} = \lfloor \frac{F_{total}}{S_r} \rfloor \tag{1}$$

$$C_i = \sum_{k=1}^{K_{max}} \sqrt{(x_i^{k_{idx}} - x_{i-1}^{k_{idx}})^2 + (y_i^{k_{idx}} - y_{i-1}^{k_{idx}})^2} \tag{2}$$

Table 3. Symbols and descriptions.

Symbols	Descriptions
F_{total}	Total number of frames in a sign language utterance
k_{idx}	Index of a keypoint
K_{max}	Maximum index for keypoints
S_r	Sampling rate
N_{change}	Total number of keypoint position change records ($N_{change} = \lfloor \frac{F_{total}}{S_r} \rfloor$)
C_i	Change record of the i-th keypoint position
$x_i^{k_{idx}}$	x-coordinate of keypoint at index k_{idx} in the i-th frame
$y_i^{k_{idx}}$	y-coordinate of keypoint at index k_{idx} in the i-th frame
K_{info}	The x and y coordinate information of keypoints for all frames
C_{total}	Sum of all keypoint position change records ($\sum_{i=1}^{N_{change}} C_i$)
C_{avg}	Average of all keypoint position change records (Threshold value)
S_{start}	Detected start frame of the sign language utterance
S_{end}	Detected end frame of the sign language utterance
$T_{utterance}$	Duration of the sign language utterance in mobile environments
N_{frames}	Total number of frames captured during $T_{utterance}$
fps_{server}	Frame rate in server environments
fps_{mobile}	Frame rate in mobile environments
SR_{server}	Optimal fixed sampling rate in server environments
SR_{mobile}	Dynamic sampling rate in mobile environments
R_{fps}	Ratio of fps_{mobile} to fps_{server}

In Equation (2), k_{idx} represents the index of the counted and measured keypoint, and the maximum index is K_{max}. The maximum value of K_{max} varies depending on the keypoints used at the start and end of the hand. $x_i^{k_{idx}}$ is the x coordinate of the keypoint with index k_{idx} in the i-th keypoint extraction frame. Similarly, $y_i^{k_{idx}}$ is the y coordinate of the keypoint with index k_{idx} in the i-th keypoint extraction frame.

C_{Total} is the sum of the position changes of each keypoint and can be calculated using Equation (3). As shown in Equation (3), C_{Total} is calculated as the sum of all C_i.

$$C_{total} = \sum_{i=1}^{N_{change}} C_i \tag{3}$$

C_{avg} is the average value of the position changes of each keypoint and can be calculated using Equation (4). C_{avg} is used as the threshold for determining the start and end of hand motion.

$$C_{avg} = \frac{C_{total}}{N_{change}} \tag{4}$$

Finally, in the detection algorithm, the start and end of hand motion are determined by Equations (5) and (6). For the start of hand motion, the frame with the smallest index i among the frames larger than C_{avg} becomes the start of hand motion. Conversely, for the end of hand motion, the frame with the largest index i among the frames larger than C_{avg} becomes the end of hand motion.

$$S_{start} = \min(\forall i, \text{where}(C_{avg} \leq C_i)) \tag{5}$$

$$S_{end} = \max(\forall i, \text{where}(C_{avg} \leq C_i)) \tag{6}$$

The key parameter is the sampling rate, SR_{server}. If SR_{server} is too small, many frames need to be counted and measured to detect the start and end points of the sign language utterance. Conversely, if it is too large, fewer frames are counted and measured. In short, this parameter determines the frequency at which keypoint changes are observed.

Applying this fixed sampling algorithm directly to mobile environments presents challenges due to mobile devices having lower computing power compared to servers. As a result, typical mobile devices, such as smartphones, extract fewer frames per unit of time, making the use of a server-optimized fixed sampling rate impractical. For example, in the case of a server, it is possible to obtain more than 30 fps (frames per second) during sign language utterances, whereas mobile devices, such as smartphones, may obtain less than 30 fps depending on their computing performance. Therefore, in mobile environments, it is necessary to dynamically adjust the sampling rate based on the obtained frames to detect the start and end points of sign language utterances.

As we explained, in mobile environments, frames are extracted in real time during sign language utterances, leading to variations in the number of frames extracted over the same period compared to servers, where all frames can be extracted from a video. This variation necessitates adjusting the sampling rate dynamically based on each mobile device's frame extraction rate. By doing so, the accuracy of detecting the start and end points of sign language utterances can be improved, ultimately leading to an improvement in the overall accuracy of the sign language recognition model. This dynamic method is crucial in resource-constrained environments like mobile devices to provide sign language recognition-based services.

A detailed explanation of our dynamic sampling rate adjustment method, based on the actual frame rate of the extracted images when detecting the start and end points of sign language utterances in a mobile environment, is as follows.

The most important aspect of the proposed method is the ability to calculate the actual frame rate when extracting images during sign language utterances, considering the performance of the mobile device. The frame rate in mobile environments (fps_{mobile}) can be calculated using Equation (7).

$$fps_{mobile} = \frac{N_{frames}}{T_{utterance}} \tag{7}$$

It is also necessary to calculate the ratio of the frame rate in mobile environments to the frame rate on the server. Based on this, the sampling rate can be adjusted dynamically. The ratio of the frame rate in mobile environments to the frame rate on the server (R_{fps}) is calculated using Equation (8).

$$R_{fps} = \frac{fps_{mobile}}{fps_{server}} \tag{8}$$

Finally, in the proposed method, the dynamic sampling rate (SR_{mobile}) is determined by adjusting the optimal fixed sampling rate used in server environments according to the frame rate ratio, as shown in Equation (9).

$$SR_{mobile} = \left\lfloor SR_{server} \times R_{fps} + 0.5 \right\rfloor \tag{9}$$

As shown in Equation (9), this ensures that the sampling rate is dynamically adjusted to maintain consistent recognition performance despite variations in frame rates in mobile environments.

Algorithm 1 illustrates the process of the proposed dynamic sampling rate adjustment for mobile devices. To clearly illustrate the method for calculating the dynamic sampling rate described in Algorithm 1, consider the following scenario in which the optimal fixed sampling rate in server environments (SR_{server}) is 5. Additionally, assume that the number of frames extracted in real time (N_{frames}) is 50, the duration of the sign utterance ($T_{utterance}$) is 10 s, and the frame rate in server environments (fps_{server}) is 30 fps.

Algorithm 1 Dynamic sampling rate adjustment for mobile devices

---

1: **Input:** SR_{server}, N_{frames}, $T_{utterance}$, and fps_{server}
2: **Output:** SR_{mobile}
3: Calculate fps_{mobile} using Equation (7):
4: $\qquad fps_{mobile} \leftarrow \frac{N_{frames}}{T_{utterance}}$
5: Calculate R_{fps}, the ratio of mobile to server frame rate, using Equation (8):
6: $\qquad R_{fps} \leftarrow \frac{fps_{mobile}}{fps_{server}}$
7: Calculate the dynamic sampling rate, SR_{mobile}, using Equation (9):
8: $\qquad SR_{mobile} \leftarrow \left\lfloor SR_{server} \times R_{fps} + 0.5 \right\rfloor$
9: **return** SR_{mobile}

Based on these assumptions, we can compute the frame rate in mobile environments (fps_{mobile}) as shown in Equation (10). The ratio of the mobile frame rate to the server frame rate (R_{fps}) is calculated as shown in Equation (11). Finally, we can obtain the dynamic sampling rate (SR_{mobile}) as shown in Equation (12).

$$fps_{mobile} = \frac{N_{frames}}{T_{utterance}} = \frac{50}{10} = 5\,\text{fps} \tag{10}$$

$$R_{fps} = \frac{fps_{mobile}}{fps_{server}} = \frac{5}{30} \approx 0.167 \tag{11}$$

$$SR_{mobile} = \left\lfloor SR_{server} \times R_{fps} + 0.5 \right\rfloor = \lfloor 5 \times 0.167 + 0.5 \rfloor = \lfloor 1.335 \rfloor = 1 \tag{12}$$

In this scenario, the dynamic sampling rate in mobile environments (SR_{mobile}) is set to 1. This means that each extracted frame is sampled and checked to detect the start and end points of a sign language utterance. In summary, the proposed method checks image frames at shorter intervals as the number of extracted frames from a mobile device decreases, in order to improve sign language recognition accuracy. Therefore, the proposed method can maintain consistent recognition performance, even with a lower frame rate compared to server environments.

Algorithm 2 provides a method for detecting sign language utterances' start and end points. Once the dynamic sampling rate for mobile environments (SR_{mobile}) is determined based on Algorithm 1, this algorithm uses the movement of hand keypoints across frames to identify the exact points where a sign language utterance starts and ends.

As shown in Algorithm 2, it leverages the calculated SR_{mobile} to compute the number of frames to be analyzed for changes, denoted as N_{change} (where S_r is replaced by SR_{mobile}). By assessing the cumulative movement of keypoints between frames (C_i) and comparing it to the average movement (C_{avg}), the algorithm detects the start point (S_{start}) and end point (S_{end}) of the utterance.

Algorithm 2 Detecting start and end points of sign language utterances through dynamic sampling rate

---

1: **Input:** SR_{mobile}, K_{info}, F_{total}, K_{max}
2: **Output:** S_{start}, S_{end}
3: **Initialize** $C_{total} \leftarrow 0$, $N_{change} \leftarrow 0$, $C_i \leftarrow 0$, $C_{avg} \leftarrow 0$, $S_{start} \leftarrow -1$, and $S_{end} \leftarrow -1$
4: Calculate N_{change} using Equation (1) (replacing S_r with SR_{mobile}):
5: $\qquad N_{change} \leftarrow \lfloor \frac{F_{total}}{SR_{mobile}} \rfloor$
6: **for** $i = 1$ to N_{change} **do**
7: $\qquad$ Calculate C_i using Equation (2):
8: $\qquad\qquad C_i \leftarrow \sum_{k=1}^{K_{max}} \sqrt{(x_i^{k_{idx}} - x_{i-1}^{k_{idx}})^2 + (y_i^{k_{idx}} - y_{i-1}^{k_{idx}})^2}$
9: $\qquad$ Accumulate C_{total}:
10: $\qquad\qquad C_{total} \leftarrow C_{total} + C_i$
11: **end for**
12: Calculate C_{avg} using Equation (4):
13: $\qquad C_{avg} = \frac{C_{total}}{N_{change}}$
14: **for** $i = 1$ to N_{change} **do** $\qquad\qquad\qquad\qquad$ ▷ Determine Start and End Points
15: $\qquad$ **if** $C_i \geq C_{avg}$ **then**
16: $\qquad\qquad$ **if** $S_{start} = -1$ **then**
17: $\qquad\qquad\qquad S_{start} \leftarrow i$ $\qquad\qquad\qquad\qquad$ ▷ First time exceeding C_{avg}
18: $\qquad\qquad$ **end if**
19: $\qquad\qquad S_{end} \leftarrow i$ $\qquad\qquad\qquad\qquad$ ▷ Update end point to current frame
20: $\qquad$ **end if**
21: **end for**
22: **return** S_{start}, S_{end}

By dynamically determining the sampling rate based on the number of frames extracted in mobile environments, we can gain the following advantages. First, we can provide adaptability to different mobile device capabilities and conditions by adjusting the sampling rate in real time based on frame rates. Second, we can ensure accuracy in detecting the start and end points of sign language utterances, leading to better overall recognition performance. Lastly, we can reduce the use of computational resources, minimizing unnecessary processing, and lowering the computational load.

Figure 1 illustrates an example of the service flow using the proposed dynamic sampling rate adjustment method in a mobile computing environment. As shown in Figure 1, the process begins with the user initiating a sign language utterance, which is captured by the mobile device's camera. The mobile device processes the captured frames in real time, applying algorithms to detect the start and end points of the utterance, thereby optimizing the input data for recognition.

After detecting these key segments, the processed images are transmitted to a sign language recognition model, typically located on a server or in the cloud. This model performs inference on the input data to identify the specific sign language gestures. Once the inference is completed, the recognition results are sent back to the mobile device for display or further use.

The mobile device focuses on capturing and preprocessing the gesture data, using dynamic sampling rate adjustments to enhance efficiency, particularly in resource-constrained environments. The recognition model then performs the computationally intensive task of interpreting the gestures. By dividing the workload in this manner, the proposed method ensures reliable and effective sign language recognition, despite the limited computational power available on mobile devices.

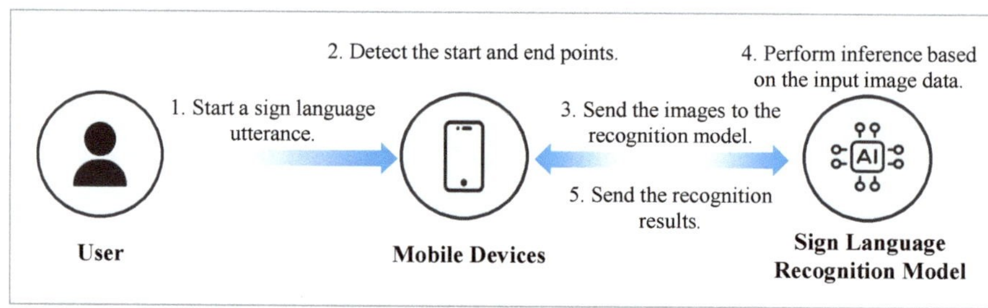

Figure 1. An example of the service flow using the proposed dynamic sampling rate adjustment method in a mobile computing environment.

4. Performance Evaluations

This section provides a detailed explanation of our performance evaluation environments and results. Ideally, the performance evaluation should be conducted in a mobile environment. However, this approach poses several challenges, including the extensive time required to evaluate a large dataset containing 510 sign language classes, the difficulty for sign language performers to master and perform all gestures accurately, and the variability in performance capabilities across different mobile devices. For these reasons, our dynamic sampling rate adjustment method was evaluated in a server environment rather than on a mobile platform. Although the performance evaluation was conducted in a server environment, it was performed with varying fps (frames per second) to account for differences in computing performance across different mobile environments, specifically by adjusting the number of image frames extracted during sign language utterances.

We compared the performance of our dynamic sampling rate adjustment method against the fixed sampling rate method proposed in previous research [17], as well as a baseline with no sampling adjustment. Table 4 shows the hardware and software configuration for the experimental environments.

Table 4. Hardware and software configuration for experimental environment.

Components	Specifications
CPU	AMD Ryzen 9 7950X3D, 16 cores
RAM	Samsung DDR4 64 GB × 2
GPU	ZOTAC GeForce RTX 4090 × 2, 24 GB GDDR6X each
OS	Ubuntu 22.04.2 LTS
CUDA Version	12.2
Python	3.10.11
MediaPipe	0.10.11
OpenCV-Python	4.7.0.72
Torch	2.0.1
Torchvision	0.15.2

4.1. Experimental Environments

4.1.1. Sign Language Recognition Model

To evaluate the performance of our dynamic sampling rate adjustment method, we used the Video Swin Transformer model [27]. This model stands out as a cutting-edge architecture in video recognition, celebrated for its efficiency and accuracy in processing video sequences. Since sign language recognition involves inferring actions from video data, the Video Swin Transformer model is highly suitable for sign language recognition and classification tasks. Specifically, we utilized the swin3d_b model with pretrained weights `torchvision.models.video.Swin3D_B_Weights.KINETICS400_IMAGENET22K_V1` from PyTorch.

The overall architecture of the Video Swin Transformer used in the performance evaluations is shown in Figure 2. The final output layer was modified to predict 510 classes to match the sign language dataset used for training and validation. The Video Swin Transformer is specifically designed to capture both spatial and temporal information from video frames, making it highly suitable for tasks such as sign language recognition. It achieves this by segmenting video frames into nonoverlapping windows and applying self-attention mechanisms within each window. This methodology enables the model to focus effectively on relevant segments of the frames, facilitating the learning of intricate patterns associated with sign language movements. Key features of the Video Swin Transformer include the following:

- Hierarchical structure: The model employs a hierarchical design that processes video data at multiple scales. This allows it to capture both fine-grained details and broader contextual information.
- Shifted window partitioning: Unlike traditional transformer models, the Video Swin Transformer uses a shifted window partitioning scheme. This approach enables connections between windows, facilitating the flow of information across different regions of the video frames.
- Relative position bias: The model incorporates relative position encoding, which helps maintain spatial relationships between different parts of the input, crucial for understanding the spatial configuration of sign language gestures.
- 3D patch embedding: Video frames are divided into 3D patches, preserving both spatial and temporal information. This approach is particularly beneficial for capturing the motion dynamics in sign language.
- Efficient computation: By limiting self-attention computation to local windows, the Video Swin Transformer achieves linear computational complexity with respect to input size, making it more efficient than global attention mechanisms.

As shown in Figure 2, in our experimental setup, the Video Swin Transformer processed sequences of images extracted from videos by first extracting keypoints of sign language utterances from these images. This approach allowed us to investigate the effectiveness of our dynamic sampling rate adjustment method in maintaining high recognition accuracy across different frame rates, reflecting real-world variability, such as background image information, in video capture conditions. Table 5 shows the hyperparameters and their settings to train the sign language recognition model. The loss function used was LabelSmoothingCrossEntropy, which is suitable for classification purposes. The batch size was set to 4, considering the computing environment used in the experiment. The learning rate was set to 1×10^{-3}, a commonly used value. The optimizer was set to SGD.

By leveraging the capabilities of the Video Swin Transformer, our performance evaluations aim to provide evidence that the recognition accuracy of the model can be improved by the proposed dynamic sampling rate adjustment method compared to methods proposed in previous research.

Table 5. Hyperparameters and their settings to train the sign language recognition model.

Hyperparameters	Configurations
Loss Function	Label smoothing cross-entropy
Learning Rate	1×10^{-3}
Optimizer	SGD
Epochs	200
Batch Size	4

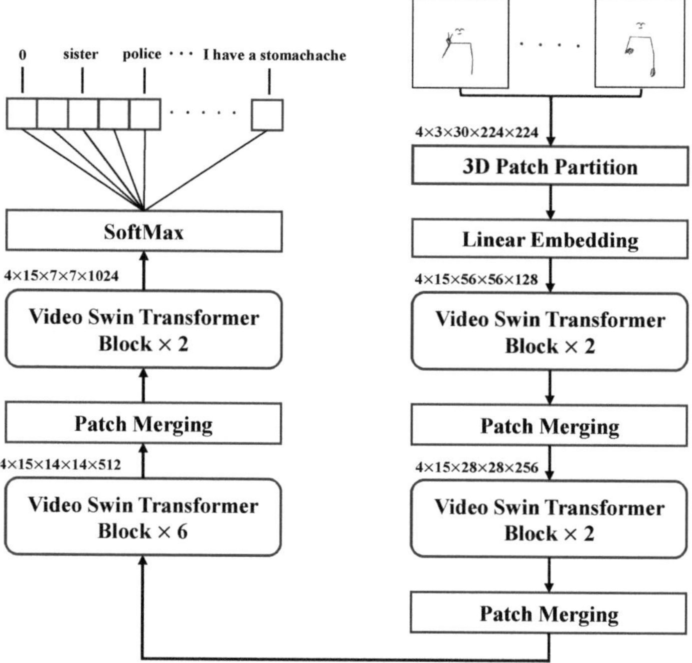

Figure 2. Overall architecture of sign language recognition model based on Video Swin Transformer.

4.1.2. Dataset and Data Preprocessing

We used a dataset provided by KETI (Korea Electronics Technology Institute, Seongnam, Republic of Korea), with some class data publicly available on AIHUB. Table 6 shows the composition of the dataset used in the experiment. Our dataset, used for model training and evaluation, comprises a total of 41,240 video samples, divided into training (28,868), testing (6186), and validation (6186) sets. It covers 510 classes, consisting of 405 individual words and 105 sentences, with each class represented by approximately 80 samples. All videos were captured in full HD (1920 × 1080) resolution, and the dataset includes recordings from multiple sign language speakers, ensuring diversity in the sign language utterances.

Table 6. Dataset composition.

Sets	Samples	Classes	Words	Sentences
Training	28,868	510	405	105
Testing	6186	510	405	105
Validation	6186	510	405	105

Each video sample in the dataset is a recording of a sign language utterance. To prepare the data for training and validation, we employed the following preprocessing steps.

1. Frame Extraction: Each video is broken down into individual frames.
2. Video Frame Centralization and Resizing:
   - Each extracted frame, originally at a resolution of 1920 × 1080 pixels, is centrally cropped to 1080 × 1080 pixels to ensure the sign language gesture is centered.
   - The centrally cropped frame is then resized to 256 × 256 pixels to standardize the input size for the model.
   - 17 keypoints (0–16) from the Pose model, shown on the left side of Figure 3.

- 21 keypoints (0–20) from each Hand model (left and right), as depicted on the right side of Figure 3.
- A total of 59 keypoints are extracted per frame.

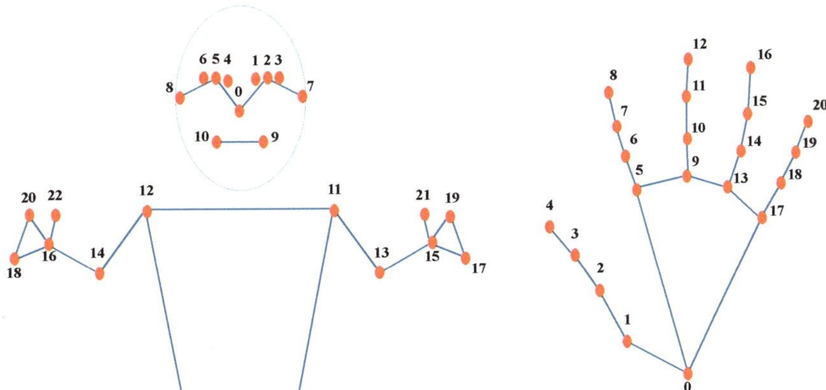

Figure 3. Illustration of the keypoints information used by the MediaPipe models. (**Left**) The Pose model handles keypoints for full-body tracking. (**Right**) The Hands model focuses on hand keypoints detection.

4.1.3. Data Augmentation and Model Training

To enhance the robustness of our sign language recognition model, we implemented several data augmentation techniques during the training process. In real-world scenarios, the form of sign language utterances can vary significantly depending on the signer. Additionally, the signer may not always perform directly in front of the camera and may even perform signs in reverse. Therefore, data augmentation techniques such as random horizontal flipping, random rotation, and random cropping were applied. Our data loading and augmentation pipeline is implemented in a custom class that is inherited from PyTorch's Dataset class. The three data augmentation techniques used in our training process are as follows.

- Random horizontal flipping: 50% probability;
- Random rotation: ±5 degrees;
- Random cropping: From 256 × 256 to 224 × 224.

The data loading process is optimized to handle varying video lengths and to apply transformations efficiently. Our implementation allows for the easy adjustment of parameters such as the number of frames, image size, and augmentation intensity. This robust data preparation and augmentation pipeline, combined with our large and diverse dataset, provides a strong foundation for training our sign language recognition model. It ensures that the model is exposed to a wide variety of input variations, promoting generalization and resilience to real-world variability in sign language performances.

4.1.4. Experimental Setup and Evaluation Scenarios

The performance evaluation was conducted using a pre-extracted test dataset consisting of 6186 samples. Each original sample in this dataset represents a sequence of images extracted from videos at 30 frames per second (fps), capturing the keypoints of sign language utterances by the signers. These keypoint images serve as the input for our sign language recognition model.

To assess the effectiveness of our dynamic sampling adjustment method, we used a modular arithmetic method to simulate lower frame rates from the original 30 fps video sequences. Specifically, we adjusted the frame rates to 5, 10, 15, 20, and 25 fps by selectively choosing frames from the original data. For example, to simulate a 15 fps mobile computing environment, we retained frames that satisfy the condition $i \mod 2 = 0$ (i.e., keep frames 0, 2, 4, 6, ... from the original frames). This approach allowed us to evaluate how effectively the dynamic sampling rate adjustment method adapts to different frame rates and maintains recognition accuracy.

The experiments were conducted using three different methods: (1) the original dataset without any modifications, (2) the proposed dynamic sampling rate adjustment method, and (3) the fixed sampling rate method proposed in previous research. We evaluated the performance of each method on the test datasets at 5, 10, 15, 20, and 25 fps. This approach allowed us to compare the recognition accuracy of the original, unmodified sequences against the proposed method and the previous fixed sampling rate method across different frame rate scenarios.

Hereafter, the methods for processing the input data for the sign language recognition model will be referred to as follows: the original dataset without any modifications will be called the Original Method, the proposed dynamic sampling rate adjustment method will be called the Dynamic Method, and, finally, the fixed sampling rate method proposed in previous research will be called the Fixed Method.

4.2. Experimental Results

4.2.1. Accuracy Results

Our experiments evaluated the performance of the Dynamic Method across various frame rates, comparing it with the Fixed Method and the Original Method. Figure 4 shows the results of the top-1 accuracy comparison according to fps. Figure 5 shows the results of the top-5 accuracy comparison according to fps. As shown in Figures 4 and 5, the Dynamic Method consistently outperforms the Fixed Method and Original Method in terms of top-1 accuracy across all tested frame rates.

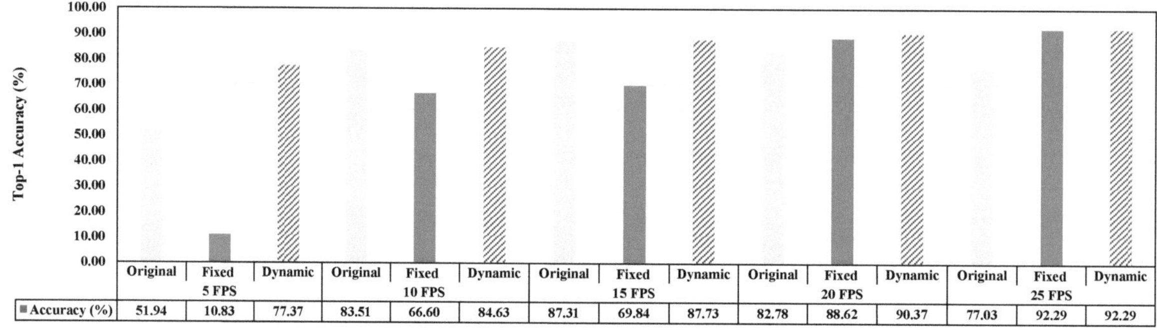

		5 FPS			10 FPS			15 FPS			20 FPS			25 FPS	
	Original	Fixed	Dynamic	Original	Fixed	Dynamic	Original	Fixed	Dynamic	Original	Fixed	Dynamic	Original	Fixed	Dynamic
■ Accuracy (%)	51.94	10.83	77.37	83.51	66.60	84.63	87.31	69.84	87.73	82.78	88.62	90.37	77.03	92.29	92.29

Figure 4. Top-1 accuracy comparison results according to FPS.

Figure 5. Top-5 accuracy comparison results according to FPS.

Figure 4 indicates that at lower frame rates of 5 fps and 10 fps, our Dynamic Method significantly outperforms the Fixed Method, achieving 77.37% accuracy at 5 fps compared to only 10.83% with the Fixed Method. As the frame rate increases, both methods generally show improved accuracy due to more frequent sampling. The Dynamic Method consistently maintains higher accuracy across all tested frame rates, reaching its highest accuracy at 25 fps with a top-1 accuracy of 92.29%, demonstrating its effectiveness in adapting to higher frame rates. Figure 5 shows the top-5 accuracy results according to the frame rate. As shown in Figure 5, the pattern of accuracy performance is similar to the top-1 results.

Based on Figures 4 and 5, it is clear that the Dynamic Method consistently outperforms the Fixed Method and Original Method across all tested frame rates, with significant improvements in accuracy at lower frame rates of 5 fps and 10 fps. This highlights its effectiveness in scenarios with computational limitations, such as mobile environments. As the frame rate increases, the performance gap between the Dynamic and Fixed Methods is reduced, but the Dynamic Method maintains a competitive edge, indicating its robustness in varying operational conditions. The observed performance enhancement validates the efficacy of the Dynamic Method in enhancing the accuracy and reliability of sign language recognition systems deployed in mobile environments.

4.2.2. Impact of FPS on Sign Language Recognition Accuracy

The detection of the start and end points of sign utterances plays a crucial role in the accuracy of sign language recognition models, and this detection is significantly influenced by the frames per second (fps) of the video input. Figure 6 shows examples of how different fps affect the quality of captured keypoint images. Figure 6a shows frames extracted at 5 fps, while Figure 6b shows frames extracted at 25 fps. Both were sourced from an original 30 fps video depicting the sign for the number '0'.

In a fixed sampling rate scenario, where the sampling rate was set to 5, keypoint information is extracted every five frames to detect the start and end points of the sign language utterance. This is because the sign language recognition model was optimized for 30 fps environments. However, when applied to a 5 fps environment, this Fixed Method becomes suboptimal.

Table 7 shows a comparison of the number of extracted images, keypoint extraction frequency, and recognition accuracy according to fps. In a 5 fps environment, during a 5-s sign utterance, only 25 frames can be extracted and made available. This scarcity results in a higher likelihood of capturing irrelevant frames that do not adequately represent the beginning or end of the gesture, thereby reducing recognition accuracy. Therefore, in summary, because the amount of keypoint movement change between each frame is much larger, it is advisable to adjust the sampling rate to a shorter interval. As shown in Figure 6, the amount of change in each keypoint between consecutive frames is greater at 5 fps than at 25 fps.

Table 7. Comparison of the number of extracted images, keypoints extraction frequency, and recognition accuracy according to FPS (when the utterance time is 5 s).

fps	Total Number of Frames	Keypoints Extraction Frequency	Recognition Accuracy
5 fps	25	5	Low
25 fps	125	25	High
30 fps	150	30	Very High

Conversely, in a 25 fps environment, 125 frames are captured in the same duration, providing the model with more opportunities to evaluate keypoint changes and accurately identify gesture dynamics and movements. This increased frame density improves the model's ability to discern the start and end points, yielding higher recognition accuracy.

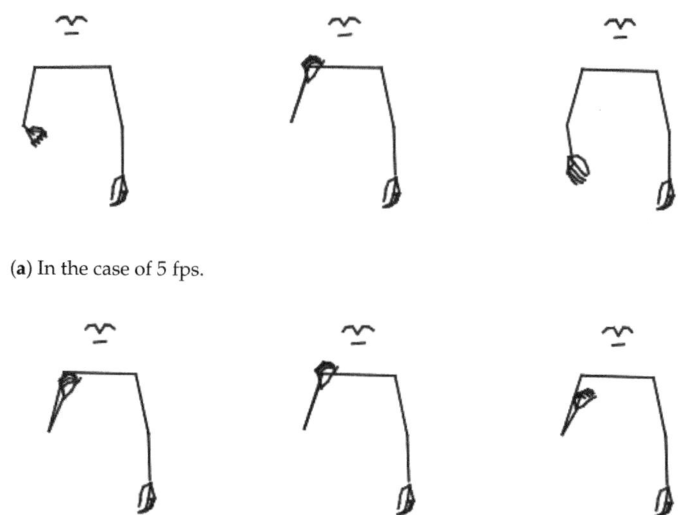

(**a**) In the case of 5 fps.

(**b**) In the case of 25 fps.

Figure 6. Examples of how different fps affect the quality of captured keypoint images.

Table 8 provides confusion matrix results for top-1 (accuracy, precision, recall, F1-score) according to frame rates and methods. As shown in Table 8, the performance of the Dynamic Method is significantly higher than the Fixed Method in lower-fps environments. In the 5 fps environment, the Dynamic Method significantly outperforms both the Original and Fixed Methods with higher precision (0.81), recall (0.77), and F1-score (0.76), while the Fixed method particularly struggles with low recall (0.11), highlighting its limitations in such environments. In the 10 fps environment, all methods show improvement, yet the Dynamic Method maintains superior performance with high precision (0.87), recall (0.85), and F1-score (0.84). The Original Method also performs well but trails slightly in F1-score (0.83). In the 15 fps environment, the Dynamic Method continues to excel with balanced precision (0.89) and recall (0.88), resulting in a strong F1-score of 0.87, while the Fixed Method, although improved, remains behind in recall (0.70). In the 20 fps and 25 fps environments, the performance of the Dynamic Method is further improved, consistently achieving the highest F1-scores of 0.90 and 0.92, compared to the Fixed and Original Methods, respectively.

As shown in Table 8, the superior performance of the Dynamic Method, particularly in low-fps environments, can be attributed to its ability to adaptively adjust the sampling rate based on real-time frame rates. Therefore, the Dynamic Method can accurately capture the start and end points of sign language utterances even when fewer frames are available. This adaptability allows the Dynamic Method to maintain high precision and recall by efficiently managing the trade-off between temporal resolution and computational resources, which is especially crucial in mobile environments where computational power is limited. Based on all of the experimental results, the Dynamic Method proves to be a reliable solution for sign language recognition in real-time applications, particularly in environments with challenging operational conditions and varying frame rates.

Table 8. Confusion matrix results for top-1 (accuracy, precision, recall, F1-score) according to frame rates and methods.

Metric	Accuracy	Precision	Recall	F1-Score
		5 fps		
Original	0.52	0.62	0.52	0.51
Fixed	0.11	0.59	0.11	0.17
Dynamic	0.77	0.81	0.77	0.76
		10 fps		
Original	0.84	0.86	0.84	0.83
Fixed	0.67	0.73	0.67	0.67
Dynamic	0.85	0.87	0.85	0.84
		15 fps		
Original	0.87	0.90	0.87	0.88
Fixed	0.70	0.73	0.70	0.69
Dynamic	0.88	0.89	0.88	0.87
		20 fps		
Original	0.83	0.90	0.83	0.85
Fixed	0.89	0.91	0.89	0.89
Dynamic	0.90	0.91	0.90	0.90
		25 fps		
Original	0.77	0.89	0.77	0.81
Fixed	0.92	0.93	0.92	0.92
Dynamic	0.92	0.93	0.92	0.92

5. Conclusions and Future Works

In this paper, we proposed a dynamic sampling rate adjustment method designed to enhance the performance of sign language recognition in mobile environments. The proposed method addresses the challenges of varying computational capacities and frame extraction rates across different mobile devices. The primary objective of the proposed scheme is to dynamically adjust the sampling rate based on the real-time frame extraction rate during sign language utterances in mobile environments. By using the dynamic sampling rate, the proposed method improves the detection of the start and end points of sign language utterances. Consequently, it improves overall recognition accuracy compared to the previous fixed method.

Based on the experimental results, the proposed dynamic sampling rate adjustment method significantly outperformed the previous fixed sampling rate method. Specifically, our approach achieved up to an 83.64% improvement in top-5 accuracy and a 66.54% enhancement in top-1 accuracy. These improvements were particularly pronounced in low-frame-rate scenarios. This shows the effectiveness of our method in environments with limited computational resources such as mobile computing environments. We expect that the proposed method can be widely applied not only to the field of sign language recognition in mobile environments but also to the preprocessing of various artificial intelligence services in mobile environments.

This paper has the following limitations. First, the effectiveness of the proposed dynamic sampling rate method was verified in an experimental setting that simulated mobile computing environments by varying the frame rate. This approach was used because performing all sign language utterances directly on a mobile device would be very time-consuming. Therefore, further performance validation on actual mobile devices may be necessary for future research. In addition, the proposed method was applied only to the Video Swin Transformer model, and it is necessary to verify its performance by applying the dynamic sampling rate method to a broader range of AI models.

Our future research work includes the following. Firstly, expanding the research to include a wider range of mobile devices and operating conditions would provide more comprehensive insights into the robustness of the proposed method. Additionally, integrating machine learning models that can predict the optimal sampling rate based on device specifications and environmental factors could further enhance performance.

Author Contributions: Conceptualization, T.K. and B.K.; methodology, T.K. and B.K.; software, T.K.; validation, T.K. and B.K.; formal analysis, T.K. and B.K.; investigation, T.K. and B.K.; resources, T.K. and B.K.; data curation, T.K. and B.K.; writing—original draft preparation, T.K. and B.K.; writing—review and editing, T.K. and B.K.; visualization, T.K. and B.K.; supervision, B.K.; project administration, B.K.; funding acquisition, B.K. All authors have read and agreed to the published version of the manuscript.

Funding: This work was supported by Institute of Information & communications Technology Planning & Evaluation (IITP) grant funded by the Korea government (MSIT) (RS-2022-II220043, Adaptive Personality for Intelligent Agents).

Institutional Review Board Statement: Not applicable.

Informed Consent Statement: Not applicable.

Data Availability Statement: The original contributions presented in the study are included in the article, further inquiries can be directed to the corresponding author.

Conflicts of Interest: The authors declare no conflicts of interest.

References

1. Nasir, H.M.; Brahin, N.M.A.; Ariffin, F.E.M.S.; Mispan, M.S.; Wahab, N.H.A. AI Educational Mobile App using Deep Learning Approach. *Int. J. Inform. Vis.* **2023**, *7*, 952–958. [CrossRef]
2. Li, Y.; Dang, X.; Tian, H.; Sun, T.; Wang, Z.; Ma, L.; Klein, J.; Bissyandé, T.F. AI-driven Mobile Apps: An Explorative Study. *arXiv* **2024**, arXiv:2212.01635. [CrossRef]
3. Karunya, S.; Jalakandeshwaran, M.; Babu, T.; Uma, R. AI-Powered Real-Time Speech-to-Speech Translation for Virtual Meetings Using Machine Learning Models. In Proceedings of the 2023 Intelligent Computing and Control for Engineering and Business Systems (ICCEBS), Chennai, India, 14–15 December 2023; pp. 1–6. [CrossRef]
4. Guo, Z.; Hou, Y.; Hou, C.; Yin, W. Locality-Aware Transformer for Video-Based Sign Language Translation. *IEEE Signal Process. Lett.* **2023**, *30*, 364–368. [CrossRef]
5. Li, W.; Pu, H.; Wang, R. Sign Language Recognition Based on Computer Vision. In Proceedings of the 2021 IEEE International Conference on Artificial Intelligence and Computer Applications (ICAICA), Dalian, China, 28–30 June 2021; pp. 919–922. [CrossRef]
6. Ko, S.K.; Kim, C.J.; Jung, H.; Cho, C. Neural Sign Language Translation Based on Human Keypoint Estimation. *Appl. Sci.* **2019**, *9*, 2683. [CrossRef]
7. Chen, Y.; Wei, F.; Sun, X.; Wu, Z.; Lin, S. A Simple Multi-Modality Transfer Learning Baseline for Sign Language Translation. *arXiv* **2023**, arXiv:2203.04287. [CrossRef]
8. Elangovan, T.; Arockia Xavier Annie, R.; Sundaresan, K.; Pradhakshya, J.D. Hand Gesture Recognition for Sign Languages Using 3DCNN for Efficient Detection. In Proceedings of the Computer Methods, Imaging and Visualization in Biomechanics and Biomedical Engineering II, Bonn, Germany, 7–9 September 2021; Tavares, J.M.R.S., Bourauel, C., Geris, L., Vander Slote, J., Eds.; Springer: Cham, Switzerland, 2023; pp. 215–233.
9. Naz, N.; Sajid, H.; Ali, S.; Hasan, O.; Ehsan, M.K. Signgraph: An Efficient and Accurate Pose-Based Graph Convolution Approach Toward Sign Language Recognition. *IEEE Access* **2023**, *11*, 19135–19147. [CrossRef]
10. Patel, B.D.; Patel, H.B.; Khanvilkar, M.A.; Patel, N.R.; Akilan, T. ES2ISL: An Advancement in Speech to Sign Language Translation using 3D Avatar Animator. In Proceedings of the 2020 IEEE Canadian Conference on Electrical and Computer Engineering (CCECE), London, ON, Canada, 30 August–2 September 2020; pp. 1–5. [CrossRef]
11. Kim, J.H.; Hwang, E.J.; Cho, S.; Lee, D.H.; Park, J. Sign Language Production with Avatar Layering: A Critical Use Case over Rare Words. In Proceedings of the Thirteenth Language Resources and Evaluation Conference, Marseille, France, 20–25 June 2022; pp. 1519–1528.
12. Moncrief, R.; Choudhury, S.; Saenz, M. Efforts to Improve Avatar Technology for Sign Language Synthesis. In Proceedings of the 15th International Conference on PErvasive Technologies Related to Assistive Environments (PETRA'22), Corfu, Greece, 29 June–1 July 2022; pp. 307–309.
13. Mondal, R. Mobile Cloud Computing. In *Emerging Trends in Cloud Computing Analytics, Scalability, and Service Models*; IGI Global: Hershey, PA, USA, 2024; pp. 170–185.

14. Mamchych, O.; Volk, M. Smartphone Based Computing Cloud and Energy Efficiency. In Proceedings of the 2022 12th International Conference on Dependable Systems, Services and Technologies (DESSERT), Athens, Greece, 9–11 December 2022; pp. 1–5. [CrossRef]
15. Silva, P.; Rocha, R. Low-Power Footprint Inference with a Deep Neural Network offloaded to a Service Robot through Edge Computing. In Proceedings of the SAC'23: 38th ACM/SIGAPP Symposium on Applied Computing, New York, NY, USA, 27–31 March 2023; pp. 800–807. [CrossRef]
16. Jayasimha, A.; Paramasivam, P. Personalizing Speech Start Point and End Point Detection in ASR Systems from Speaker Embeddings. In Proceedings of the 2021 IEEE Spoken Language Technology Workshop (SLT), Shenzhen, China, 19–22 January 2021; pp. 771–777. [CrossRef]
17. Kim, G.; Cho, J.; Kim, B. A Keypoint-based Sign Language Start and End Point Detection Scheme. *KIISE Trans. Comput. Pract.* **2023**, *29*, 184–189. [CrossRef]
18. Waheed, T.; Qazi, I.A.; Akhtar, Z.; Qazi, Z.A. Coal not diamonds: How memory pressure falters mobile video QoE. In Proceedings of the 18th International Conference on Emerging Networking EXperiments and Technologies, New York, NY, USA, 6–9 December 2022; CoNEXT '22, pp. 307–320. [CrossRef]
19. Ekbote, J.; Joshi, M. Indian sign language recognition using ANN and SVM classifiers. In Proceedings of the 2017 International Conference on Innovations in Information, Embedded and Communication Systems (ICIIECS), Coimbatore, India, 17–18 March 2017; pp. 1–5. [CrossRef]
20. Pathan, R.; Biswas, M.; Yasmin, S.; Khandaker, M.U.; Salman, M.; Youssef, A.A.F. Sign language recognition using the fusion of image and hand landmarks through multi-headed convolutional neural network. *Sci. Rep.* **2023**, *13*, 16975. [CrossRef] [PubMed]
21. Katoch, S.; Singh, V.; Tiwary, U.S. Indian Sign Language recognition system using SURF with SVM and CNN. *Array* **2022**, *14*, 100141. [CrossRef]
22. Kothadiya, D.; Bhatt, C.; Sapariya, K.; Patel, K.R.; Gil-González, A.B.; Corchado, J.M. Deepsign: Sign Language Detection and Recognition Using Deep Learning. *Electronics* **2022**, *11*, 1780. [CrossRef]
23. Kothadiya, D.R.; Bhatt, C.M.; Saba, T.; Rehman, A.; Bahaj, S.A. SIGNFORMER: DeepVision Transformer for Sign Language Recognition. *IEEE Access* **2023**, *11*, 4730–4739. [CrossRef]
24. Alharthi, N.M.; Alzahrani, S.M. Vision Transformers and Transfer Learning Approaches for Arabic Sign Language Recognition. *Appl. Sci.* **2023**, *13*, 11625. [CrossRef]
25. Tripathi, S.; Ranade, S.; Tyagi, A.; Agrawal, A. PoseNet3D: Learning Temporally Consistent 3D Human Pose via Knowledge Distillation. In Proceedings of the 2020 International Conference on 3D Vision (3DV), Fukuoka, Japan, 25–28 November 2020; pp. 311–321. [CrossRef]
26. Bird, J.J.; Ihianle, I.K.; Machado, P.; Brown, D.J.; Lotfi, A. A Neuroevolution Approach to Keypoint-Based Sign Language Fingerspelling Classification. In Proceedings of the 2023 15th International Congress on Advanced Applied Informatics Winter (IIAI-AAI-Winter), Bali, Indonesia, 11–13 December 2023; pp. 215–220. [CrossRef]
27. Liu, Z.; Ning, J.; Cao, Y.; Wei, Y.; Zhang, Z.; Lin, S.; Hu, H. Video Swin Transformer. *arXiv* **2021**, arXiv:2106.13230. [CrossRef]

MDPI AG
Grosspeteranlage 5
4052 Basel
Switzerland
Tel.: +41 61 683 77 34

Applied Sciences Editorial Office
E-mail: applsci@mdpi.com
www.mdpi.com/journal/applsci

www.ingramcontent.com/pod-product-compliance
Lightning Source LLC
LaVergne TN
LVHW071936160726
843515LV00011B/2627